# Metastasis Research Protocols

# METHODS IN MOLECULAR MEDICINE™

## *John M. Walker,* Series Editor

69. **Gene Therapy Protocols, 2nd ed.,** edited by *Jeffrey R. Morgan, 2002*
68. **Molecular Analysis of Cancer,** edited by *Jacqueline Boultwood and Carrie Fidler,* 2002
67. **Meningococcal Disease:** *Methods and Protocols,* edited by *Andrew J. Pollard and Martin C. J. Maiden,* 2001
66. **Meningococcal Vaccines:** *Methods and Protocols,* edited by *Andrew J. Pollard and Martin C. J. Maiden,* 2001
65. **Nonviral Vectors for Gene Therapy:** *Methods and Protocols,* edited by *Mark A, Findeis,* 2001
64. **Dendritic Cell Protocols,** edited by *Stephen P. Robinson and Andrew J. Stagg,* 2001
63. **Hematopoietic Stem Cell Protocols,** edited by *Christopher A. Klug and Craig T. Jordan,* 2001
62. **Parkinson's Disease:** *Methods and Protocols,* edited by *M. Maral Mouradian,* 2001
61. **Melanoma Techniques and Protocols:** *Molecular Diagnosis, Treatment, and Monitoring,* edited by *Brian J. Nickoloff,* 2001
60. **Interleukin Protocols,** edited by *Luke A. J. O'Neill and Andrew Bowie,* 2001
59. **Molecular and Cellular Pathology in Prion Disease,** edited by *Harry F. Baker,* 2001
58. **Metastasis Research Protocols:** *Volume 2, Cell Behavior In Vitro and In Vivo,* edited by *Susan A. Brooks and Udo Schumacher,* 2001
57. **Metastasis Research Protocols:** *Volume 1, Analysis of Cells and Tissues,* edited by *Susan A. Brooks and Udo Schumacher,* 2001
56. **Human Airway Inflammation:** *Sampling Techniques and Analytical Protocols,* edited by *Duncan F. Rogers and Louise E. Donnelly,* 2001
55. **Hematologic Malignancies:** *Methods and Protocols,* edited by *Guy B. Faguet,* 2001
54. ***Mycobacterium tuberculosis* Protocols,** edited by *Tanya Parish and Neil G. Stoker,* 2001
53. **Renal Cancer:** *Methods and Protocols,* edited by *Jack H. Mydlo,* 2001
52. **Atherosclerosis:** *Experimental Methods and Protocols,* edited by *Angela F. Drew,* 2001
51. **Angiotensin Protocols**, edited by *Donna H. Wang,* 2001
50. **Colorectal Cancer:** *Methods and Protocols,* edited by *Steven M. Powell,* 2001
49. **Molecular Pathology Protocols**, edited by *Anthony A. Killeen,* 2001
48. **Antibiotic Resistance Methods and Protocols,** edited by *Stephen H. Gillespie,* 2001
47. **Vision Research Protocols,** edited by *P. Elizabeth Rakoczy,* 2001
46. **Angiogenesis Protocols,** edited by *J. Clifford Murray*, 2001
45. **Hepatocellular Carcinoma:** *Methods and Protocols,* edited by *Nagy A. Habib*, 2000
44. **Asthma:** *Mechanisms and Protocols,* edited by *K. Fan Chung and Ian Adcock*, 2001
43. **Muscular Dystrophy:** *Methods and Protocols,* edited by *Katherine B. Bushby and Louise Anderson*, 2001
42. **Vaccine Adjuvants:** *Preparation Methods and Research Protocols,* edited by *Derek T. O'Hagan*, 2000
41. **Celiac Disease:** *Methods and Protocols,* edited by *Michael N. Marsh,* 2000
40. **Diagnostic and Therapeutic Antibodies,** edited by *Andrew J. T. George and Catherine E. Urch*, 2000
39. **Ovarian Cancer:** *Methods and Protocols,* edited by *John M. S. Bartlett,* 2000
38. **Aging Methods and Protocols,** edited by *Yvonne A. Barnett and Christopher R. Barnett*, 2000
37. **Electrochemotherapy, Electrogenetherapy, and Transdermal Drug Delivery:** *Electrically Mediated Delivery of Molecules to Cells*, edited by *Mark J. Jaroszeski, Richard Heller, and Richard Gilbert,* 2000
36. **Septic Shock Methods and Protocols**, edited by *Thomas J. Evans*, 2000
35. **Gene Therapy of Cancer:** *Methods and Protocols,* edited by *Wolfgang Walther and Ulrike Stein*, 2000
34. **Rotaviruses:** *Methods and Protocols,* edited by *James Gray and Ulrich*

METHODS IN MOLECULAR MEDICINE™

# Metastasis Research Protocols

## Volume I: Analysis of Cells and Tissues

Edited by

**Susan A. Brooks**

*Research School of Biological and Molecular Sciences, Oxford Brookes University, Oxford, UK*

and

**Udo Schumacher**

*Institute for Anatomy, Department of Neuroanatomy, University Hospital Hamburg-Eppendorf, Hamburg, Germany*

Humana Press Totowa, New Jersey

999 Riverview Drive, Suite 208
Totowa, New Jersey 07512



This publication is printed on acid-free paper. ∞
ANSI Z39.48-1984 (American Standards Institute) Permanence of Paper for Printed Library Materials.

Cover Illustration: Fig. 2 from Chapter 7, "Surrogate Markers of Angiogenesis and Metastasis" by Girolamo Ranieri and Giampetro Gasparini.

Production Editor: Jessica Jannicelli.

Cover design by Patricia Cleary.

For additional copies, pricing for bulk purchases, and/or information about other Humana titles, contact Humana at the above address or at any of the following numbers: Tel.: 973-256-1699; Fax: 973-256-8341; E-mail: humana@humanapr.com; Website: http://humanapress.com



Printed in the United States of America. 10 9 8 7 6 5 4 3 2 1

Library of Congress Cataloging in Publication Data

Main entry under title:

Methods in molecular medicine™.

Metastasis research protocols/edited by Susan A. Brooks and Udo Schumacher.
p. cm. -- (Methods in molecular medicine ; 57-58)
Includes bibliographical references and index.
Contents: v. 1. Analysis of cells and tissues -- v. 2. Analysis of cell behavior in vitro and in vivo.
ISBN 0-89603-610-3 (v. 1 : alk. paper) -- ISBN 0-89603-615-4 (v. 2 : alk. paper)
1. Metastasis--Laboratory manuals. 2. Cancer cells--Laboratory manuals. 3. Cytology--Laboratory manuals. I. Brooks, Susan A. II. Schumacher, Udo. III. Series.

RC269.5 .M478 2001
616.99'407--dc21 2001016931

# Preface

The process of metastasis formation is hugely complex, as described in the introductory chapter of this book, and this complexity has led us to compile two volumes of methods, from a vastly divergent background that attempts to encompass the whole spectrum of cancer biology. This first volume, *Metastasis Research Protocols: Analysis of Cells and Tissues*, concentrates on analysis and mapping of molecules produced by cells and tissues and analysis of the molecular biology underlying their expression, whereas the second volume, *Metastasis Research Protocols: Cell Behavior In Vitro and In Vivo*, focuses sharply on the determination of cell behavior in vitro and in vivo. We have deliberately included chapters describing well-established and familiar techniques (for example, SDS-PAGE and Western blotting [Chapter 11], and immunocytochemistry [Chapter 2]) in addition to the newer and more specialized approaches and specific examples of their application, because—although the methodology is readily available in the published literature and established in many laboratories—we wished these volumes to "stand alone" and to make accessible here the standard techniques that underpin much metastasis research for both the newcomer to the field and the seasoned researcher. Undoubtedly, owing to the complexity of the metastatic cascade and the wealth of research techniques involved in scientific approaches to its unraveling, and despite our best efforts to make these volumes as comprehensive as seems feasible, this is a tall order, and there will inevitably be omissions. For these we apologize.

Part I of the present volume, *Detection Methods for Cellular Markers of Metastatic Potential*, begins with a chapter on the evaluation of the presence or absence of metastatic spread by "traditional" histopathological assessment. It seemed to us that this topic was the best way to launch a book of metastasis research protocols because it is essential to so much of clinically based metastasis research. Old fashioned hitology and its ramifications, such as immunocytochemistry, remain the gold standard for the diagnosis of the metastatic spread of a tumor. The methodology for basic immunocytochemistry is described, as are variations such as multiple labeling and lectin cytochemistry, and some examples of clinically relevant applications of immunocytochemistry to metastasis research are given. Several approaches to the assessment of cellular proliferation, including immunocytochemistry, calculation of mitotic index, and flow cytometry, are described. Another classical method of enormously wide applicability in metastasis research, SDS-PAGE and Western blotting, is detailed. This technique allows at least initial analysis of protein extracts from cells and tissues and is inevitably referred to in numerous other protocols in this and the following volume. The last chapter in this section describes enzyme zymography for assessment of levels of gelatinases.

Having outlined techniques for mapping cellular expression of gene products of potential significance in the metastatic cascade, Part II extends the level of analysis further with a range of techniques for analysis of *Genetic Aspects of Metastasis*. A range of molecular biological methods applied to metastasis research are presented, facilitating a focused and specific single gene approach. They include *in situ* hybridization to localize mRNAs, a variation on this approach, fluorescence *in situ* hybridization (FISH), and comparative genomic hybridization (CGH). RNA purification and quantification methods are also presented. The classical approaches of Northern and Southern blotting and PCR are described. Methylation analysis of CpG islands is covered, as is detection of metastatic tumor cells in blood by the classical technique of reverse transcriptase polymerase chain reaction (RT-PCR). The final chapter in this section presents the technique of differential display.

The last section of the book deals with the *Mathematical Modeling of Metastasis*. This aspect might seem unfamiliar to many cancer scientists, but it may well be that mathematical models will help to direct further research efforts in metastasis research or may help to evaluate whether a particular strategy in the treatment of metastases will have any influence on the overall outcome of the disease. Complex and nonlinear biological systems will need the help of mathematicians to adequately model their effects.

Cancer mortality will only drop if metastases can be treated successfully. To do so, it is probably necessary to understand the metastatic cascade first at the cellular and molecular level, then at the tissue level, and finally—the subject of Volume II—to appreciate the subtleties of cellular behavior in both in vitro assays and at the level of the whole organism. Based on this comprehensive knowledge, rational strategies to combat metastatic disease might be developed. We hope that this collection of protocols will help to work toward this goal.

***Susan A. Brooks***
***Udo Schumacher***

# Introduction to Volumes I and II

## Why a Collection of Metastasis Research Protocols?

A metastasis is a secondary tumor derived from a primary tumor at a distant site. The process leading to the formation of metastases is complex and owing to its complexity many researchers have shunned tackling this problem. "Science is to do the doable" and many excellent minds have tried to tackle the problem of metastases unsuccessfully, leading to a general frustration in addressing this important research topic. Metastases formation is by far the overwhelming cause of death in cancer patients and should therefore be a major issue to be addressed in cancer research. In 1971, President Nixon declared war on cancer with the aim of eradicating it as a major cause of death in the United States. Yet, at the beginning of the new millennium, cancer is still undefeated and account for approximately 25% of the deaths in the developed world. In contrast, mortality because of cardiovascular diseases in the age groups below 60 years has declined impressively *(1,2)*. Though surgical techniques and anesthesia have made progress during the last decades, cancer mortality remains high. The primary tumor can often be removed surgically, but no surgical cure for generalized metastases can be offered to the patient. Metastatic disease is frequently resistant to other forms of treatment and may not be eradicated by even the most aggressive chemotherapy or radiotherapy. Metastatic disease is the most common cause of death in cancer patients. And even here there remain unanswered questions. If metastases erode a large blood vessel resulting in a massive bleeding or if they destroy a vital brain center, then the cause of death is clear. However, at post-mortem examinations, no such clear signs for the cause of death are detected for the vast majority of cancer victims and so the ultimate reason why cancer patients die from metastatic disease remains an unsolved problem.

Metastasis formation is a multistep process (for review, *see* **refs.** ***3,4)***. The following simple description will focus on the formation of systemic metastases and will not deal with local lymph node metastasis. The process of metastasis formation starts when the transformed malignant cell starts to grow. Once the primary tumor reaches a certain size, it needs blood vessel penetration in order to grow further. Once tumor neoangiogenesis has started, tumor cells have to loosen themselves from the primary tumor, penetrate the surrounding connective tissue and basal lamina of the blood vessel. In order to proceed further, tumor cells have to invade through the endothelium into the bloodstream. Once tumor cells have entered the systemic circulation, they have to survive in this environment and have to be transported to the site of the future metastasis. Here, they have to attach to the endothelium, penetrate it, and complete the reverse of the process observed at the original site of invasion: Cross the endothelium and its basal lamina and establish themselves in the connective tissue surrounding the blood vessel. Once this has been achieved, the metastatic cancer cell has to start to grow in order to form a clinically detectable metastasis.

The process of metastasis formation is often referred to as a cascade, meaning that every single step has to be successfully completed in order to proceed to the next. The cascade nature of the process makes it so difficult to analyze, and, hence, the rate-limiting step of metastasis formation is still unknown. During the last decades, a shift has occurred of what is thought to be important in this process. In the 1970s, much attention was paid to the degradation of the basal lamina as the initial step of metastasis formation *(5)*. The focus of the attention shifted then toward cell adhesion molecules. Cell adhesion molecules have a dual role to perform: First of all, their expression at the site of the primary tumor make them act as an anti-metastatic factor because they ensure that a cell with a metastatic potential remains at the site of the primary tumor. On the other hand, expression of cell adhesion molecules at the site of the future metastasis are thought to be essential for the interaction between the tumor cell in the bloodstream with the endothelial cell of the target organ. At this site, the cell adhesion molecule would act as a pro-metastatic molecule. The apparent paradox can be solved in two ways: Either two different cell adhesion molecules act at the two different sites or the behavior of one cell adhesion molecule is subject to down- and upregulation at the invasion and at the evasion sites, respectively (for dual role of a cell adhesion molecule, *see* **ref.** *6)*.

While the importance of the processes occurring at the evasion site has been the focus of attention in the 1970s and 1980s, processes in the bloodstream and at the invasion site are of particular interest at present. That viable tumor cells are frequently detectable within the blood of cancer patients has been known for many years. Although many tumor cells are shed into the blood, only very few tumor cells form distant metastases. Only recently, however, the molecular basis of this phenomenon has been elucidated. If tumor cells lose the contact to their basement membrane at the site of the primary tumor, they switch toward apoptosis, and only if the signaling pathway leading to apoptosis, such as the p53 mediated one, is abrogated, can tumor cells survive within the circulation in order to form metastases *(7)*. The clinical relevance of circulating tumor cells has been reinvestigated using PCR analysis for the presence of genetic material derived from tumor cells. In addition to the problem of the viability of the circulating tumor cells mentioned above, further problems such as the presence of pseudogenes *(8)* and methodological problems that occur when working at the maximum sensitivity needed for single cell detection and preanalytical and statistical influences have to be considered *(9)*. Hence, the clinical significance of detecting pieces of circulating-tumor-derived DNA by PCR amplification remains unclear and has to be assessed more vigorously.

Another issue of fundamental importance is the question how the metastatic process is initiated and regulated. At the molecular level, the best understood model of carcinogens is the colorectal one. This famous model of colorectal carcinogenesis worked out by Vogelstein and his coworkers implies that a specific mutation occurs prior to the formation of metastasis *(10)*. This theory would imply that metastasis is primarily a somatic mutation, which is not or to a lesser extent subject to regulatory processes. Using the serial analysis of gene expression (SAGE) technique, Vogelstein's group has screened the expression of at least 45,000 different genes in colorectal cancer, colon cancer cell lines, and colonic mucosa *(11)*. The result of this extensive survey was surprising: Few genes were differently expressed and no specific metastasis associated gene was described. Hence, the opposite proposition that the formation of metastases is subject to cell regulatory mechanisms may at least in some cancers be true.

That regulatory mechanisms are indeed governing metastasis are best illustrated in solid neoplasms. Cancers are malignant tumors of epithelial origin. In simple epithelia such as the ones lining the gut lumen, lining cells are characterized by their attachment to the basement membrane. This attachment to the basement membrane mediated by hemidesmosomes and cell adhesion molecules such as integrins renders the cells to be immotile. This cellular behavior is in contrast to the mesenchymal cells: These are the cells of the embryonic connective tissue that are characterized by their ability to migrate. Tumor cells that are to become metastatic are undergoing the epithelial mesenchymal transition: This implies that regulatory mechanisms are present that enable tumor cells to change their phenotype from an epithelial to a mesenchymal one, which represents the migrating tumor cell in the bloodstream *(12)*. At the evasion site, the tumor cells have to reverse their behavior and form epithelial formations. Evidence for the shift of antigen expression during the sequence epithelial phenotype–mesenchymal phenotype–epithelial phenotype in the mature large metastasis has been shown for the cell adhesion molecule epithelial glycoprotein-2 *(13)*.

In vitro systems by their very nature can only mimic limited steps of the metastatic cascade, which can, however, be studied in depth. Because the rate-limiting step of the metastatic cascade is not known, it is often difficult to meaningfully interpret the results of an in vitro assay for the metastatic cascade as a whole. Hence, old fashioned animal models of metastasis, for which there is an increased demand *(14,15)*, are included in these volumes. While the in depth analysis is better done in an in vitro system, animal models of metastases offer not only the great advantage of analyzing the process as a whole, but also reflect the biological diversity of tumors much better than in vitro systems. From cell culture conditions in general and the two-dimensional growth pattern in particular, the heterogeneity of antigen expression as seen in vivo is most often not fully reflected in cell culture systems, but is much better observed in vivo in experimental animals *(16)*. However, the clinical significance of animal models has to be assessed critically as well. In murine breast cancer, the mammary tumor virus induces hormone-independent breast cancer. This does not reflect the clinical situation, where many breast cancers are hormone sensitive and where no evidence for a viral cause exists *(17)*.

Despite this complexity, the problem of metastasis formation can be resolved. It may well be that relatively few mechanisms are rate-limiting in this complex process and, hence, simple ways of predicting metastasis formation can be found that ultimately lead to novel therapies once a through road has been established. We hope that these two volumes provide some tools wih which to do this.

## References

1. Bailar, J. C. and Gornik, H. L. (1997) Cancer undefeated. *N. Engl. J. Med.* **336,** 1969–1574.
2. Sporn, M. B. (1996) The war against cancer. *Lancet* **347,** 1377–1381.
3. Hart, I. R., Goode, N. R., and WIlson, R. E. (1989) Molecular aspects of the metastatic cascade. *Biochim. Biophys. Acta* **989,** 65–84.
4. Hart, I. R. and Saini, A. (1992) Biology of tumour metastasis. *Lancet* **339,** 1453–1457.
5. Liotta, L. A., Rao, N. C., Barsky, S., and G. Bryant (1984) The laminin receptor and basement membrane dissolution: role in tumour metastasis. *Ciba Foundation Symposium* **108,** 146–154.
6. Takeishi, M. (1993) Cadherins in cancer: implications for invasion and metastasis. *Current Opin. Cell Biol.* **5,** 806–811.
7. Nikiforov, M. A., Hagen, K., Ossovaskaya, V. S., Connor, T. M. F., et al. (1996) p53 modulation of anchorage independent growth and experimental metastasis. *Oncogene* **13,** 1709–1719.

8. Nollau, P., Jung, R., Neumaier, M., and Wagener, C. (1995) Tumour diagnosis by PCR-based detection of tumour cells. *Scand. J. Clin. Lab. Invest. Suppl.* **221,** 116–121.
9. Jung, R., Ahmad-Nejad, P., Wimmer, M., Gerhard, M., Wagener, C., and Neumaier, M. (1997) Quality management and influential factors for the detection of single metastatic cancer cells by reverse transcriptase polymerase chain reaction. *Eur. J. Clin. Chem. Clin. Biochem.* **35,** 3–10.
10. Fearon, E. R. and Vogelstein, B. (1990) A genetic model for colorectal tumorigenesis. *Cell* **61,** 759–767.
11. Zhang, L. Zhou, W., Velculescu, V. E., Kern, S. E., Hruban, R. H., Hamilton, S. R., Vogelstein, B., and Kinzler, K. W. (1997) Gene expression profiles in normal and cancer cells. *Science* **276,** 1268–1272.
12. Birchmeier, C., Birchmeier, W., and Brand-Saberi, B. (1997) Epithelial-mesenchymal transitions in cancer progression. *Acta Anat.* **156,** 217–226.
13. Jojovic, M., Adam E., Zangemeister-Wittke, U., and Schumacher, U. (1998) Epithelial glycoprotein-2 (EGP-2) expression is subject to regulatory processes in epithelial-mesenchymal transitions during metastases: an investigation of human cancers transplanted into severe combined immunodeficient mice. *Histochem. J.* **30,** 723–729.
14. Tannock, I. F. (1998) Conventional cancer therapy: promise broken or promise delayed? *Lancet* **351** (Suppl. II), 9–16.
15. Lane, D. (1998) The promise of molecular oncology. *Lancet* **351,** (Suppl II), 17–20.
16. Schumacher, U., Mohamed, M., and Mitchell, B. S. (1996) Differential expression of carbohydrate residues in human breast and colon cancer cell lines grown in vitro and in vivo in SCID mice. *Cancer J.* **9,** 247–254.
17. Van de Vijver, M. J. and Nusse, R. (1991) The molecular biology of breast cancer. *Biochem. Biophys. Acta* **1072,** 33–50.

# Contents

## Contents of the Companion Volume

### Metastasis Research Protocols

### *Volume II: Analysis of Cell Behavior In Vitro and In Vivo*

# Contributors

AMANDA J. ATHERTON • *LICR/UCL Breast Cancer Laboratory, Department of Surgery, Royal Free and University College Medical School, London, UK*

CHRISTINE BLANCHER • *Institute of Molecular Medicine, John Radcliffe Hospital, Oxford, UK*

STEPHAN BRAUN • *Department of Obstetrics and Gynaecology, Technical University, Munich, Germany*

SUSAN A. BROOKS • *Research School of Biological and Molecular Sciences, Oxford Brookes University, Oxford, UK*

IAN D. BULEY • *Department of Cellular Pathology, John Radcliffe Hospital, Oxford, UK*

RICHARD S. CAMPLEJOHN • *Richard Dimbleby Department of Cancer Research, Rayne Institute, St. Thomas Hospital, London, UK*

MARINA CHEKMAREVA • *University of Chicago, Chicago, IL*

CATHERINE CLARKE • *LICR/UCL Breast Cancer Laboratory, Department of Surgery, Royal Free and University College Medical School, London, UK*

GIULIO FRANCIA • *Richard Dimbleby Department of Cancer Research/ICRF Laboratory, Rayne Institute, St. Thomas Hospital, London, UK*

RAFAEL FRIDMAN • *Department of Pathology, Wayne State University, Detroit, MI*

CHISATO FUJIYAMA • *Molecular Angiogenesis Laboratory, Imperial Cancer Research Fund, University of Oxford, Institute of Molecular Medicine, John Radcliffe Hospital, Oxford, UK*

GIAMPIETRO GASPARINI • *Division of Medical Oncology, Azienda Complesso Ospedaliero "S. Filippo Neri," Rome, Italy*

CHERYL E. GILLETT • *Hedley Atkins/ICRF Breast Pathology Laboratory, Guy's Hospital, London, UK*

HAKAN GOKER • *Gene Function and Regulation, Institute of Cancer Research, Chester Beatty Laboratories, London, UK*

KATJA HAACK • *Medical Clinic, Department of Clinical Chemistry, University Hospital Eppendorf, Hamburg, Germany*

DEBBIE M. S. HALL • *Research School of Biological and Molecular Sciences, Oxford Brookes University, Oxford, UK*

ADRIAN L. HARRIS • *Molecular Angiogenesis Group, Imperial Cancer Research Fund, Institute of Molecular Medicine, John Radcliffe Hospital, Oxford, UK*

LAWRENCE JOHN • *Division of Histopathology, Department of Pathology and Microbiology, University of Bristol, Bristol Royal Infirmary, Bristol, UK*

ADAM JONES • *Imperial Cancer Research Fund, Molecular Angiogenesis Laboratory, University of Oxford, Institute of Molecular Medicine, John Radcliffe Hospital, Oxford, UK*

RUEY HO KAO • *Richard Dimbleby Department of Cancer Research/ICRF Laboratory, Rayne Institute, St. Thomas Hospital, London, UK*

HERMANN KNEITZ • *Institute of Pathology, Hannover Medical School, Hannover, Germany*

HANS KREIPE • *Institute of Pathology, Hannover Medical School, Hannover, Germany*

RICHARD LILISCHKIS • *Institute of Pathology, Hannover Medical School, Hannover, Germany*

MICHAEL NEUMAIER • *Medical Clinic, Department of Clinical Chemistry, University Clinic Hamburg-Eppendorf, Hamburg, Germany*

KLAUS PANTEL • *Department of Obstetrics and Gynaecology, University Hospital Humburg-Eppendorf, Hamburg, Germany*

MASSIMO PIGNATELLI • *Division of Histopathology, Department of Pathology and Microbiology, University of Bristol, Bristol Royal Infirmary, Bristol, UK*

RICHARD POULSOM • *In Situ Hybridisation Service, Imperial Cancer Research Fund, London, UK*

GIROLAMO RANIERI • *Division of Medical Oncology, Azienda Ospedali Riuniti, Reggio Calabria, Italy*

CARRIE W. RINKER-SCHAEFFER • *University of Chicago, Chicago, IL*

DEREK E. ROSKELL • *Department of Cellular Pathology, John Radcliffe Hospital, Oxford, UK*

REBECCA ROYLANCE • *Molecular and Population Genetics Laboratory, Imperial Cancer Research Fund, London, UK*

UDO SCHUMACHER • *Institute for Anatomy, Department of Neuroanatomy, University Hospital Hamburg-Eppendorf, Hamburg, Germany*

JONATHAN A. SHERRATT • *Department of Mathematics, Heriot Watt University, Edinburgh, UK*

Janet Shipley • *Molecular Cytogenetics, Haddow Laboratories, Institute of Cancer Research, Surrey, UK*

Kenneth Smith • *Imperial Cancer Research Fund, Molecular Angiogenesis Laboratory, University of Oxford, Institute of Molecular Medicine, John Radcliffe Hospital, Oxford, UK*

Krishna Soondrum • *Medical Clinic, Department of Clinical Chemistry, University Hospital Eppendorf, Hamburg, Germany*

Thelma Tennant • *University of Chicago, Chicago, IL*

Marta Toth • *Department of Pathology, Wayne State University, Detroit, MI*

Helen E. Turner • *Department of Endocrinology Radcliffe Infirmary, Oxford, UK*

# I

# Detection Methods for Cellular Markers of Metastatic Potential

# 1

# Histopathological Assessment of Metastasis

**Derek E. Roskell and Ian D. Buley**

## 1. Introduction

In spite of advances in the fields of immunohistochemistry and molecular biology, in clinical practice much of the assessment of metastases still relies on light microscopy using conventional histological stains. This is not so much a reflection of a reluctance by histopathologists to adopt new techniques, but more an indication that for most malignancies an enormous amount of useful prognostic data can be gained from relatively unsophisticated assessment of tissues, and that many of the strongest studies of prognostic factors in malignancy predate the era of molecular diagnostics. Although it is undoubtedly true that newer techniques have added prognostic information in the assessment of many tumors, and many, such as the measurement of estrogen receptor status in breast cancer, could be considered routine, a skilled assessment of the morphology of the tissues still provides the fundamental basis of assessing prognosis in the vast majority of cases.

## 2. Metastatic Potential of Primary Tumors

Probably the most important factor in determining the metastatic potential of a primary tumor is the correct identification of the tumor type. Some tumors, such as basal cell carcinoma of the skin, are very unlikely to metastasize in any circumstance, whereas others, such as small cell carcinoma of the lung, metastasize in almost every case. For these tumors, simply making the diagnosis usually offers an adequate assessment of metastatic potential. However, the majority of common malignancies fall into a group in which the incidence of metastasis varies considerably, and a more detailed assessment must be undertaken in order to establish the prognosis in each individual case.

From: *Methods in Molecular Medicine, vol. 57:*
*Metastasis Research Protocols, Vol. 1: Analysis of Cells and Tissues*
Edited by: S. A. Brooks and U. Schumacher © Humana Press Inc., Totowa, NJ

Although for some of these malignant tumors the presence of metastatic disease may be obvious at presentation, it is well known that even therapeutic interventions that seem to remove the whole primary tumor in the absence of overt metastatic disease do not always, and in many cancers seldom result in long-term cure.

### 2.1. Stage and Grade

Histopathological assessment of the excised primary tumor to predict the likelihood of recurrence or metastasis generally involves assessing two aspects of the tumor's growth: stage and grade. Although often confused, these simple terms are quite different. The stage simply refers to how far the tumor has spread, whereas the grade refers to its perceived aggressiveness, regardless of how far it has gone.

Staging a malignancy from assessment of the primary site has its limitations as it will not detect the presence of tumor in tissues left in the patient. However, there are clues that can point to the likelihood of tumor cells having escaped. These include the proximity of the surgical margin, the size of the tumor, the presence of lymphatic or vascular invasion, and the invasion of the tumor through structures that are known to provide physical barriers to keep the tumor from areas rich in lymphatics. Thus, the invasion of a colonic carcinoma into and through the muscle of the bowel wall confers a worse prognosis, as does the invasion of a malignant melanoma of the skin into the deeper layers of the dermis which contain the lymphatics. These are aspects of staging that can be assessed from the resected tissue. Numerous clinical and pathological staging proformas have been developed, the most well known of which is the TNM (tumor, nodes, metastasis) system. In this case, numerical values are given to the aspects of the tumor that measure its stage in terms of the primary tumor (size, spread into surrounding tissues), lymph nodes (number and site of affected nodes), and presence or absence of distant metastases, so that, for example, a colonic carcinoma that has invaded the submucosa and has spread to four pericolic nodes with no evidence of distant metastasis would be staged as T1, N2, M0. The score for the individual TNM components can be used to place the tumor into a reliable prognostic group, in this case stage III out of IV. As would be expected, most weight in this numerical "stage grouping" is given to scores for distant (M) and lymph node (N) metastasis (*see* **Table 1**).

The morphological grading of tumors relies on assessing their rate of growth and their degree of differentiation. Growth rate may be assessed by counting mitotic figures in the histological sections, usually expressed as the number of mitoses per 10 standard high power fields on the microscope, but other surrogate markers of growth rate also contribute to assessment of grade. Necrosis is thought to be a reflection of a fast growing tumor outgrowing its blood supply, so the presence of necrosis usually points to a higher grade.

**Table 1**
**Summary of TNM Staging of Colonic Carcinoma**

| | |
|---|---|
| *Primary tumor T* | |
| Tis | Carcinoma *in situ* or invasion of *lamina propria* only |
| T1 | Invasion of submucosa |
| T2 | Invasion of external muscle (*muscularis propria*) |
| T3 | Invasion through *muscularis propria* into subserosa or pericolic tissue |
| T4 | Invasion of other organs or structures or through visceral peritoneum |
| *Lymph node N* | |
| N0 | No lymph node metastasis |
| N1 | Metastasis in one to three pericolic nodes |
| N2 | Metastasis in four or more pericolic nodes |
| N3 | Metastasis to any node on a major (named) vascular trunk |
| *Distant metastasis M* | |
| MX | Cannot be assessed |
| M0 | No distant metastasis |
| M1 | Distant metastasis |

| *Stage grouping* | | | |
|---|---|---|---|
| 0 | Tis | N0 | M0 |
| I | T1 | N0 | M0 |
| | T2 | N0 | M0 |
| II | T3 | N1 | M0 |
| | T4 | N0 | M0 |
| III | Any T | N2 | M0 |
| | Any T | N2 | M0 |
| | Any T | N3 | M0 |
| IV | Any T | Any N | M1 |

The degree of differentiation of a malignancy refers to how much it resembles normal tissue. A well-differentiated tumor is of a lower grade than a poorly differentiated one, and the less the tissue resembles normal, the more likely it is to have lost the factors that allow organization and adhesion of the cells, and these changes correlate with metastatic potential. An anaplastic tumor is so poorly differentiated that no recognizable feature is present to point to its tissue of origin. Pleomorphism among the nuclei of tumor cells refers to the degree of difference seen within the population of tumor cells. In a normal tissue, the cell nuclei largely resemble each other, and severe pleomorphism in a tumor may be a reflection of genetic diversity and mutations in the tumor cells. The more pleomorphism is present, the more likely there is to be a subset of the tumor cells that will develop the genetic capability to metastasise.

The mitotic count, degree of differentiation, and nuclear pleomorphism can all be used to place a malignant tumor into a grade. For many more common malignancies, such as breast carcinoma, these features are scored numerically and combined to give an assessment of grade that is as reproducible and standardized as possible (*see* **Table 2**).

Other factors that are useful in the grading of some tumors are the degree of inflammatory response, which implies a better prognosis in colonic cancer, and the pattern of the invading margin, in which a worse prognosis may be associated with an infiltrative, rather than a compressive, expansile margin. In many tumors, differentiation toward a particular tissue type may be as significant as lack of differentiation (anaplasia) in establishing a poor prognosis. A good example of this is differentiation towards trophoblastic tissue. The normal function of trophoblast is to form the part of the placenta that invades the wall of the uterus to gain nutrients from the mother's blood. Indeed, in normal pregnancy trophoblastic cells can be found circulating in the maternal bloodstream. Trophoblastic differentiation in a malignant tumor can confer an ability for highly aggressive invasion and bloodborne spread.

## 3. Metastasis to Lymph Nodes

Metastasis to lymph nodes forms part of the staging of malignancies. Prognosis may be affected not just by the presence of lymph node metastasis, but also by the number of nodes involved, their site, the size of the deposits, and the extension of the tumor through the capsule of a node. Detailed assessment of lymph nodes is therefore important in establishing the prognosis and the possible need for further treatment, such as chemotherapy.

The assessment of lymph nodes in an excised surgical specimen is relatively straightforward, although there are some diagnostic problems that may face the pathologist.

### *3.1. Morphology of Lymph Node Metastasis*

Metastatic tumor arrives at a lymph node via the afferent lymphatics which empty into the subcapsular sinus, and it is here that the majority of "early" metastases are first seen. The pattern of growth usually resembles the primary tumor quite closely, and very well differentiated tumors may appear histologically benign but for the fact that they have spread to a node. Diagnosis is difficult, however, when poorly cohesive individual cells spread into the substance of the node and epithelial structures are not produced. This is frequently the case with lobular carcinoma of the breast in which the malignant cells may resemble macrophages and lymphoid cells. In such situations histochemical stains, in this case for mucin, or immunohistochemical stains (for cytokeratins) may reveal a far greater metastatic burden than would otherwise be suspected.

**Table 2**
**Histological Grading of Breast Carcinoma**

| Tubule formation (score 1–3) | Nuclear pleomorphism (score 1–3) | Mitotic count (using nomogram, score 1–3) |
|---|---|---|
| 1 Majority of tumor | 1 Mild | 1 Low |
| 2 Moderate | 2 Moderate | 2 Moderate |
| 3 Little or none | 3 Severe | 3 High |

Overall scores: Grade 1, 3–5 points; Grade 2, 6–7 points; Grade 3, 8–9 points.

Equally, it is important to recognize that proliferation of epithelioid cells in the subcapsular sinus does not imply metastasis. A frequent finding in lymph nodes, including those draining sites of malignancy, is sinus histiocytosis, a totally benign proliferation of macrophages that can be mistaken for malignancy. The diffuse expansion of the sinuses throughout the node suggests this reaction pattern, but in difficult cases immunohistochemistry for a macrophage marker such as CD68, or cytokeratins allows any difficulty to be resolved. Techniques for immunocytochemistry are described in Chapter 2 by Brooks.

A pathological curiosity that may be mistaken for metastasis is the presence of ectopic normal tissue within a lymph node structure. Thus, normal salivary gland tissue may be mistaken for metastatic carcinoma in a neck node, and nests of melanocytic nevus cells can be found both within lymph nodes and lymphatic vessels, raising the diagnostic question of malignant melanoma. Malignancies of these cell types may present as lymph node metastasis with an occult primary, so a recognition of the benign nature of these ectopic tissues will prevent damaging and unnecessary investigations or therapy.

Prognostically, the detection of small or occult metastases is an area of some controversy. Clearly, because conventional histological sections are just four thousandths of a millimeter thick, a single section will miss a proportion of small deposits, and the more sections that are examined, and the more special techniques used to detect tumor cells, the greater the number of metastases detected. However, because most prognostic data have been collected using simple histological methods, these simple methods can be used to give useful information, even if more involved techniques would detect more metastatic cells. Indeed, for some malignancies, it is unclear if the detection of micrometastases confers a worse prognosis.

The staging of malignancies of lymphoid origin, lymphomas, can present particular difficulties when assessing lymph nodes for the presence of disease, as often the malignant cells resemble a phase of the normal differentiation of lymphoid cells. Distinguishing malignancy in more obvious cases depends on assessing the growth pattern of the cells within the node, but in more subtle cases may involve establishing the presence of immunohistochemical markers either of a particular stage in lymphoid differentiation, or of genetic damage such as a chromosomal translocation.

## 4. Diagnosis of Metastasis to Other Sites

Much of what has been discussed regarding lymph node metastasis applies to metastasis to other sites. Particular problems may be encountered, however, in distinguishing metastasis in an organ from a primary malignancy at that site. This is obviously only a difficulty if the metastasis is of a type that could occur there as a primary, so, for example, the only problem when finding a deposit of adenocarcinoma in bone is in determining the primary site, as primary adenocarcinoma does not arise in bone, but an adenocarcinoma involving the ovary could be primary or metastatic. Methodology for detection of individual micrometastatic tumor cells in bone marrow is given in Chapter 5 by Braun and Pantel.

Clues to a deposit of carcinoma in a lung, liver, or elsewhere being a primary include the absence of an overt alternative primary site, and the presence of a single tumor deposit, as in many cases multiple deposits imply metastasis within the organ. However, primary carcinomas, for example, a primary liver carcinoma, may seed metastases within the same organ. This is not particularly surprising as tumors frequently demonstrate tissue tropism in the distribution of metastasis. Certain cell types prefer to grow in a particular environment, and malignant liver cells might be expected to settle preferentially and establish metastatic deposits in the liver. Another situation in which multiple primary tumors may be seen are some of the genetic "cancer syndromes," in which predisposed individuals are at high risk of developing certain malignancies, and frequently suffer multiple primary tumors in one or more organs at the same time. Probably the best indicator of primary, rather than metastatic disease, is the presence of *in situ* carcinoma adjacent to the invasive tumor. *In situ* carcinoma or severe dysplasia is essentially a preinvasive "malignancy," which is thought in many cases to be an important stage in the development of clinical disease.

## 5. Establishing the Site of an Unknown Primary Tumor

An enlarged lymph node may be the first presentation of malignancy. In such cases, the major distinction to be made is between a reactive enlargement, a primary tumor of the lymph node (almost always a lymphoma), and a metastatic deposit. Fine needle aspiration cytology is a rapid and minimally invasive investigation that allows such a distinction to be made in many cases. In clinical practice, a standard procedure following a fine-needle aspiration diagnosis of probable lymphoma would be to excise the node and submit it to detailed histological and immunohistochemical analysis to place the lymphoma into a precise diagnostic category. A diagnosis of metastatic carcinoma, however, is not generally followed by excision of the node, unless the enlarged node is causing, or threatens, local complications. Instead a detailed consideration of the patient's clinical history and examination, followed by investigations, particularly radiological imaging, directed at identifying the primary site is undertaken. Sometimes the pattern of metastatic disease strongly suggests a particular primary site, reflecting tissue tropism. Small cell carcinoma of the lung, for example, is a frequent cause of metastasis to the adrenal gland, and may present in the adrenal before a lung primary is detected. Nevertheless, in a proportion of cases the pathologist is faced with an excised lymph node or a biopsy containing a metastasis from an unknown primary. Even when considering only a few cells from a fine-needle aspirate, some suggestion of likely primary sites can usually be made.

A few malignancies are immediately recognizable from their characteristic morphology in tissue sections. These are, however, the exceptions, and more often the pathologist is able to place the tumor into a category by its type of differentiation, for example, squamous or adenocarcinoma, which narrows down the likely primary sites. Within these groups there are additional features, such as calcified psammoma bodies which are associated with papillary carcinomas of thyroid and ovary, or the clear cell morphology associated with renal carcinoma, the identification of which narrows the broad category of adenocarcinoma to a more focused area. However, few morphological features are absolutely specific, and increasingly the use of tissue markers such as prostate specific antigen detected either by immunohistochemistry or as a raised level in the patient's blood can be expected to offer a more directed suggestion of primary site in patients presenting with metastatic disease.

## 6. Histopathology and the Clinical Detection of Metastasis

Clinical or radiological (X-rays and scans) detection of new mass lesions in a patient with a history of malignancy is likely to indicate metastatic disease. However, as the diagnosis of metastasis has considerable implications for the future treatment and life expectancy of the patient it is important to remember that a conclusive tissue diagnosis should be made wherever possible. Enlarged lymph nodes in a patient with malignancy do not in themselves imply metastasis, although they may be very suggestive. The presence of inflammation associated with a primary tumor, even a totally benign one, may lead to reactive enlargement of regional lymph nodes. This is particularly likely following biopsy or surgery to the primary tumor, as tissue damage from these procedures can lead to lymphadenopathy.

Cytology is a diagnostically useful and minimally invasive means of diagnosing metastasis at most sites, often using radiological (ultrasound or CT scan) imaging as guidance for fine-needle aspiration, if the lesion is in a deep location where it cannot be felt. Pleural effusions, urine, sputum, and other fluids are easily collected and can also be examined for the presence of malignant cells, although multiple samples may be necessary. Core biopsy is a widely used alternative to fine-needle aspiration that removes a small cylinder of intact tissue about 1–2 mm in diameter. This has the advantage of keeping the morphology of the tissue intact, but the disadvantage of an increased risk of hemorrhage or perforation of organs by the larger needle. Occasionally, the difficult location of a putative metastatic deposit (e.g., the brain), and overwhelmingly supportive radiological evidence, reduce the requirement for a tissue diagnosis, but there will always be cases in which the multiple liver deposits seen on ultrasound, the "hot spot" on a radioisotopic bone scan, and the shadowing on the chest X-ray turn out not to be the metastases that seemed so obvious.

## 7. Intraoperative Diagnosis of Metastasis

Occasionally, particularly rapid diagnosis of metastasis is required if an unexpected deposit is found during surgery. The progress of the operation and the type of surgery performed may depend on whether or not the lesion found distant from the main tumor is a metastasis, or whether the surgical resection margin is free of tumor. In such cases, cytology is sometimes an option, but the usual diagnostic method where possible is to remove a small piece of tissue and submit it for frozen section histology.

Conventional processing of tissues for histology involves fixation in formalin followed by chemical processing to embed the dehydrated tissue in a paraffin wax block, from which sections are cut with a microtome, dewaxed, and stained. This processing takes considerable time, particularly for larger pieces of tissue, and even the most rapid processing takes several hours. Frozen section avoids the fixation, processing, and wax embedding stages, shortening the process to a few minutes. The fresh tissue is rendered solid and therefore able to be cut into thin slices by freezing in liquid nitrogen. Sections are cut on a cryostat, which is essentially a microtome in a refrigerated cabinet, and the sections can be dried and stained immediately. Although the technique is relatively fast, the major disadvantages are that the morphology of the tissues is poorly displayed compared to paraffin sections, and that only relatively small pieces of tissue can be successfully cut in this way. In many cases metastatic tumor can be reliably identified from a frozen section, but more subtle examples may be missed and there is also a significant risk of false-positive diagnosis.

Apart from the occasional need for immediate diagnosis, surgical access at the time of primary tumor resection offers opportunities for assessing the spread of tumor and therefore contributing to accurate staging of the disease. Needle core biopsy of local or distant lymph nodes or other lesions may be possible if they can be seen or felt, and isotonic fluid washings from, for example, the peritoneal cavity, can be examined for the presence of malignant cells.

## 8. When Metastasis Is Not Malignant

Up to this point, the discussion has assumed that a tumor that has spread to lymph nodes or a distant site is by definition malignant. Although this is almost universally correct, there are some interesting exceptions that may cause diagnostic difficulty. These fall into two categories: benign tumors that spread and tumors that were once malignant but no longer pose a threat to the patient.

The first group of benign tumors that spread is largely made up of tumors that grow inside blood vessels, and can break off and be carried to a distant site. A good example is atrial myxoma, which is a benign tumor growing inside the heart. This passive embolization of a tumor is not equivalent to the active invasion of blood vessels by a malignancy, and although tumor emboli can cause disease by obstructing blood vessels, and they can continue to grow within vessels at the distant site, they do not break out of the blood vessel and invade surrounding tissues. Thus, the detection of such a tumor away from its primary site does not constitute metastasis in the malignant sense, and does not have the prognostic or therapeutic implications of malignancy.

The second unusual group is made up largely of tumors originating from germ cells or from precursor tissues of developing organs in children (so-called "blastomas"). The primitive, poorly differentiated nature of these malignancies is to some extent different from conventional poorly differentiated malignancies. In this case proliferation of the primitive cells from which organs originate produces a tumor with little visible differentiation, but the cells may retain their normal capacity to mature into fully differentiated adult tissues. Apparent maturation of these tumors may be seen following chemotherapy, which kills the primitive cells but has little effect on the differentiated ones, so that subsequent biopsy of a metastatic lesion may demonstrate only the mature, fully differentiated tissue which, although it undoubtedly represents metastasis, is no longer a threat to the patient.

## Further Reading

Cotran, R. S., Kumar V., and Collins, T., eds. (1999) *Robbins Pathologic Basis of Disease*, 6th ed. W.B. Saunders, Philadelphia.

Rosai, J., ed. (1996) *Ackerman's Surgical Pathology*, 8th ed. Mosby, St. Louis, MO.

2

# Basic Immunocytochemistry for Light Microscopy

**Susan A. Brooks**

## 1. Introduction

### 1.1. What is Immunocytochemistry?

Immunocytochemistry may be defined as the identification of a cell- or tissue-bound antigen *in situ*, by means of a specific antibody–antigen reaction, tagged microscopically by a visible label. Successful immunocytochemistry therefore requires (1) preservation of the antigen in a form that is recognizable by the antibody, (2) a suitable antibody, and (3) an appropriate label. The basic technique was first described by Coons and colleagues ***(1–3)***, who employed antibody directly labeled with a fluorescent tag to identify antigen in tissue sections. Since that time, the technique has been refined and expanded enormously. Some significant developments include the use of horseradish peroxidase ***(4)*** and alkaline phosphatase ***(5)*** as label molecules; the development of many, increasingly sensitive, multilayer detection methods; and exploitation of the strong binding between avidin and biotin in detection techniques ***(6,7)***.

### 1.2. Range of Applications

Immunocytochemistry is appropriate for a remarkably wide range of applications. Any cell- or tissue-bound immunogenic molecule can, theoretically, be detected *in situ* using the technique. It is a technique of particular interest in metastasis research, as it facilitates the detection of virtually any molecule of interest to the researcher in samples of tumor or normal tissues or cells. Of particular interest in this field is the heterogeneity in expression by cells within a morphologically homogeneous tumor mass, or between normal vs cancer cells. The gain or loss of expression of certain antigens by tumor cells at different stages in the natural history of the disease and in relation to metastatic potential is also of great relevance. For example, loss of expression of cell

From: *Methods in Molecular Medicine, vol. 57: Metastasis Research Protocols, Vol. 1: Analysis of Cells and Tissues*
Edited by: S. A. Brooks and U. Schumacher © Humana Press Inc., Totowa, NJ

adhesion molecules by tumor cells may be instrumental in their breaking away from the primary tumor mass. Immunocytochemistry is the only technique that allows detection of such molecules *in situ*. Examples of immunocytochemistry as applied to metastasis research applications are explored in more depth in a number of chapters in this volume, including Chapter 5 by Braun and Pantel on immunocytochemical detection and characterization of individual micrometastatic tumor cells, Chapter 6 by John and Pignatelli on assessment of integrin expression, Chapter 8 by Turner and Harris on the measurement of microvessel density in primary tumor, Chapter 9 by Gillett on assessment of cellular proliferation, Chapter 13 by Kilic and Ergün in the companion volume on methods to evaluate the formation and stabilization of blood vessels and their role for tumor growth and metastasis, and Chapter 14 in the companion volume on galectin-3 binding and metastasis by Nangia-Makker et al.

### 1.3. Types of Cell and Tissue Preparations

As the first requirement for successful immunocytochemistry is preservation of the antigen, the primary consideration must be what type of cell or tissue preparation to employ. Immunocytochemistry can be performed on a range of different cell and tissue preparations of interest in metastasis research, including cell suspensions, cell smears, frozen (cryostat) sections and fixed, paraffin wax-embedded sections. Some of the advantages and disadvantages of these types of preparation are summarized in **Table 1**.

Of particular interest may be the application of immunocytochemistry to routinely formalin-fixed, paraffin wax-embedded archival tissues to facilitate mapping of expression of molecules of interest retrospectively with the benefit of long-term patient follow-up. Many antigens are well preserved for many years in such tissues, and retrospective analysis can be successfully carried out after years (possibly decades) of tissue storage, making this a powerful and informative approach. The only limitations are, first, that lipids are dissolved out and lost during processing to paraffin wax, and therefore their antigenic structures are not present. Second, an antibody is required that successfully recognizes antigen preserved in this manner. Some antigens may be damaged, sequestered, or altered by fixation and processing to paraffin wax, and many antibodies will therefore no longer recognize them and will give successful results only on fresh, frozen (cryostat) tissue sections or fresh cell preparations. Enzyme and heat-mediated antigen retrieval techniques are described later in this chapter and may often be very successful in partially, or fully, reversing the alterations caused by fixation and processing and facilitate successful detection of otherwise undetectable antigens. The relatively recent popularity of heat-mediated antigen retrieval (e.g., *see* **refs. *8–11***) has vastly expanded the repertoire of antibodies that can be used successfully on fixed

**Table 1**
**Advantages and Disadvantages of Different Tissue Preparations**

| Preparation | Suitable for | Advantages | Disadvantages |
|---|---|---|---|
| Cell suspensions | Living cells, e.g., blood cells, cultured cells, cells released from solid tissue masses.<br><br>Direct method using fluorescent-labeled antibodies most suitable. | Unaltered antigen expression in the living cell seen. Excellent for cell surface antigens. | Not suitable for demonstration of cytoplasmic antigens. Cells seen in isolation; no indication of tissue distribution of antigen. |
| Cell smears | Any living cells in suspension, e.g., blood, cultured cells.<br><br>Any staining method is suitable. | Quick and easy.<br><br>Good for cytoplasmic antigens. | Morphology sometimes indistinct. |
| Frozen sections | Any fresh solid animal or human tissue.<br><br>Any staining method is suitable. | Relatively quick.<br><br>Fairly good morphology; spatial relationships of cells within tissues seen. Good for cytoplasmic and cell surface antigens. | Technically more demanding than suspensions or smears. |
| Paraffin sections | Any solid animal or human tissue.<br><br>Any staining method is suitable. | Tissue preserved indefinitely. Excellent morphology. Relationships between cells in tissues seen.<br><br>Cell surface and cytoplasmic antigens seen. | More time consuming than other methods.<br><br>Glycolipids lost.<br><br>Fixation and processing may damage some antigens. |

and processed tissues. Expression of molecules by preparations of cultured cell lines or of cell suspensions from body fluids such as ascites, blood, or pleural effusions may also be of particular interest in metastasis research, and immunocytochemistry on such preparations in the form of cytospins, cell smears, or cells cultured on coverslips is generally very successful.

For the simpler, quicker immunocytochemical methods, cell and tissue preparations will adhere well to clean, dry glass microscope slides. For longer, multistep techniques, use of an adhesive is recommended. Silane treatment, described in this chapter, is probably one of the most effective adhesives available and is cheap and simple to use. Its use is essential if heat-mediated antigen retrieval methods are going to be used subsequently.

### 1.4. Choice of Antibody

The second requirement for successful immunocytochemistry is the availability of a suitable antibody directed against the antigen of interest. Detailed description of the raising and production of antibodies lies beyond the scope of this volume, but, in brief, the choice lies between monoclonal and polyclonal antibodies. A huge range of both types of antibodies, directed against thousands of antigens of potential interest, are available commercially, and both types of antibodies can be produced ''in house'' if the appropriate facilities and expertise are available. Commercial companies exist that will produce ''tailor made'' antibodies directed against peptide sequences requested by the customer at (relatively) modest cost.

It is essential to choose an antibody appropriate for immunocytochemistry specifically, as antibodies developed for other applications, for example, enzyme-linked immunosorbent assay (ELISA), may simply not work well. It is also important to realize that all commercially available antibodies directed against the same molecule, or epitope on a particular molecule, may not be equally effective in immunocytochemistry and some ''shopping around'' may be helpful. Many commercial companies will provide small samples of antibodies free of charge for researchers to evaluate.

The choice of monoclonal or polyclonal antibody depends largely on what antibodies directed against the antigen of interest are available. Each type of antibody has specific advantages and disadvantages and it cannot be assumed that either polyclonal or monoclonal antibodies are invariably superior. In brief, polyclonal antibodies contain a cocktail of immunoglobulins directed against different epitopes of the antigen of interest, and other, irrelevant antigens also. They can therefore sometimes crossreact with molecules other than the one of interest and give spurious results or ''dirty'' background staining. The cocktail of immunoglobulins present may, however, react with multiple epitopes on the antigen molecule of interest, resulting in stronger and more effective labeling than achieved with a comparable monoclonal antibody. They are usually raised in rabbit (or sometimes other large animals such as goat or sheep; chicken antibodies are also gaining in popularity), and tend to be cheaper to buy than monoclonal antibodies. Monoclonal antibodies are usually raised in mice or rats, and, as the name suggests, represent immunoglobulins produced by a single immortalized clone of cells, and therefore directed against a single epitope. They tend to be more expensive than polyclonal antisera, but can sometimes be used at extremely high working dilutions. The great advantage of monoclonal antibodies is their absolute specificity, which means that labeling results are often very clean. Many modern monoclonal antibodies are raised to synthetic peptide sequences, which has the advantage that the precise epitope they recognise is known. Crossreactivity can sometimes occur even with monoclonal antibodies if the epitope they are directed against is shared by other, irrelevant molecules.

There are no hard and fast rules as to choice of antibody —for some applications, a particular monoclonal antibody may be ideal; for others a polyclonal antiserum may give better results. It is important to note what class of antibody is being used. Most monoclonal antibodies used in immunocytochemistry are of the immunoglobulin G (IgG) class, but some may be immunoglobulin M (IgM). This is an important consideration in many immunocytochemical methods, as detection of antibody binding to antigen may be achieved by subsequent reaction with a secondary antibody directed against the first—for example, to detect a monoclonal mouse IgG binding, a labeled secondary antibody raised in, for example rabbit, against mouse IgG may be applied. A secondary antibody directed against mouse IgM would not be appropriate in this example.

For any application, the appropriate dilution of antibody must be determined. This can usually be ascertained only by performing a range a of dilutions and checking which gives optimum results in terms of strong specific labeling coupled with clean background. When using polyclonal antibodies, doubling dilutions are convenient, ranging from, perhaps, as a rough guide, 1:50–1:3200. Monoclonal antibodies may be tested in the range of, possibly, 1–50 μg/mL. In the more complex, multistep techniques, different dilutions of primary antibody and secondary labeling reagents may need to be titrated against each other in a ''checkerboard of dilutions'' to determine optimum working dilutions. An example of a typical ''checkerboard of dilutions'' is given in **Table 2**. One would expect the more concentrated solutions of primary and/or secondary antibody to give strong staining but unacceptably high background; too high a dilution of either primary or secondary antibody will yield low intensity, but probably very clean labeling. The optimum dilution of both in combination should yield deep, intense specific labeling allied to clean background and absence of nonspecific labeling. It is worth taking time and care over titration experiments to achieve optimal experimental results.

**Table 2**
**Determination of Optimum Dilution of Primary and Secondary Antibody[a]**

| Dilution of secondary antibody, e.g., swine antisera raised against rabbit immunoglobulins | Dilution of primary antibody, e.g., polyclonal rabbit antisera raised against the molecule of interest | | | | | | |
|---|---|---|---|---|---|---|---|
| | 1:50 | 1:100 | 1:200 | 1:400 | 1:800 | 1:1600 | 1:3200 |
| 1:50 | | | | | | | |
| 1:100 | | | | | | | |
| 1:200 | | | | | | | |
| 1:400 | | | | | | | |
| 1:800 | | | | | | | |

[a]Sample "checkerboard of dilutions" used to determine the optimum dilution of primary and secondary antibody for use in, for example, an indirect detection method. The optimum combination of both will yield strong, specific labeling with clean background and no nonspecific labeling.

### *1.5. Choice of Labels for Immunocytochemistry*

The third requirement for successful immunocytochemistry is the presence of a visible label. The choice of label usually lies between a fluorescent label or the colored product of an enzyme reaction, although other labels such as colloidal gold, silver, or ferritin can also be employed (usually in immunocytochemistry for electron microscopy, which is beyond the scope of this chapter).

Traditionally, the most commonly used fluorescent label is fluorescein isothiocyanate (FITC), which fluoresces a bright yellow-green ***(12)***. Alternatives include tetrarhodamine isothiocyanate (TRITC) and Texas red, which fluoresce red *(13)*. Many primary and secondary antibodies labeled with these compounds are available commercially. There is also an ever increasing range of other fluorescent labels that open up the possibility of multiple labeling, as described in Chapter 3 by Atherton and Clarke.

The most commonly used enzyme labels are horseradish peroxidase and alkaline phosphatase. Many primary and secondary antibodies and other immunocytochemical reagents labeled with these compounds are available. Other, less commonly used, enzyme labels include glucose oxidase and β-galactosidase. The principle of using any enzyme label is that its reaction with substrate plus a soluble chromagen yields a precipitated or insoluble colored product visible by light microscopy. For horseradish peroxidase reaction with hydrogen peroxide plus the chromogen diaminobenzidine (DAB) ***(14)*** yields a granular, brown, alcohol insoluble product or with 3-amino-9-ethylcarbazole (AEC) ***(15)*** yields a granular, red, alcohol-soluble product. Other chromagens are also available, but are less commonly employed. For alkaline phosphatase, reaction with naphthol phosphate as a substrate and a diazonium salt can yield a variety of colored—most typically red or blue—alcohol-soluble azo dyes as products ***(16)***.

Enzyme label detection kits, usually in the form of dropper bottles of concentrated reagents ready to be diluted in water or buffer, are commercially available as a convenient alternative to preparation of the necessary solutions "in house."

When enzyme labels are employed, the issue of endogenous, cell- or tissue-bound enzyme becomes an issue, and steps often need to be incorporated into the detection method to block endogenous enzyme prior to development of the final colored label product. It is worth noting that endogenous alkaline phosphatase is usually destroyed in processing to paraffin wax.

### 1.6. Range of Detection Methods Available

A number of fairly standard immunocytochemical techniques exist, which vary in terms of complexity and sensitivity. They range from the simple "direct" technique in which antigen is detected by the binding of a directly labeled antibody, through to much more complex, but highly sensitive multilayer techniques. Examples of a range of detection methods are outlined in this chapter and their relative advantages and disadvantages are summarized in **Table 3**. They are also represented diagramatically in **Figs. 1–5**. For any particular application, the choice of technique must usually be determined largely by trial and error. The simpler "direct" techniques are often employed for labeling of living cells—for example, cultured cells or cells from body fluids—as they are least likely to damage delicate cells. They are commonly used in conjunction with fluorescent labels, although enzyme labels can also be used. The more complex and sensitive, multilayer techniques are usually used in conjunction with more robust cell and tissue preparations and are particularly appropriate where antibody titres are low, or where antigen expression is scanty as the "layering" of reagent results in amplification of the final signal. Amplification is achieved because at every step, multiple reagent molecules have the opportunity to bind to the previous "layer," resulting eventually in a much amplified "cloud" of label molecules marking the initial binding of antibody molecule to antigen. This important point is not shown in the figures illustrating the methods as, for the sake of clarity, the "layers" are represented in a simplified, linear manner.

In addition to the methods listed in this chapter, similar approaches are described in detail in Chapter 4 by Brooks and Hall on the related technique of lectin histochemistry. Lectin histochemistry facilitates the detection of carbohydrate structures, as part of, for example, glycoproteins, glycolipids, or glycosaminoglycans, *in situ*, by means of their recognition by a lectin.

### 1.7. Controls

The incorporation of appropriate positive and negative controls is, naturally, of paramount importance. The most appropriate positive control is a cell or tissue preparation that is known to express high levels of the antigen of inter-

**Table 3**
**Advantages and Disadvantages of Different Detection Methods**

| | Advantages | Disadvantages |
|---|---|---|
| Direct method | Simplest method available. Quick. Limited number of reagents required. Works particularly well using fluorescent labelled antibody and cell suspensions. | Lacks sensitivity; therefore may not be appropriate for scantily expressed antigens. May suffer from high background. |
| Simple indirect method | Increased sensitivity over direct method (~20× more sensitive). Relatively quick and straightforward. | Requires more reagents than the direct method; therefore potentially mor expensive. Extra step, therefore takes longer. |
| Simple avidin-biotin method | Relatively quick and simple, but highly sensitive and yields clean labeling. | Endogenous biotin may confuse interpretation in some cases. Glycosylated avidin may be recognized by tissue bound lectins or bind charged sites non-specifically. |
| ABC method | Highly sensitive (at least 100× more sensitive than the direct method). Clean results. | Large ABC complex sometimes causes steric hindrance. Tissue lectins may bind glycosylated avidin; avidin may attach nonspecifically by charge Endogenous biotin may confuse interpretation. Cost implications of reagents. |
| PAP or APAAP methods | Highly sensitive (~50× more sensitive than direct method). | Very time consuming Cost implications of extra reagents. Largely superseded by ABC method. |

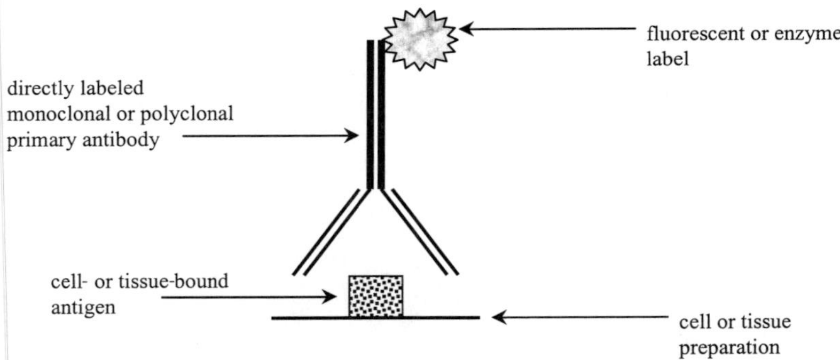


Fig. 1. Direct method. Cell- or tissue-bound antigen is detected by binding of drectly labeled primary antibody,

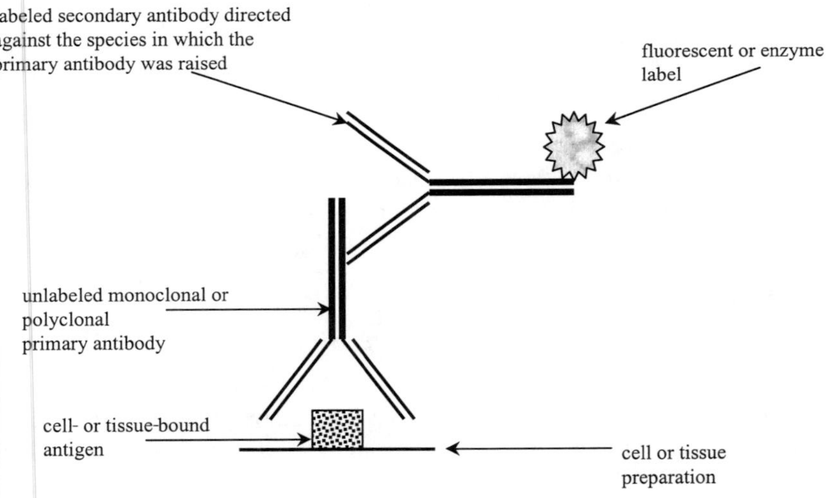


Fig. 2. Simple indirect method. Cell- or tissue-bound antigen is detected by binding of unlabeled primary antibody, then labeled secondary antibody directed against the species in which the primary antibody was raised.

est. The simplest negative control is to omit the primary antisera, and replace it with either buffer, preimmune serum at the same working dilution as the antibody, or an antibody directed against an irrelevant antigen. Specificity of binding can be confirmed by competitive inhibition in the presence of, or preabsorption of the antisera with, the antigen of interest.

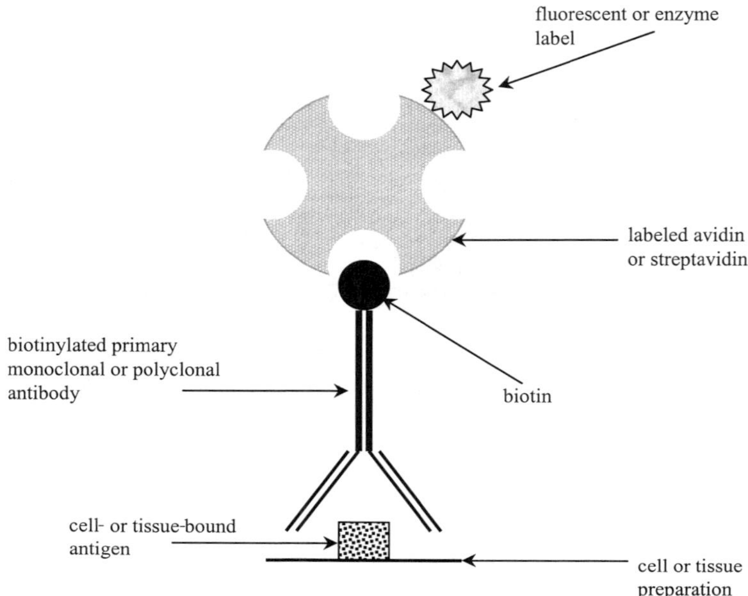


Fig. 3. Simple avidin–biotin method. Cell- or tissue-bound antigen is detected by binding of a biotinylated primary antibody, then labeled avidin or streptavidin.

## 2. Materials

### 2.1. Silane Treatment of Microscope Slides (see Note 1)

1. Acetone.
2. Acetone/silane solution: 2% v/v aminopropyltriethoxysilane or 3-(triethoxysilyl)-propylamine in acetone; make fresh on the day of use. Discard after treating a maximum of 1000 slides.
3. Distilled water: Discard and refresh after every five racks of slides have passed through.

### 2.2. Preparation of Cells Cultured on Coverslips

1. Cells cultured under standard conditions.
2. Fetal calf serum (FCS) cell free culture medium.
3. Alcohol- or autoclave-sterilized round glass coverslips (13 mm diameter, thickness 0).
4. Dental wax or Parafilm.
5. 0.1 *M* PIPES buffer, pH 6.9: Stir 12.1 g of PIPES (piperazine-*N*,*N'*-*bis*-[2-ethane-sulfonic acid]) into 50 mL of ultrapure water to give a cloudy solution. Add approx 40 mL of 1 *M* NaOH and the solution should clear. Check pH and adjust to 6.9, if necessary, using 1 *M* NaOH. Add ultrapure water to give a final volume of 400 mL.

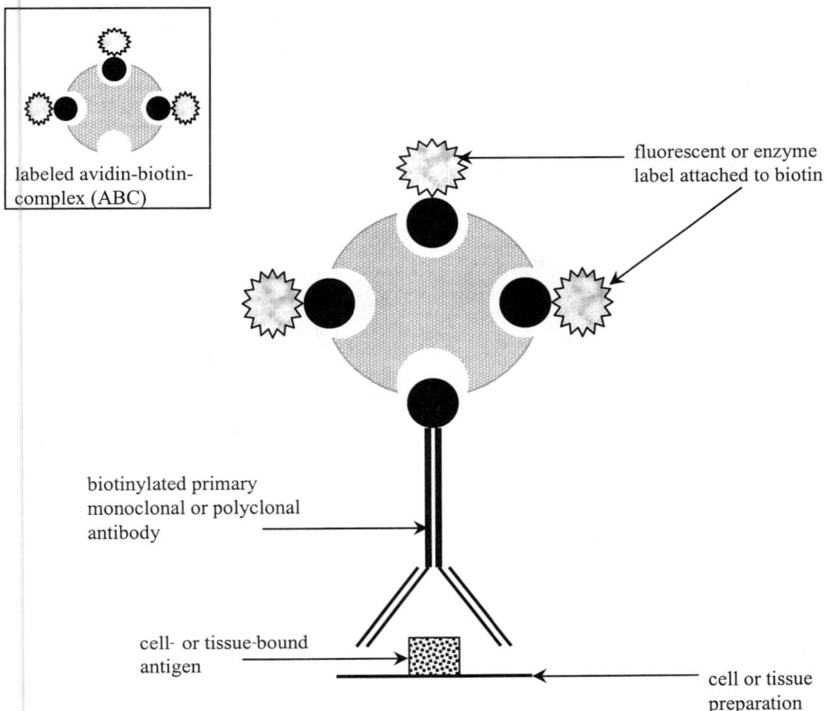


Fig. 4. Avidin–biotin complex (ABC) method. Cell- or tissue-bound antigen is detected by binding of a biotinylated primary antibody, then labeled ABC.

6. 3% v/v paraformaldehyde in 0.1 *M* PIPES buffer, pH 6.9: Place 3 g of paraformaldehyde in a 250 mL conical flask, add 30 mL ultrapure water, loosely stopper, and heat on a 60°C hotplate, in a fume cupboard for about 30 min to give a cloudy solution. Add 1 *M* NaOH (the purest grade), with continual stirring until the solution clears. Add ultrapure water to give a total volume of 50 mL, then add 50 mL of 0.2 *M* PIPES buffer, pH 6.9 (*see* **step 5**). Divide into 10-mL aliquots and store frozen. Defrost in a warm water bath for use.
7. 0.1% v/v Triton X-100 or Saponin in 0.1 *M* PIPES buffer, pH 6.9.

### 2.3 Smears Prepared from Cells in Suspension

1. Cells in suspension (*see* **Note 2**).
2. Silane-treated glass microscope slides (*see* **Subheading 2.1.** and **Note 1**).
3. Aluminum foil or "cling film."
4. Acetone.

### 2.4 Frozen (Cryostat) Sections

1. Chunk of fresh tissue >0.5 $cm^3$ in size.
2. Cryostat embedding medium, for example, OCT or similar.
3. Isopentane or hexane.

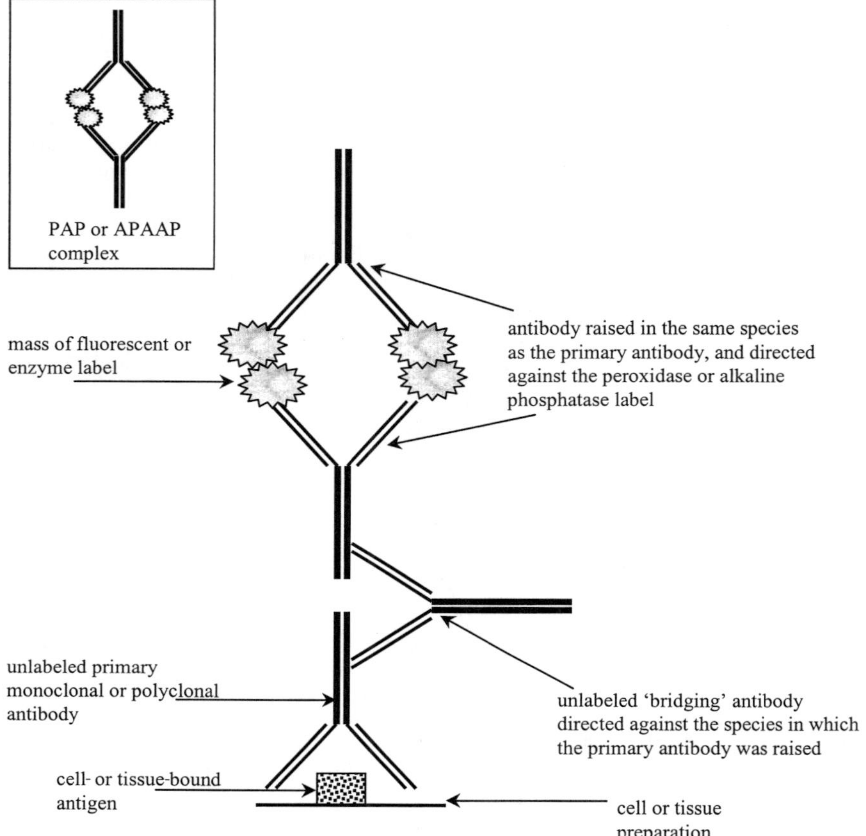


Fig. 5. Peroxidase–antiperoxidase (PAP) or alkaline phosphatase–antialkaline phosphatase (APAAP) method. Cell- or tissue-bound antigen is detected by, first, unlabeled primary monoclonal or polyclonal antibody, then a "bridging" antibody directed against the species in which the primary antibody was raised, and finally a labeled PAP or APAAP complex raised in the same species as the primary antibody. The labeled PAP or APAAP consists of a mass of enzyme label complexed with antibodies directed against it.

4. Liquid nitrogen.
5. Silane-coated clean glass microscope slides (*see* **Subheading 2.1.** and **Note 1**).
6. Acetone.
7. Aluminum foil or "cling film."

### 2.5. Fixed, Paraffin Wax-Embedded Sections

1. Paraffin wax-embedded tissue blocks.
2. 20% v/v ethanol or industrial methylated spirit in distilled water.

3. Silane-treated glass microscope slides (*see* **Subheading 2.1.** and **Note 1**).
4. Xylene (*see* **Note 3**).
5. Absolute ethanol or industrial methylated spirit.
6. 70% v/v ethanol or industrial methylated spirit in distilled water.
7. Distilled water.

### 2.6. Buffers for Blocking, Dilutions, and Washes

1. Washing buffer: Tris-buffered saline (TBS), pH 7.4–7.6: 60.57 g of Tris, 87.0 g of NaCl dissolved in 1 L of distilled water. Adjust pH to 7.4–7.6 using concentrated HCl. Make up to total vol 10 L using distilled water. This buffer is recommended for all washes, unless otherwise stated.
2. Blocking buffer: 5% v/v normal horse or goat serum in washing buffer. All immunocytochemical methods (*see* **Subheading 3.9.**) incorporate a step in which cell and tissue preparations are incubated with a blocking buffer to reduce nonspecific binding of antibodies (*see* **Note 4**).
3. Dilution buffer for antibodies: 3% v/v normal horse or goat serum in blocking buffer. Antibodies are diluted to their working concentration in buffer containing a low percentage of normal serum. This, again, reduces nonspecific binding of antibodies and minimizes ''dirty'' background staining (*see* **Note 4**).

### 2.7. Enzyme-Based Antigen Retrieval Methods (see Note 5)

1. Trypsin solution: 1 mg/mL of crude, type II trypsin, from porcine pancreas (*see* **Note 6**) and 1 mg/mL of calcium chloride in washing buffer (*see* **Subheading 2.6.**) warmed to 37°C.
2. Protease solution: Protease XXIV, bacterial, 7–14 U/mg in washing buffer (*see* **Subheading 2.6.**) prewarmed to 37°C.
3. Pepsin solution: pepsin from porcine stomach, 1:2500, 600–1000 U/mg in 0.01 *M* hydrochloric acid, prewarmed to 37°C.
4. Neuraminidase solution: neuraminidase (sialidase), type V from *Clostridium perfringens* at a concentration of 0.1 U/mg in 0.1 *M* sodium acetate pH to 5.5 with citric acid, containing 0.01% w/v calcium chloride, prewarmed to 37°C.

### 2.8. Microwave Oven Heat-Mediated Antigen Retrieval Method (see Note 7)

1. Citrate buffer, pH 6.0: 2.1 g of citric acid dissolved in 1 L of distilled water. Adjust pH to 6.0 using concentrated NaCl.
2. Distilled water.

### 2.9. Quenching Endogenous Enzyme

1. Methanol–hydrogen peroxide solution: 3% v/v hydrogen peroxide in methanol. Make up fresh every 2–3 d.

## 2.10. Examples of Some Histochemical Staining Techniques

### 2.10.1. Direct Method

1. Blocking buffer (*see* **Subheading 2.6.**).
2. Monoclonal or polyclonal antibody labeled with a fluorescent or enzyme label made up at optimum working dilution in dilution buffer (*see* **Subheading 2.6.**).
3. Washing buffer (*see* **Subheading 2.6.**).

### 2.10.2. Simple Indirect Method

1. Blocking buffer (*see* **Subheading 2.6.**).
2. Unlabeled monoclonal or polyclonal antibody made up at optimum working dilution in dilution buffer (*see* **Subheading 2.6.**).
3. Washing buffer (*see* **Subheading 2.6.**).
4. Fluorescent- or enzyme-labeled secondary antibody directed against the immunoglobulins of the species in which the primary antibody was raised, made up at optimum working dilution in dilution buffer (*see* **Subheading 2.6.**).

### 2.10.3. Simple Avidin–Biotin Method

1. Blocking buffer (*see* **Subheading 2.6.**).
2. Biotin-labeled monoclonal or polyclonal primary antibody made up at optimum working dilution in dilution buffer (*see* **Subheading 2.6.**).
3. Washing buffer (*see* **Subheading 2.6.**).
4. Avidin or streptavidin labeled with a fluorescent or enzyme label made up at optimum working dilution in dilution buffer (*see* **Subheading 2.6.** and **Note 8**).

### 2.10.4. Avidin–Biotin-Complex Method

1. Blocking buffer (*see* **Subheading 2.6.**).
2. Biotin-labeled monoclonal or polyclonal primary antibody made up at optimum working dilution in dilution buffer (*see* **Subheading 2.6.**).
3. Washing buffer (*see* **Subheading 2.6.**).
4. Fluorescent- or enzyme-labeled avidin–biotin complex (ABC) made up according to the manufacturer's instructions (*see* **Note 9**).

### 2.10.5. Peroxidase–Anti-Peroxidase or Alkaline Phosphatase–Anti-Alkaline Phosphatase Methods (see **Note 10**)

1. Blocking buffer (*see* **Subheading 2.6.**).
2. Unlabeled monoclonal or polyclonal antibody made up at optimum working dilution in dilution buffer (*see* **Subheading 2.6.**).
3. Washing buffer (*see* **Subheading 2.6.**).
4. Unlabeled "bridging" antibody directed against the species in which the primary antibody was raised made up in excess concentration (*see* **Note 11**) in dilution buffer (*see* **Subheading 2.6.**).
5. Peroxidase–anti-peroxidase (PAP) or alkaline phosphatase–anti-alkaline phosphatase (APAAP) raised in the same species as the primary antibody (e.g.,

when using a mouse monoclonal primary, use *mouse* PAP or APAAP; when using rabbit polyclonal primary antibody, use *rabbit* PAP or APAAP), made up at optimum working dilution in dilution buffer (*see* **Subheading 2.6.**).

## 2.11. Enzyme Development Methods

### 2.11.1. DAB for Horseradish Peroxidase Label

1. Washing buffer (*see* **Subheading 2.6.**).
2. DAB–$H_2O_2$: 0.5 mg/mL of 3,3-diaminobenzidine tetrahydrochloride (DAB) in washing buffer (*see* **Subheading 2.6.**). Add $H_2O_2$ to give a concentration of 0.03% v/v immediately before use. This substance is potentially carcinogenic (*see* **Note 12**).

### 2.11.2. Fast Red for Alkaline Phosphatase Label

1. TBS, pH 8.2–9.0: 6.57 g of Tris, 8.7 g of NaCl dissolved in a total volume of 1 L of distilled water. Adjust pH to 8.2–9.0 using concentrated HCl.
2. Stock solution of naphthol phosphate: Dissolve 20 mg of naphthol AS-MX phosphate sodium salt in 500 µL of *N,N*-dimethylformamide in a small glass vessel (*see* **Note 13**).
3. Stock solution fast red salt: Dissolve 20 mg of fast red salt in 1 mL of TBS, pH 8.2–9.0.
4. Levamisole hydrochloride.

## 2.12. Counterstaining

1. Mayer's hematoxylin solution (*see* **Note 14**).
2. 1% v/v ammonia in tap water.

## 2.13. Mounting

1. For fluorescently labeled preparations: An antifade mounting medium, for example, Citifluor or similar.
2. For alkaline phosphatase/fast red labeled, or other preparations labeled with an alcohol-soluble chromogenic product: An aqueous mountant, for example, Aquamount or similar.
3. For horseradish peroxidase/DAB labeled, or other preparations labeled with an alcohol-insoluble chromogenic product: 70% v/v ethanol or industrial methylated spirit in distilled water; 95% v/v ethanol or industrial methylated spirit in distilled water; absolute alcohol; xylene (*see* **Note 3**); and an appropriate xylene-based mounting medium, for example, Depex or similar.

# 3. Methods

## 3.1. Silane Treatment of Microscope Slides (see Note 1)

1. Place slides in a slide carrier and immerse in acetone for 5 min.
2. Immerse in acetone/silane solution for 5 min.
3. Immerse in two consecutive baths of either acetone or distilled water for 5 min each.
4. Drain slides, dry either at room temperature or in a warm oven, and store in closed boxes at room temperature indefinitely.

### 3.2. Preparation of Cells Cultured on Coverslips

1. Wash cultured cells in fresh FCS free culture medium.
2. Aspirate and discard the medium.
3. Scrape cells from the flask using a rubber policeman and resuspend in fresh FCS cell free culture medium.
4. Count cells and subculture $1 \times 10^5$ cells into Petri dishes.
5. Place sterile coverslips in Petri dishes and allow cells to proliferate for 24 h at 37°C under 5% $CO_2$.
6. Carefully remove coverslips using fine forceps or the edge of a scalpel blade.
7. Place coverslips, cell side up, onto a piece of dental wax or Parafilm for support, and cover each with 100 µL of cold 3% v/v paraformaldehyde in 0.1 *M* PIPES buffer, pH 6.9, for 15 min.
8. Wash thoroughly in 0.1 *M* PIPES buffer, pH 6.9.
9. Permeabilize in 0.1% v/v Triton X-100 or 0.1% v/v Saponin in 0.1 *M* PIPES buffer, pH 6.9, for 10 min.
10. Wash thoroughly in 0.1 *M* PIPES buffer, pH 6.9.

### 3.3. Smears Prepared from Cells in Suspension

1. Place a drop of cells in suspension approx 5 mm from one end of a silane-treated glass microscope slide.
2. Place a second microscope slide on top of the first, allowing approx 1 cm of glass to protrude at either end, and allowing the drop to spread between the two.
3. Drag one slide over the other in a rapid, smooth movement, spreading the cells in a thin smear over the surface of both slides.
4. Air-dry the slides for approx 5 min. They may then be used at once, or wrapped individually in foil or "cling film" and stored in the freezer until required. If stored frozen, allow to thaw to room temperature before use.
5. When ready for use, fix by dipping in acetone for 1 min and air-dry.

### 3.4. Frozen (Cryostat) Sections

1. Using a sharp, clean blade, cut a solid tissue block of fresh tissue of approx 0.5 $cm^3$ (*see* **Note 15**).
2. Place the tissue on cryostat chuck thickly coated in cryostat embedding medium (*see* **Note 16**).
3. Using long-handled tongs, pick up the chuck and immerse chuck and tissue in isopentane or hexane precooled in liquid nitrogen for approx 1–2 min (*see* **Note 17**).
4. Place the frozen chuck in the cabinet of the cryostat and leave to equilibrate for approx 30 min.
5. Using the cryostat, cut 5–10 µm thick sections and pick them up on clean, dry silane-treated microscope slides. Allow to air-dry for between 1 h and overnight.
6. Sections may then either be stored until required by wrapping individually or back-to-back in aluminum foil or "cling film," sealing in polythene bags or boxes containing desiccant, and storing in the freezer, or may be used at once. If stored frozen, allow to thaw and equilibrate to room temperature before opening.
7. Immediately before use, dip slides in acetone for 1–10 min and air-dry for approx 5 min.

### 3.5. Fixed, Paraffin Wax-Embedded Sections

1. Cool wax-embedded tissue blocks on ice for approx 15 min.
2. Cut 4–7 μm thick sections by microtome (*see* **Note 18**).
3. Float sections out on a pool of 20% v/v ethanol or industrial methylated spirit in distilled water on a clean glass plate supported by a suitable receptacle such as a glass beaker or jar (*see* **Note 19**).
4. Carefully transfer the sections, floating on the alcohol, onto the surface of a water bath heated to 40°C — they should puff out and become flat (*see* **Note 20**).
5. Separate out individual sections very gently using the tips of fine, bent forceps.
6. Pick up individual sections on silane-treated microscope slides.
7. Allow slides to drain by up-ending them on a sheet of absorbent paper for 5–10 min.
8. Dry slides either in a 37°C incubator overnight or in a 60°C oven for 15–20 min (*see* **Note 21**). Slides may then be cooled, stacked or boxed, and stored at room temperature in a dust-tight container until required.
9. When required, soak slides in xylene for approx 15 min to remove the paraffin wax.
10. Transfer through two changes of absolute ethanol or industrial methylated spirit, then through one change of 70% v/v ethanol or industrial methylated spirit in distilled water, then distilled water, agitating the slides vigorously for 1–2 min at each stage to equilibrate (*see* **Note 22**).

### 3.6. Enzyme-Based Antigen Retrieval Methods (see Note 5)

When using fixed, paraffin embedded tissues *only*, methods to retrieve antigens damaged or sequestered by harsh fixation and processing procedures may be necessary, or may significantly enhance results. These methods are *not* appropriate for other types of cell or tissue preparation (*see* **Note 23**).

1. Trypsinization: Immerse slides in a bath of trypsin solution at 37°C, in an incubator or water bath, for 5–30 min, then wash in running tap water for 5 min.
2. Protease treatment: Place slides face up in a suitable chamber (*see* **Note 24**) and apply a few drops of protease solution to cover the tissue preparation. Incubate in an incubator at 37°C, for 5–30 min, then wash in running tap water for 5 min.
3. Pepsin treatment: Place slides face up in a suitable chamber (*see* **Note 24**) and apply a few drops of pepsin solution to cover the tissue preparation, or immerse slides in a bath of pepsin solution. Allow to digest in an incubator at 37°C for 5–30 min, then wash in running tap water for 5 min.
4. Neuraminidase treatment: Place slides face up in a suitable chamber (*see* **Note 24**) and apply a few drops of neuraminidase solution to cover the tissue preparation. Incubate in an incubator at 37°C for 5–30 min, then wash in running tap water for 5 min.

### 3.7. Microwave Oven Heat-Mediated Antigen Retrieval Method

An alternative to enzyme mediated antigen retrieval is antigen retrieval mediated by heat. The method given in the following is for heat-mediated treatment using a microwave oven, but other methods exist using, for example, pressure cooking or autoclaving (*see* **Note 7**).

1. Immerse slides in citrate buffer, pH 6.0, in any suitable microwave-safe container, such as a plastic sandwich box (*see* **Note 25**).
2. Place in a conventional microwave oven and heat on full power until the buffer boils.
3. Reduce the power to "simmer" or "defrost" for 5 min, so that the buffer boils gently. After 5 min, check the level of the buffer, and top up with hot distilled water if necessary. Heat on "simmer" or "defrost" for another 5 min.
4. Allow slides to cool at room temperature for 30 min (*see* **Note 26**).
5. Wash under running tap water for 5 min.

### 3.8. Quenching Endogenous Enzyme

If horseradish peroxidase is to be employed as the label molecule, then endogenous peroxidase must be quenched as follows. This step is most conventionally performed immediately prior to the addition of the primary antisera.

1. Immerse slides in methanol–hydrogen peroxide solution for 20 min.
2. Wash under running tap water for approx 5 min.

If alkaline phosphatase is to be employed as the label molecule it may be necessary to quench endogenous alkaline phosphatase, although it is usually destroyed by processing to paraffin wax rendering this procedure unnecessary. If required, 1 m*M* levamisole is added to the final enzyme development medium (*see* **Subheadings 2.11.2.** and **3.11.2.**).

### 3.9. Examples of Some Histochemical Staining Techniques

As described in the Introduction, a number of basic immunocytochemical techniques are available that vary in their relative complexity and sensitivity. Illustrative examples are listed here that should give good results, but the researcher is urged to experiment and adapt these basic technique to give optimum results in his or her experimental system. Other techniques also exist. The methods outlined here are also illustrated diagramatically in **Figs. 1–5**.

#### 3.9.1. Direct Method (see *Fig. 1*)

1. Incubate slides with blocking buffer for 30 min.
2. Drain slides and wipe around cell or tissue preparation using a clean, dry tissue.
3. Incubate slides with directly (fluorescent or enzyme) labeled monoclonal or polyclonal antibody in a humid chamber (*see* **Note 24**), for 30 min to 2 h at room temperature, or at 4°C overnight.
4. Wash in three changes of washing buffer (*see* **Note 27**).
5. If fluorescently-labeled antibody is employed, mount and view directly using a fluorescent microscope. If enzyme-labeled antibody is used, proceed to enzyme development as described in **Subheading 3.10.** onwards.

#### 3.9.2. Simple Indirect Method (see *Fig. 2*)

1. Incubate slides with blocking buffer for 30 min.
2. Drain slides and wipe around cell or tissue preparation using a clean, dry tissue.

3. Incubate slides with unlabeled monoclonal or polyclonal antibody in a humid chamber (*see* **Note 24**), for 30 min to 2 h at room temperature, or at 4°C overnight.
4. Wash in three changes of washing buffer (*see* **Note 27**).
5. Incubate slides with either fluorescent- or enzyme-labeled secondary antibody directed against the immunoglobulins of the species in which the primary antibody was raised (*see* **Note 28**) in a humid chamber (*see* **Note 24**) for 1 h.
6. Wash in three changes of washing buffer (*see* **Note 27**).
7. If fluorescently-labeled secondary antibody is employed, mount and view directly using a fluorescent microscope. If enzyme-labeled secondary antibody is used, proceed to enzyme development as described in **Subheading 3.10.** onwards.

#### *3.9.3. Simple Avidin–Biotin Method (*see ***Fig. 3****)*

1. Incubate slides with blocking buffer for 30 min.
2. Drain slides and wipe around cell or tissue preparation using a clean, dry tissue.
3. Incubate slides with biotin-labeled monoclonal or polyclonal primary antibody in a humid chamber (*see* **Note 24**), for 30 min to 2 h at room temperature, or at 4°C overnight.
4. Wash in three changes of washing buffer (*see* **Note 27**).
5. Incubate with avidin or streptavidin labeled with fluorescent or enzyme label, in a humid chamber (*see* **Note 24**), for 30 min.
6. Wash in three changes of washing buffer (*see* **Note 27**).
7. If fluorescently-labeled avidin or streptavidin is employed, mount and view directly using a fluorescent microscope. If enzyme-labeled avidin or streptavidin is used, proceed to enzyme development as described in **Subheading 3.10.** onwards.

#### *3.9.4. ABC Method (*see ***Fig. 4****)*

1. Incubate slides with blocking buffer for 30 min.
2. Drain slides and wipe around cell or tissue preparation using a clean, dry tissue.
3. Incubate slides with biotin-labeled monoclonal or polyclonal primary antibody in a humid chamber (*see* **Note 24**), for 30 min to 2 h at room temperature, or at 4°C overnight.
4. Wash in three changes of washing buffer (*see* **Note 27**).
5. Incubate with fluorescent- or enzyme-labeled ABC in a humid chamber (*see* **Note 24**) for 30 min.
6. Wash in three changes of washing buffer (*see* **Note 27**).
7. If fluorescently-labeled ABC is employed, mount and view directly using a fluorescent microscope. If enzyme-labeled ABC is used, proceed to enzyme development as described in **Subheading 3.10.** onwards.

#### *3.9.5. PAP or APAAP Methods (*see ***Fig. 5****)*

1. Incubate slides with blocking buffer for 30 min.
2. Drain slides and wipe around cell or tissue preparation using a clean, dry tissue.
3. Incubate slides with unlabeled monoclonal or polyclonal primary antibody in a humid chamber (*see* **Note 24**), for 30 min to 2 h at room temperature, or at 4°C overnight.

4. Wash in three changes of washing buffer (*see* **Note 27**).
5. Incubate with an unlabeled "bridging" antibody (*see* **Note 11**) for 1 h in a humid chamber (*see* **Note 24**).
6. Wash in three changes of washing buffer (*see* **Note 27**).
7. Incubate with PAP or APAAP for 1 h in a humid chamber (*see* **Note 24**).
8. Wash in three changes of buffer (*see* **Note 27**).
9. Proceed to enzyme development as described in **Subheading 3.10.** onwards.

### 3.10. Enzyme Development Methods

#### 3.10.1. DAB for Horseradish Peroxidase Label

1. Wash in three changes of washing buffer (*see* **Note 27**).
2. Incubate with DAB–$H_2O_2$ for 10 min (*see* **Note 12**).
3. Wash under running tap water for 5 min.
4. Proceed to counterstaining (*see* **Subheading 3.11.**) and mounting (*see* **Subheading 3.12.**).

#### 3.10.2. Fast Red for Alkaline Phosphatase Label

1. Wash slides briefly in TBS, pH 8.2–9.0.
2. Take 18.5 mL of TBS pH 8.2–9.0, add 500 µL of stock solution of naphthol phosphate and mix, then add levamisole to give a 1 m*M* solution and mix, then add 1 mL of fast red solution and mix. Filter and apply to slides immediately.
3. Immerse the slides in fast red solution for 5–30 min (*see* **Note 29**).
4. Wash under running tap water for 5 min.
5. Proceed to counterstaining (*see* **Subheading 3.11.**) and mounting (*see* **Subheading 3.12.**).

### 3.11. Counterstaining

1. Immerse in Mayer's hematoxylin solution for 3–5 min.
2. "Blue" by immersing in running tap water for 5 min, or dip briefly in 1% v/v ammonia in tap water, then wash in tap water (*see* **Note 30**).
3. Proceed to mounting (*see* **Subheading 3.12.**).

### 3.12. Mounting

1. Preparations labeled using fluorescent tags should be mounted directly in an antifade fluorescent mountant such as Citifluor or similar.
2. Preparations labeled using alcohol-soluble chromagens such as fast red for alkaline phosphatase should be mounted directly in an aqueous mountant such as Aquamount or similar.
3. Preparations labeled using alcohol-insoluble chromagens such as DAB for horseradish peroxidase should be dehydrated by immersing, with agitation, for 1 min each in 70% v/v, 95% v/v ethanol or industrial methylated spirit, then two changes of absolute ethanol, cleared by immersion, with a agitation, in two changes of xylene, then mounted in a xylene-based mountant such as Depex or similar (*see* **Note 31**).

### 3.13. Viewing, Interpretation, and Quantification of Labeling Results

Slides should be viewed by light or fluorescence microscope, as appropriate. Fluorescently labeled preparations should be should be stored in the dark and in the refrigerator until viewing. They should be viewed as soon as possible after labeling, as fluorescence will fade over time, and photographic images should be made for permanent record. Enzyme labels should be permanent, and slides may therefore be stored for longer. Some aqueous mountants deteriorate over time. Good labeling is indicated by a strong specific label and low, or preferably nonexistent, background and nonspecific labeling.

It is often helpful to score labeling on an arbitrary scale where the observer estimates the percentage of cells, for example cancer cells, labeled (10%, 50%, 95%, etc.) and the intensity of labeling on a scale of – (no labeling at all), + – very weak labeling, + (weak but definite labeling) to ++++ (extremely intense labeling) to give results ranging from completely negative to 100% ++++. Preferably, this should be carried out by at least two independent observers and results compared.

Many attempts have been made to quantify immunocytochemistry results using automated, computer-based approaches, but the author is unaware of any truly satisfactory and reproducible system.

### 3.14. Some Common Problems and the Most Likely Suggested Solutions to Them

#### 3.14.1. High Nonspecific Background Staining

Possibly the most common problem, this usually can be caused by a number of different factors. The most usual is probably insufficient washing between steps (*see* **Note 27**). The second most likely cause is employment of too high a concentration of one or more of the reagents. All reagents (primary antibody, secondary antibody, avidin and biotin products, PAP or APAAP, etc.) should be titrated carefully to give optimum results (*see* **Subheading 1.4.** and **Table 2**). When using horseradish peroxidase as a label, residual endogenous peroxidase may sometimes be a problem — check by incubating a slide that has been treated simply with the standard methanol–hydrogen peroxide solution (*see* **Subheading 2.9.**) with the chromogenic substrate DAB–$H_2O_2$ (*see* **Subheading 2.11.1.**) — there should be no brown staining present; if there is, this indicates the presence of unquenched endogenous peroxidase. If this is the case, try freshly made methanol–hydrogen peroxide, increase the concentration of hydrogen peroxide or the incubation time in this step, or try an alternative label, for example, alkaline phosphatase.

If the problematic high background staining is absent in a negative control where primary antibody is omitted, this would indicate a crossreaction between the primary antibody and some cell or tissue component. This may be remedied by one or more of the following: increasing the concentration of blocking serum or protein in blocking buffer (*see* **Subheading 2.6.** and **Note 4**), incorporating more sodium chloride (up to 0.1 *M*) into blocking, dilution, and washing buffers, or adding a small amount of detergent (e.g., 0.05% v/v Tween-20) to washing buffers.

In avidin–biotin based methods, endogenous biotin can sometimes cause confusing results. It can be blocked by applying unconjugated avidin (which binds to tissue-bound biotin), then saturating with further free, excess, unlabeled biotin. Avidin may also sometimes attach to charged cell/tissue sites: This may be most easily remedied by increasing the pH of washing, dilution, and blocking buffer to 9.0, or may be avoided by using the more costly streptavidin products instead of avidin.

### 3.14.2. Weak or Absent Labeling

Obviously, the positive control—a preparation known to express the antigen of interest—should be checked. If satisfactory labeling is achieved here, it would suggest that the antigen is present in only low levels, or absent, in the test slides. If low levels only are present, perhaps indicated by weak labeling, a more sensitive detection technique should be employed. If fixed, paraffin-embedded material is being examined, antigen retrieval methods should be tried. If positive controls show inadequate labeling, all reagents should be systematically checked for reactivity.

## 4. Notes

1. We use silane-treated slides for *all* cell and tissue preparations in immunocytochemistry. Alternative, commercial, brand-named preparations are also available, but tend to be more expensive. Silane, or equivalent, treatment is *essential* if a heat-mediated antigen retrieval method is to be employed subsequently. Slides should not be agitated in the baths of reagents, as air bubbles will prevent the silane solution reaching the glass surface and will result in patchy and inadequate treatment.
2. Any cells in suspension are suitable—for example, blood, cancer cells in ascites or in pleural effusions taken from patients or from animal models, cells derived from solid tissue tumors and released into suspension, or cultured cells in suspension.
3. Xylene is potentially hazardous and should be handled with care in a fume cupboard. Modern, safer chemical alternatives are commercially available, but it is our experience that they give slightly less satisfactory results.
4. Blocking buffer and buffer for dilution of antibodies: The use of normal (nonimmune) animal serum and its incorporation into working solutions of antibodies effectively reduces non-specific background staining. Goat or horse serum is recommended here for simplicity as it is unlikely that the detection methods employed will involve specific recognition of antibodies raised in these species.

It is also appropriate to employ normal (nonimmune) serum from the final antibody-producing species in any detection method — for example, in an indirect method in which a labeled rabbit secondary antibody is employed to detect binding of an unlabeled mouse monoclonal antibody, nonimmune rabbit serum would be entirely appropriate. Solutions of an inert protein or protein mixture such as bovine serum albumin, casein, or commercially available dried skimmed milk powder are also routinely used, and are cheaper than nonimmune serum. As a guide, a solution of 1–5% w/v solution in buffer is appropriate.

5. It is not possible to predict if enzyme-based antigen retrieval methods will be effective, or which to choose; try them out and select conditions which work best for your application. Test a range of treatment times, for example, 0, 5,10, 20, and 30 min. Trypsinisation is probably most widely used, probably because it is relatively cheap, and can be extremely effective, if, sometimes a little brutal! The other enzymes tend to be more expensive (especially neuraminidase), but are more selective. Try heat-mediated antigen retrieval (*see* **Subheadings 2.8.** and **3.8.**) also. These methods should be used only in conjunction with fixed, paraffin-embedded tissue preparations mounted on silane (or equivalent; *see* **Note 1**)-treated glass slides.
6. Use crude, type II trypsin, from porcine pancreas. Impurities (e.g., chymotrypsin) enhance its effect. Do not use purer (and more expensive!) products.
7. It is not possible to predict if heat-mediated antigen retrieval methods will be effective in any particular case. This can be determined only by trial and error. Other heat-mediated antigen retrieval methods, such as autoclaving and pressure cooking also exist. Try enzyme-based antigen retrieval methods (*see* **Subheadings 2.7.** and **3.7.**) also. These methods should be used only in conjunction with fixed, paraffin-embedded tissue preparations mounted on silane (or equivalent, *see* **Note 1**)-treated glass slides.
8. Avidin is a large glycoprotein extracted from egg white that has four binding sites for the vitamin biotin. Streptavidin is a protein, similar in structure to avidin, and is derived from the bacterium *Streptomyces avidinii*. Avidin products are generally significantly cheaper than streptavidin products. Streptavidin is said to give a cleaner result, but avidin may be perfectly acceptable for most applications.
9. ABC is available commercially as convenient dropper bottle kits. Follow kit instructions, which usually require that avidin and labeled biotin are combined 30 min before use. Avidin and labeled biotin are mixed together in such a ratio that three of the four possible biotin-binding sites are saturated, leaving one free to combine with the biotin label attached to the primary antibody.
10. The PAP and APAAP techniques were extremely popular some years ago owing to their sensitivity, resulting from the multilayering. They are less commonly used today and have been superseded to some extent by the avidin–biotin and ABC methods.
11. The "bridging" antibody forms a "bridge" by linking the primary antibody with the PAP or APAAP complex. So, for example, if a monoclonal mouse primary antibody is used, the "bridging" antibody will be antisera raised against mouse

immunoglobulins; if a rabbit polyclonal primary antibody is used, the "bridging" antibody will be antisera directed against rabbit immunoglobulins. This antisera should be applied in excess. The "bridging" antibody needs to be present in excess so that only one of the two possible antigen binding sites of each "bridging" antibody molecule are occupied by primary antibody, leaving the second binding site free to bind to the PAP or APAAP complex. As a rough guide, in our experience, most commercially available secondary antibodies will work best in this context when diluted at approx 1:50.

12. DAB is potentially carcinogenic. It should be handled with care, using gloves. Avoid spillages and aerosols. Work in a fume cupboard. After use, soak all glassware etc. in a dilute solution of bleach overnight before washing. Swab down working surfaces with dilute bleach after use. Clean up spillages with excess water, then swab with dilute bleach. We usually make a concentrated DAB stock solution at 5 mg/mL in distilled water and freeze in 1-mL aliquots in 10-mL plastic screw top tubes until required. This minimizes the risk of aerosols from weighing out powder when required. It is also available in convenient tablet or dropper bottle kit form, which minimizes hazards, but is more expensive.
13. Use glass, as plastic will dissolve in the dimethylformamide.
14. A number of different hematoxylin solutions are commercially available. Mayer's, a progressive stain, is particularly convenient, but other types are equally effective.
15. When cutting tissue, use a very sharp blade and use single, firm, swift, downward stokes. Avoid hacking and crushing, which will result in poor morphology. Cut the tissue into a straight edged, geometrical shape, most conveniently a cube, as this will make sectioning easier. Clearly, when handling potentially infective tissue, appropriate health and safety guidelines should be adhered to.
16. The cryostat embedding medium acts as a support for the tissue. Apply it generously to the chuck and immerse the tissue block into a pool of it. Align the tissue block with a straight edge parallel with the cutting edge of the chuck.
17. To obtain optimum morphological integrity, the tissue should be frozen as rapidly as possible, avoiding the formation of morphology-destroying ice crystals. The most effective method is to immerse the tissue in isopentane or hexane precooled in liquid nitrogen. These solvents conduct heat away from the tissue more rapidly that liquid nitrogen alone. Other methods of freezing the block—for example, by using a commercially available freezing spray, by blasting with $CO_2$ gas, or even simply placing it in the chamber of the cryostat until frozen are less effective.
18. Most tissues cut most best when chilled. An ice cube or handful of crushed ice should be applied to the surface of the tissue block every few minutes during cutting. This is particularly important when working in a warm room.
19. Small creases or wrinkles in the section should begin to flatten out; the effect may be enhanced by gentle manipulation using, for example, a soft paintbrush and/or forceps.
20. If the wax begins to melt, it is too hot. If sections remain wrinkled, it may indicate that the water is slightly too cold.
21. Do not heat directly on a hotplate, as this may damage some antigens.

22. Take careful note of the appearance of the slides during this process. When slides are transferred from one solvent to the next, they initially appear smeary as the two solvents begin to mix. Vigorous agitation, that is "sloshing them up and down" ensures that the slides equilibrate effectively with the new solvent. When they are equilibrated, the surface smearing will disappear. During the rehydration process, if white flecks or patches become visible around the sections, this indicates that the wax has not been adequately removed—return the sections to xylene for a further 10–15 min. A common cause of poor immunocytochemistry results is inadequate removal of paraffin wax.
23. Make enzyme solutions fresh immediately before use. Use glassware, solutions, etc. that have been prewarmed to 37°C before use. Initially try a range of digestion times, for example, 0, 5, 10, 20, 30 min. Digestion times of >30 min are not recommended, as visible damage to tissue morphology becomes apparent. This is especially true when using trypsin.
24. The idea is to have a flat platform on which to place the slides, in a lidded, humid chamber. Humidity is important so that small volumes of solution placed onto the surface of the slides do not evaporate and therefore either dry out completely or become more concentrated. Drying of cell/tissue preparations will result in high nonspecific background staining. For small numbers of slides, it may be convenient to place them, face up, in a lidded Petri dish lined with a disc of dampened filter paper. For larger numbers of slides, specially designed incubation chambers—usually fashioned in Perspex, and containing raised ridges to support slides over troughs that may be partially filled with water to maintain a humid atmosphere—are commercially available. These tend to be expensive to buy. They can be made "in house" if appropriate facilities exist. It is possible to make a perfectly functional incubation chamber very simply using a large sandwich box with supports for slides formed from, for example, glass or wooden rods supported in "plasticine" or "blu-tack." Again, a small amount of water or dampened filter paper may be added to the base of the chamber to maintain humidity.
25. Slides can conveniently be placed in commercially available slide carriers, which typically hold up to about 12 or 25 slides and may be housed in, for example, appropriately sized plastic sandwich boxes, or, alternatively, upright in plastic Coplin jars. Space slides out evenly in the buffer. Do not overcrowd slides, as this results in "hot spots" and uneven antigen retrieval.
26. This cooling down period is part of the retrieval method and should not be skipped.
27. We recommend vigorous washing in three changes of washing buffer. Each wash should consist of vigorous "sloshing up and down" of slides in buffer for about 30 s to 1 min, then allowing slides to stand in the buffer for about 4 min. Insufficient washing, in particular omitting one or more changes of buffer, can result in unacceptably high levels of dirty background staining.
28. For example, if a monoclonal mouse primary antibody was used, incubate with labeled secondary antisera raised against *mouse* immunoglobulins; if a rabbit polyclonal primary antibody was used, incubate with a labeled secondary antibody directed against *rabbit* immunoglobulins.

29. Monitor the progress of color development by periodic examination using a microscope. Stop development when specifically labeled structures show deep red and before nonspecific background staining begins to occur.
30. Cell/tissue preparations will initially stain deep purple-red after immersion in Mayer's hematoxylin. The stain changes to navy blue when exposed to mildly alkaline conditions (known as "bluing"). This is most commonly achieved by washing in the slightly alkaline tap water that is available in most areas. If "bluing" is unsuccessful owing to unusually acidic tap water, dip slides briefly in 1% v/v ammonia in tap water, then wash in tap water.
31. Take careful note of the appearance of the slides during this process. When slides are transferred from one solvent to the next, they initially appear smeary as the two solvents begin to mix. Vigorous agitation, that is, "sloshing them up and down," ensures that the slides equilibrate effectively with the new solvent. When they are equilibrated, the surface smearing will disappear. White clouding of the xylene (it appears "milky") indicates contamination with water. Slides should be passed back through graded alcohols (absolute ethanol, 95% ethanol, 70% ethanol) to tap water, solvents should be discarded and replaced, and the process repeated.

## References


1. Coons, A. H., Creech, H. J., and Jones, R. M. (1941) Immunological properties of an antibody containing a fluorescent group. *Proc. Soc. Exp. Biol. Med.* **47,** 200–202.
2. Coons, A. H., Leduc, E. H., and Connolly, J. M. (1955) Studies on antibody production. I: A method for the histochemical demonstration of specific antibody and its application to a study of the hyperimmune rabbit. *J. Exp. Med.* **102,** 49–60.
3. Coons, A. H. and Kaplan, M. H. (1950) Localisation of antigen in tissue cells. *J. Exp. Med.* **91,** 1–13.
4. Nakane, P. K. and Pierce, G. B., Jr. (1966) Enzyme-labeled antibodies: preparation and application for the localization of antigens. *J. Histochem. Cytochem.* **14,** 929–931.
5. Mason, D. Y. and Sammons, R. E. (1978) Alkaline phosphatase and peroxidase for double immunoenzymic labeling of cellular constituents. *J. Clin. Pathol.* **31,** 454–462.
6. Guesdon, J. L., Ternynck, T., and Avrameas, S. (1979) The uses of avidin-biotin interaction in immunoenzymatic techniques. *J. Histochem. Cytochem.* **27,** 1131–1139.
7. Hsu, S. M., Raine, L., and Fanger, H. (1981) Use of avidin-biotin-peroxidase complex (ABC) in immunoperoxidase techniques: a comparison between ABC and unlabeled antibody (PAP) procedures. *J. Histochem. Cytochem.* **29,** 577–580.
8. Cattoretti, G., Pileri, S., Parravicini, C., Becker, M. H., Poggi, S., Bifulco, C., et al. (1993) Antigen unmasking on formalin fixed, paraffin embedded tissue sections. *J. Pathol.* **171,** 83–98.

9. Cattoretti, G. and Suurjmeijer, A. J. H.(1995) Antigen unmasking on formalin-fixed paraffin embedded tissues using microwaves: a review. *Adv. Anat. Pathol.* **2,** 2–9.
10. Norton, A. J., Jordan, S., and Yeomans, P. (1994) Brief, high temperature heat denaturation (pressure cooking): a simple and effective method of antigen retrieval for routinely processed tissues. *J. Pathol.* **173,** 371–379.
11. Bankfalvi, A., Navabi, H., Bier, B., Bocker, W., Jasani, B., and Schmid, K. W. (1994) Wet autoclave pre-treatment for antigen retrieval in diagnostic immunocytochemistry. *J. Pathol.* **174,** 223–228.
12. Riggs, J. L., Seiwald, R. J., Burkhalter, J. H., Downs, C. M., and Metcalf, T. (1958) Isothiocyanate compounds as fluorescent labeling agents for immune serum. *Am. J. Pathol.* **34,** 1081–1097.
13. Titus, J. A., Haughland, R., Sharrows, S. O., and Segal, D. M. (1982) Texas red, a hydrophilic, red-emitting fluorophore for use with fluorescein in dual parameter flow microfluorometric and fluorescence microscopic studies. *J. Immunol. Methods* **50,** 193–204.
14. Graham, R. C. and Karnovsky, M. J. (1966) The early stages of absorption of injected horseradish peroxidase in the proximal tubules of mouse kidney: ultrastructural cytochemistry by a new technique. *J. Histochem. Cytochem.* **14,** 291–302.
15. Graham, R. C., Ludholm, U., and Karnovsky, M. J. (1965) Cytochemical demonstration of peroxidase activity with 3-amino-9-ethylcarbazole. *J. Histochem. Cytochem.* **13,** 150–152.
16. Burstone, M. S. (1961) Histochemical demonstration of phosphatases in frozen sections with naphthol AS-phosphates. *J. Histochem. Cytochem.* **9,** 146–153.

## Suggested Further Reading

Polak, J. M. and van Noorden, S. (1997) *Introduction to Immunocytochemistry*, 2nd ed. Royal Microscopical Society Handbooks, No. 37. Bios Scientific Publishers, Oxford, UK.

# 3

# Multiple Labeling Techniques for Fluorescence Microscopy

**Amanda J. Atherton and Catherine Clarke**

## 1. Introduction

As described in Chapter 2 by Brooks, it has long been possible to localize antigens immunocytochemically using specific antibodies in conjunction with a label that is visible microscopically. Although much information can be derived by localizing a single protein/peptide, it is often useful to label simultaneously for two or more antigens within the same cells or tissue sections. There are a number of occasions when such multiple labeling techniques can be used: (1) to phenotype cells, for which no specific marker is available, using an appropriate panel of antibodies; (2) to identify which cells in a tissue or culture express an antigen of interest, by simultaneously labeling with antibodies to both this antigen and to a known phenotype marker; (3) to identify the distribution of an antigen at the subcellular level by simultaneously labeling with antibodies to both this and a known organelle marker; (4) to investigate whether several antigens of interest are colocalized, either at the cellular or the subcellular level.

Although it is possible to directly label a primary antibody with a fluorochrome (direct immunofluorescence), the overall fluorescence signal achieved using this technique is often weak *(1)*. Indirect immunofluorescence involves the use of secondary antibodies conjugated to different fluorochromes *(2)*. This approach has the advantage that multiple secondary antibodies can bind to each primary antibody, resulting in an amplification of the signal.

The most basic form of multiple labeling involves the simultaneous use of two or more primary antibodies that have been raised in different species of animals. This is generally successful, although crossreactions may occur where the secondary antibodies have been raised in the same species as one of the primary

From: *Methods in Molecular Medicine, vol. 57: Metastasis Research Protocols, Vol. 1: Analysis of Cells and Tissues*
Edited by: S. A. Brooks and U. Schumacher © Humana Press Inc., Totowa, NJ

antibodies. For example, although Ab1-mouse–antirabbit + Ab2-rabbit–antimouse may crossreact, Ab1-goat–antirabbit + Ab2-sheep–antimouse is probably all right. Crossreactions may also occur when antibodies raised in rats are recognized by antimouse secondary antibodies and vice versa. For this reason, it is important to select secondary antibody conjugates that are known not to crossreact with other species.

If the primary antibodies for study are of the same species, it is still feasible to multiple label using the class and subclass specific secondary antibodies that are now widely available from a number of suppliers. Where primary antibodies are of the same subclass, it is often preferable to directly conjugate the primary antibodies of interest to different fluorophores and localize them by direct immunofluorescence. However, if this is not possible, they can be localized by sequential application to the cells/tissue sections. Although this approach can be successful, it does require great skill and careful handling to avoid crossreactions. Furthermore, it is important that a blocking step is included in between the two sets of reactions.

When planning multiple labeling experiments it is essential to consider not only which antibodies are most appropriate, but also how to label and visualize them. In general, a combination of fluorochromes, visualized either by standard fluorescence or confocal microscopy, is preferable to the use of enzymic reaction products such as diaminobenzidine (DAB) (horseradish peroxidase) and Nitro Blue Tetrazolium (NBT) (alkaline phosphatase) viewed by light microscopy. Although it is possible to use such enzyme-linked secondary antibodies for multiple labeling, the developing steps must be carried out separately for each enzyme and require a degree of skill to ensure that each color is visible. Furthermore, where two antigens are colocalized it is likely that one color may swamp the other, whereas fluorescently-labeled antibodies remain independently detectable even where colocalized.

A wide range of fluorochromes conjugated to secondary antibodies, and therefore suitable for multiple labeling, are now readily available (*see* **Table 1**). Such fluorochromes, when excited, emit colored light which is either visible to the human eye or detectable by a CCD camera. To choose the best fluorochromes for each multiple labeling experiment it is necessary to consider a number of points.

First, the number of colors that are required needs to be determined. For a basic double-labeling experiment, it is common practice to use a combination of a green-emitting (fluorescein isothiocyanate, FITC) and a red-emitting (tetramethyl rhodamine isothiocyanate, TRITC) fluorochrome ***(3)***. If a third color is necessary, aminomethylcoumarin acetate (AMCA), which emits blue fluorescence, is commonly chosen. To increase the number of fluorochromes that can be used at any one time to four, it is possible to use one that emits in the far red, for example, indodicarbocyanine (Cy5). This fluorescence is not

**Table 1**
**Examples of Fluorochromes Commonly Used for Multiple Labeling**

| Fluorochrome | Excitation peak (nm) | Emission peak (nm) | Color |
|---|---|---|---|
| Aminoacetylcoumarin, AMCA | 350 | 450 | Blue |
| 4',6'-Diamidino-2-phenylindole | 360 | 460 | Blue |
| Fluorescein, FITC | 492 | 520 | Green |
| Propidium iodide, PI | 535 | 615 | Red |
| Tetramethyl rhodamine, TRITC | 550 | 570 | Red |
| Indocarbocyanine, Cy3 | 550 | 570 | Red |
| Lissamine rhodamine, LRSC | 570 | 590 | Red |
| Texas red, TR | 596 | 620 | Deep red |
| Indodicarbocyanine, Cy5 | 650 | 670 | Far red |

visible to the human eye, however, it can be detected by a CCD camera. Alternatively, where this equipment is not available, such fluorescence can be captured on far-red-sensitive film, but the image must be focused using one of the other fluorophores.

It is also important to consider the degree of spectral overlap that will occur when using several fluorochromes at one time. By studying the emission spectrum (*see* **Fig. 1**) for each fluorochrome, it is possible to ensure that the peak emission of those chosen is as far apart as possible. Having chosen the preferred fluorochromes, it is usual to equip the fluorescence microscope with a filter set designed to allow maximum emission with minimum overlap for each fluorochrome *(4)*. For multiple labeling purposes it is preferable, therefore, that the filters chosen transmit light over a narrow band of wavelengths. This is in contrast to the long-pass filters which are often employed for viewing single labeling. These filters are designed to maximize detection of emission over a wide range of wavelengths and would therefore result in spectral overlap when used for multiple fluorescence *(4)*.

Filter sets that allow more than one fluorochrome to be excited and detected simultaneously (multipass filters) are also available. The advantage of multipass filters is that the distribution of more than one antigen can be viewed in real time, without any risk of the slight image shift that can occur between images from single band filters of different wavelengths. This image shift results from lack of coincidence, in sets of single band filters, and can be overcome by careful filter selection, although a perfectly coincident filter set will be expensive.

Multipass filters do have some disadvantages when compared to single-pass filters. They can result in a loss of brightness, although this would be significant only when viewing very dim fluorescence. Furthermore, if two coincident

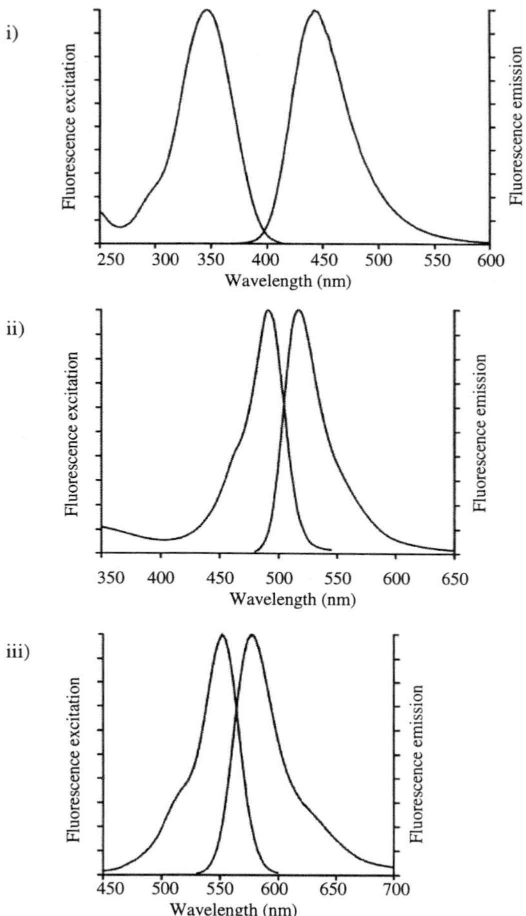


Fig. 1. Excitation and emission spectra of (i) AMCA, (ii) FITC, and (iii) TRITC.

signals are viewed, where one is very dim compared to the other, the dimmer signal may be swamped and thus not be visible using a multipass filter, whereas the two signals may be viewed independently using single-pass filters.

When equipping a microscope for multiple immunofluorescence labeling, it is also important to choose an appropriate light source *(4)*. The most commonly used is the mercury vapor arc lamp, which has an intense emission over a few wavelengths that can be easily selected with filters. Although the output in the blue/green is relatively low, it is quite adequate for immunofluorescence. Tungsten–halogen lamps have lower overall emission levels, and are not suitable for ultraviolet (UV) excitation, but can be used for green or red excitation. In contrast, xenon arc lamps have a broad band of emission in both the visible spectrum and UV, which allows a wider choice of excitation wavelengths, but requires more complex filter systems than mercury arc lamps. Although xenon arc lamps require a greater capital outlay than mercury vapor lamps, the life of the lamp is proportionally greater so that the running cost per hour over the life of the lamp may actually be less.

Prior to selecting fluorochromes for a multiple labeling experiment, it is also important to consider the relative abundance of the proteins to be localized. Some fluorochromes, particularly FITC, are susceptible to photobleaching and will not be compatible with the long exposure times required to photograph low levels of expression. It is best, therefore, to label the least abundant proteins with the more photostable fluorochromes such as the carbocyanines. Antifade reagents such as Citifluor can also be applied to the sample to help reduce photobleaching *(5)*.

Where a particular antigen of interest is at very low levels within cells or tissues, it is possible to increase the number of fluorochrome molecules available for detection and thus amplify the fluorescent signals. This can be achieved by using biotinylated secondary antibodies in combination with a fluorochrome conjugated to streptavidin *(6)*. Alternatively, enzyme-labeled secondary detection reagents in conjunction with fluorogenic substrates, for example, the tyramide signal amplification kit produced by NEN (Hounslow, UK; Boston, MA), can be used *(7)*.

A particular problem with multiple immunofluorescence is that nonspecific fluorescence can occur, either as autofluorescence of the tissue itself or as secondary fluorescence, caused by fixatives such as glutaraldehyde which form aromatic fluorescing compounds when reacting with cells *(8)*. Use of nonfluorescent fixatives such as acetone, ethanol, and methanol helps to reduce the problem of secondary fluorescence. Autofluorescence can occur in a number of human and animal tissues including connective tissue fibers (including collagen and elastin), ocular lens tissue, bone, and teeth. Although autofluorescence can occur over a number of different wavelengths of light, it is most commonly seen in response to UV excitation, resulting in blue fluorescence. This can be a problem when using multiple immunofluorescence, as AMCA, which also fluoresces in the blue, is frequently employed in this technique.

## 2. Materials

1. Fixatives: When using ice-cold methanol or acetone it should be precooled and stored in a spark-proof freezer.
2. PBS/BSA: 0.5% (w/v) bovine serum albumin (BSA) is dissolved in phosphate-buffered saline (PBS) and stored at 4°C for no more than 2–3 d.
3. Primary antibodies: Dilution should be determined using single labeling, prior to their use in multiple-labeling experiments.

4. Secondary conjugates: Should be stored according to manufacturer's instructions, and dilutions determined as for primary antibodies (*see also* **Note 1**).
5. Nuclear counterstain: It is often helpful to label the nucleus to give an idea of the overall tissue structure. DAPI (4',6-diamidino-2-phenylindole; 1.5 μg/mL) can be added to the mounting medium. Alternatively, propidium iodide (PI) will label nuclei red and contrast well with FITC labeling. If the cells are methanol fixed, label prior to mounting by incubating for 15 min in a solution of PI (10 mg/mL) in PBS.
6. Antifadent/mountant: Citifluor can be purchased commercially (Agar Scientific, Stanstead, Essex) and used neat. It will keep well at 4°C for extended periods of time.

## 3. Methods

### *3.1. Concurrent Multiple Labeling of Cells in Culture*

1. Grow cells to appropriate density on glass coverslips (13 mm) in 24-well plates. Alternatively, grow cells in tissue culture chamber slides and label the cells *in situ*. Keep cells in their individual wells throughout the labeling procedure.
2. Remove medium from the wells and wash cells twice in PBS.

3a. For cytoplasmic antigens: Fix cells in ice-cold methanol or acetone for 2 min, then wash well in PBS to remove all the fixative. Carry out all subsequent labeling steps at room temperature.

3b. For membrane antigens: Leave cells unfixed and keep on ice throughout the labeling procedure. Allow 10 min for the cells to cool down before starting and keep all reagents ice-cold to prevent internalization of antibodies.

4. Incubate cells with 200 μL of primary antibodies, diluted appropriately in PBS/BSA, for 1–2 h (*see* **Notes 2** and **3**). Where the primary antibody is in short supply, it is possible to minimize the amount used by inverting the coverslips onto just 50 μL of the antibody. In the instance it is important to place the coverslips in a humid chamber to prevent the antibody from drying out.
5. Wash cells in PBS 3×.
6. Incubate cells with appropriate combination of secondary antibodies conjugated to fluorochromes and/or biotin. Dilute each secondary antibody as recommended in the manufacturer's data sheet, and incubate cells with 200 μL of the mixture for 1 h (*see* **Note 1**).
7. Wash cells in PBS 3×.
8. If using a biotinylated secondary antibody, follow with a 30 min incubation in streptavidin conjugated to a fluorochrome. This is a chemical reaction that reaches saturation after 20 min.
9. Wash cells in PBS 3×.
10. If using a nuclear counterstain, for example, DAPI or propidium iodide, add it at this stage and incubate for 1 min or 15 min respectively.
11. Wash cells again in PBS 3×.
12. Carefully remove each coverslip from the well using a pair of angled forceps and invert it onto a drop of Citifluor. Seal the edges of the coverslip using clear nail varnish and keep slides at 4°C.

13. View under an epifluorescence microscope fitted with a suitable lamp and narrow band pass filters appropriate to the fluorochromes used (*see* **Note 4**).
14. Photograph or store images for processing (*see* **Note 5**).

### 3.2. Concurrent Labeling of Tissue Sections

1a. Wax sections: Cut 5-μm sections onto Vectabond or poly-L-lysine coated slides and allow to dry. Dewax sections and rehydrate following normal procedures.
1b. Frozen sections: Cut 5-μm sections onto Vectabond coated slides and allow to air-dry overnight. Fix sections in chloroform–acetone (1:1) at 4°C for 5 min and wash carefully under running tap water.
2. Carefully dry around sections using a tissue and draw a water-resistant barrier around them using a pen designed for immunocytochemistry, such as the Dako pen.
3. Place slides in a humid chamber and add 100 μL of combined antibodies, diluted appropriately in PBS/BSA to each section. Incubate for 1–2 h at room temperature or overnight at 4°C.
4. Process as for cells (*see* **Subheading 3.1., steps 5–14**).

## 4. Notes

1. If the background persists despite appropriate blocking of the tissue and careful washing of the tissues/cells it is often possible that the secondary antibody is too concentrated. Generally, it is possible to dilute the secondary antibody further than suggested by the manufacturers.
2. When an antigen has been localized previously using immunoperoxidase, it is possible that a decrease in sensitivity may occur when attempting to localize this antigen using immunofluorescence. Although this problem may be solved by using an amplification step, as described earlier, it is sometimes possible to solve the problem simply by using a higher concentration of the primary antibody.
3. Problems with nonspecific binding of antibodies to the cells or tissues can be often solved by preblocking the tissues/cells in bovine serum albumin. However, where this method fails to work it is possible to preblock using 5% normal serum from the same host species as the labeled antibody. The immunoglobulins in the serum should bind to any sticky sites on the samples, thus preventing nonspecific binding of the labeled antibody. It is worth noting that tissue should never be blocked using serum from the same animal host as the primary animal, as this could result in increased background.
4. If autofluorescence is suspected and labeling is equivocal as a result, it is worth repeating the multiple labeling experiment and changing the fluorochromes. If the labeling pattern remains the same it is likely to represent the localization of the primary antibody. However, if the labeling pattern changes as a result of swapping the fluorochromes the original labeling pattern was probably a result of autofluorescence.
5. It is important to remember that it is difficult to illustrate quadruple labeling, as only three colors can be represented at anyone time; red, green and blue. To

demonstrate four separate antigens it is important to ensure that at least two of the antigens have a completely separate distribution pattern, so that they can be displayed in the same color. Alternatively, the four antigens can be displayed as two separate pictures.

## References

1. Coons, A. H. and Kaplan, M. H. (1941) Immunological properties of an antibody containing a fluorescent group. *Proc. Soc. Exp. Biol. Med.* **47,** 200–205.
2. Coons, A. H. and Kaplan, M. H (1950) Localization of antigen in tissue cells. II. Improvements in method for the detection of antigen by means of fluorescent antibody. *J. Exp. Med.* **91,** 1–13.
3. Atherton, A. J., O'Hare, M. J., Buluwela, L., Titley, J., Monaghan, P., Paterson, H. F., et al. (1994) Ectoenzyme regulation by phenotypically distinct fibroblast sub-populations isolated from the human mammary gland. *J. Cell Sci.* **107,** 2931–2939.
4. Rost, F. W. D. (1992) The optical system of the fluorescence microscope, in *Fluorescence Microscopy*, vol. 1. Cambridge University Press, Cambridge, UK, pp. 35–63.
5. Johnson, G. D., Davidson, R. S., McNamee, K. G., Russel, G., Goodwin, D., and Holborow, E. J. (1982) Fading of immunofluorescence during microscopy: a study of the phenomena and its remedy. *J. Immunol. Methods* **55,** 231–242.
6. Rost, F. W. D. (1995) Indirect fluorochromy: immunofluorescence, hybridization, lectins, in *Fluorescence Microscopy*, vol. 2. Cambridge University Press, Cambridge, UK, pp. 108–134.
7. Hunyady, B., Krempels, K., Harta, G., and Mazcy, E. (1996) Immunohistochemical signal amplification by catalyzed reporter deposition and its application in double immunostaining. *J. Histochem. Cytochem.* **44,** 1353–1362.
8. Rost, F. W. D. (1995) Autofluorescence in human and animal tissues, in *Fluorescence Microscopy*, vol. 2. Cambridge University Press, Cambridge, UK, pp. 1–15.

# 4

# Lectin Histochemistry to Detect Altered Glycosylation in Cells and Tissues

**Susan A. Brooks and Debbie M. S. Hall**

## 1. Introduction

### *1.1. What Are Lectins? Why Use Them?*

A lectin is "a protein or glycoprotein of non-immune origin, not an enzyme, that binds to carbohydrates and agglutinates cells" ***(1)***. Lectins are naturally occurring substances, most commonly derived from plant or sometimes invertebrate sources, that can be exploited in the laboratory to detect and reveal carbohydrate structures in or on the surface of cells in very much the same way that antibodies can be used to reveal specific antigens. Lectins can detect very subtle alterations in cellular glycosylation. This is of interest in metastasis research as there is increasing evidence that marked glycosylation changes can attend both transformation to malignancy and tumor progression.

### *1.2. Nomenclature of Lectins*

Lectins are named after the source from which they are derived—sometimes the Latin binomial (e.g., *Bandeirea simplicifolia* lectin or *Dolichos biflorus* lectin), sometimes the common name (e.g., peanut lectin or wheatgerm lectin), or sometimes by a slightly obscure historical term (e.g., Concanavalin A [Con A] for the lectin from *Canavalia ensiformis*, the jack bean). The term lectin is used fairly interchangeably with the older term "agglutinin," as in peanut agglutinin or *Helix pomatia* agglutinin. Lectins are often referred to by an abbreviation for their names, for example, PNA (peanut agglutinin) or DBA (*Dolichos biflorus* agglutinin); obscurely, PHA, which actually stands for phytohaemagglutinin is, for historical reasons, the abbreviation usually employed for the lectin derived from *Phaseolus vulgaris*. Many sources yield more than one lectin, termed isolectins, which may

From: *Methods in Molecular Medicine, vol. 57:*
*Metastasis Research Protocols, Vol. 1: Analysis of Cells and Tissues*
Edited by: S. A. Brooks and U. Schumacher © Humana Press Inc., Totowa, NJ

have quite different carbohydrate binding specificities (e.g., the gorse *Ulex europaeus* yields two major isolectins, *Ulex europaeus* agglutinin I [UEA-I], which has a strong binding preference for fucose, and *Ulex europaeus* agglutinin II [UEA-II], which has a strong binding preference for *N*-acetylglucosamine).

### 1.3. Monosaccharide Binding Specificity and Complex Natural Binding Partners

The sugar specificity of a lectin is usually quoted in terms of the monosaccharide that most effectively inhibits its binding. For example, the lectin from the Roman snail *Helix pomatia* (HPA) has a monosaccharide binding specificity for α-D-*N*-acetylgalactosamine, as does the lectin from the horse gram *Dolichos biflorus* (DBA). It is very important, however, to bear in mind that the combining site of the lectin commonly encompasses more than the terminal monosaccharide (e.g., α-D-*N*-acetylgalactosamine) in an oligosaccharide chain, and that lectins with nominally identical monosaccharide binding specificities will actually recognize subtly different complex binding partners and will give quite different results in lectin histochemistry. In the example given above, HPA and DBA will actually give quite distinct staining patterns in lectin histochemistry in spite of their identical nominal binding specificities.

### 1.4. Range of Lectins Available Commercially

Many hundreds of lectins are readily available, at modest cost, from commercial sources. Very few are well characterized with well-defined binding characteristics, and little or nothing is known about some. In most cases, limited information, such as monosaccharide binding preference, will be available. It is therefore often the case that the experimenter may have to test several lectins before the tool most ideally suited to the experiment is determined.

Although the literature abounds with lectin histochemical studies, surprisingly little useful data have yet been revealed by their use. This may be in the most part owing to the conservative nature of many lectin histochemical studies, which often concentrate on a small panel of much-used and well-characterized lectins, probably chosen for their nominal coverage of the major monosaccharides expressed by human tissues, and often purchased as a kit from one of the major suppliers. Progress seems more likely if experimenters are more adventurous and use a broader spectrum of the lectins available.

### 1.5. Range of Lectin Histochemistry Methods

Lectin histochemistry can be carried out on a range of different cell and tissue preparations including, as described in this chapter, frozen or cryostat sections of fresh tumor tissue; formalin-fixed, paraffin wax-embedded surgical specimens; cells in suspension (either cultured cell lines or clinical samples such as blood samples, ascites, or pleural effusions); and cultured cancer cell lines either grown on coverslips or pelleted, fixed and processed to paraffin wax blocks. If cells or tissues are fixed and processed to paraffin wax, carbohydrate retrieval methods (trypsinization or microwave treatment) may assist recovery of sequestered carbohydrates and improve labeling results significantly.

A very large number of histochemical labeling techniques are available for histochemical detection of lectin binding to cells and tissues. They are basically variations of the techniques used for immunocytochemistry, as described in Chapter 2 by Brooks. Each carry their own advantages and disadvantages, but the several simple and basic techniques described here should all work well and give good results. Lectin binding may be visualized by either a fluorescent (especially suitable for fresh cell smears or cells cultured on coverslips) or an enzyme label which yields a colored product (the enzyme label in the methods given here is horseradish peroxidase, which yields a deep chocolate brown product with the chromogenic substrate diaminobenzidine–$H_2O_2$, but alternative enzyme labels such as alkaline phosphatase are also commonly used). Detection methods range from the very straightforward, but insensitive, direct method to the more complex, multistep, but sensitive avidin–biotin complex method. Other, even more sensitive, but more complex methods also exist.

### 1.6. Some Examples of Glycosylation Changes, Detectable by Lectin Histochemistry, of Interest in Metastasis Research

There are a number of major alterations in cellular glycosylation, detectable by lectin histochemistry, that appear to be of importance in metastasis research. Some examples, to be briefly discussed here, include altered sialylation, overexpression of complex β1–6 branched *N*-acetylglucosamine oligosaccharides (detected by the lectin PHA), and overexpression *N*-acetylgalactosamine sugars (detected by the lectin HPA). It seems likely that in the rapidly advancing field of glycobiology, as the functions of complex sugars and their interactions with their receptors become better understood, their relevance and interactions in the complex processes of metastasis will increasingly be revealed.

#### 1.6.1. Alterations in Sialylation

Sialic acids are commonly found at the nonreducing termini of oligosaccharide chains, and disturbances in their expression associated with malignancy and tumor progression have been reported. Increased expression of sialylated structures, for example, the sialylated forms of the Tn epitope, sialyl Lewis a and sialyl Lewis x, have been associated with malignancy and with metastatic competence, as has overexpression of sialyltransferase enzymes *(2–5)*. Conversely, other studies have correlated failure in sialylation mechanisms with experimental metastatic capacity *(6)*.

#### 1.6.2. PHA Binding to Detect Complex β1–6 Branched Oligosaccharides

Increased expression of complex β1–6 branched oligosaccharides, detected by the lectin PHA, has been associated to increased metastatic potential of tumor cells in a number of studies ***(7–9)***.

#### 1.6.3. HPA Binding to Detect Overexpression of N-Acetylgalactosamine

HPA recognizes oligosaccharides expressing terminal α-D-*N*-acetylgalactosamine. A number of studies have reported that positive binding of HPA by lectin histochemistry, detecting overexpression of oligosaccharides expressing terminal *N*-acetylgalactosamine, is associated with metastatic competence and poor patient survival in a range of human adenocarcinomas including breast, colon, gastric, prostate, and esophageal ***(10–13)***.

## 2. Materials

### 2.1. Silane Treatment of Microscope Slides (see Note 1)

1. Acetone.
2. Acetone/silane: 7 mL of 3-(triethoxysilyl)-propylamine (APES) in 400 mL of acetone; make fresh on the day of use. Discard after treating a maximum of 1000 slides.
3. Distilled water.

### 2.2. Different Types of Cell and Tissue Preparations

#### 2.2.1. Frozen or Cryostat Sections

1. Chunk of fresh tissue >0.5 $cm^3$ in size.
2. OCT embedding medium.
3. Isopentane.
4. Liquid nitrogen.
5. Silane-coated clean glass microscope slides (*see* **Note 1**).
6. Acetone.
7. Aluminum foil.

#### 2.2.2. Fixed, Paraffin Wax-Embedded Sections

1. Paraffin wax-embedded tissue blocks (*see* **Note 2**).
2. 20% v/v ethanol in distilled water.
3. Silane-coated glass microscope slides (*see* **Note 1**).
4. Xylene.
5. Absolute ethanol or industrial methylated spirit.
6. 70% v/v ethanol or industrial methylated spirit in distilled water.
7. Distilled water.

#### 2.2.2.1. Carbohydrate Retrieval by Trypsinisation

1. Tris-buffered saline (TBS), pH 7.6: 60.57 g of Tris, 87.0 g of NaCl dissolved in 1 L of distilled water. Adjust pH to 7.6 using concentrated HCl. Make up to a total volume of 10 L using distilled water.
2. Trypsin solution: 1 mg/mL of crude, type II trypsin, from porcine pancreas (*see* **Note 3**) in Tris-buffered saline, pH 7.6.

#### 2.2.2.2. Carbohydrate Retrieval by Microwave Treatment

1. Citrate buffer, pH 6.0: 2.1 g of citric acid dissolved in 1 L of distilled water. Adjust pH to 6.0 using concentrated NaCl.
2. Distilled water.

### 2.2.3. Cell Smears

1. Cells in suspension (*see* **Note 4**).
2. Silane-coated glass microscope slides (*see* **Note 1**).
3. Aluminum foil.
4. Acetone.

### 2.2.4. Preparation of Cultured Cells

#### 2.2.4.1. Grown on Coverslips

1. Cells cultured under standard conditions to 70% confluence.
2. Cell culture medium, without fetal calf serum.
3. Alcohol- or autoclave-sterilized round glass coverslips (13 mm diameter, thickness 0).
4. 0.2 *M* PIPES buffer, pH 6.9: Stir 12.1 g of PIPES (piperazine-*N,N'*-*bis*-[2-ethane-sulfonic acid]) into 50 mL of ultrapure water to give a cloudy solution. Add approx 40 mL of 1 *M* NaOH and the solution should clear. Check pH and adjust to 6.9, if necessary, using 1 *M* NaOH. Add ultrapure water to give a final volume of 200 mL.
5. 0.1 *M* PIPES buffer, pH 6.9: Prepare 0.2 *M* PIPES buffer, as in **step 4**, but add ultrapure water to give a total volume of 400 mL, instead of 200 mL, or take a small volume, for example, 50 mL of 0.2 *M* PIPES buffer and dilute 1:1 with ultrapure water.
6. 3% v/v paraformaldehyde in 0.1 *M* PIPES buffer, pH 6.9: Place 3 g of para-formaldehyde in a 250-mL conical flask, add 30 mL of ultrapure water, loosely stopper, and heat on a 60°C hotplate in a fume cupboard for about 30 min to give a cloudy solution. Add 1 *M* NaOH (the purest grade) with continual stirring until the solution clears. Add ultrapure water to give a total volume of 50 mL, then add 50 mL of 0.2 *M* PIPES buffer, pH 6.9 (*see* **step 4**). Divide into 10-mL aliquots and store frozen. Defrost in warm a water bath for use.
7. 0.1% v/v Triton X-100 in 0.1 *M* PIPES buffer, pH 6.9.

#### 2.2.4.2. Fixed and Processed for Paraffin Wax Embedding

1. Cells cultured to 70% confluence under standard conditions.
2. Cell culture medium, without fetal calf serum.
3. 4% v/v formol saline in distilled water, at 4°C: Dissolve 4.25 g of NaCl in 500 mL of a 4% v/v formaldehyde solution in distilled water.
4. Lectin buffer: 60.57 g of Tris, 87.0 g of NaCl, 2.03 g of $MgCl_2$, 1.11g $CaCl_2$ dissolved in 1 L of distilled water. Adjust pH to 7.6 using concentrated HCl. Make up to a total volume of 10 L using distilled water (*see* **Note 5**).
5. 4% w/v agarose in lectin buffer: Suspend 4 g of low gelling temperature agarose in 100 mL of lectin buffer and heat gently (in a water bath or microwave oven, with regular stirring) until the agarose dissolves (at approx 60°C). Cool to approx 37–40°C (the agar should be thickening slightly as it begins to gel) for use.

## 2.3. Quenching Endogenous Peroxidase

1. Methanol–hydrogen peroxide solution: 3% v/v hydrogen peroxide in methanol. Make up fresh every 2–3 d.

## 2.4. Examples of Histochemical Staining Techniques

### 2.4.1. Direct Fluorescent Label Method

1. Fluorescently-labeled lectin, for example, fluorescein isothiocyanate (FITC)-labeled lectin, tetramethyl rhodamine (TRITC)-labeled lectin, or Texas red-labeled lectin.
2. Lectin buffer: 60.57 g of Tris, 87.0 g of NaCl, 2.03 g of $MgCl_2$, 1.11 g of $CaCl_2$ dissolved in 1 L of distilled water. Adjust pH to 7.6 using concentrated HCl. Make up to a total volume of 10 L using distilled water (*see* **Note 5**).

### 2.4.2. Direct Horseradish Peroxidase Label Method

1. Horseradish peroxidase labeled lectin.
2. Lectin buffer: 60.57 g of Tris, 87.0 g of NaCl, 2.03 g of $MgCl_2$, 1.11 g of $CaCl_2$ dissolved in 1 L of distilled water. Adjust pH to 7.6 using concentrated HCl. Make up to a total volume of 10 L using distilled water (*see* **Note 5**).
3. DAB–$H_2O_2$: 0.5 mg/mL of 3,3-diaminobenzidine tetrahydrochloride (DAB) in lectin buffer. Add $H_2O_2$ to give a concentration of 5% v/v immediately before use (*see* **Notes 6–8**).
4. Mayer's hematoxylin (*see* **Note 9**).

### 2.4.3. Biotin-Labeled Lectin: Simple Avidin–Biotin Method

1. Biotin-labeled lectin.
2. Lectin buffer: 60.57 g of Tris, 87.0 g of NaCl, 2.03 g of $MgCl_2$, 1.11 g of $CaCl_2$ dissolved in 1 L of distilled water. Adjust pH to 7.6 using concentrated HCl. Make up to a total volume of 10 L using distilled water (*see* **Note 5**).
3. Streptavidin labeled with horseradish peroxidase.
4. DAB–$H_2O_2$: 0.5 mg/mL of DAB in lectin buffer. Add $H_2O_2$ to give a concentration of 5% v/v immediately before use (*see* **Notes 6–8**).
5. Mayer's hematoxylin (*see* **Note 9**).

### *2.4.4. Indirect Antibody Method*

1. Native, unlabeled lectin.
2. Lectin buffer: 60.57 g of Tris, 87.0 g of NaCl, 2.03 g of $MgCl_2$, 1.11 g of $CaCl_2$ dissolved in 1 L of distilled water. Adjust pH to 7.6 using concentrated HCl. Make up to a total volume of 10 L using distilled water (*see* **Note 5**).
3. Peroxidase-labeled rabbit polyclonal antibody directed against the lectin (*see* **Note 10**).
4. DAB–$H_2O_2$: 0.5 mg/mL of DAB in lectin buffer. Add $H_2O_2$ to give a concentration of 5% v/v immediately before use (*see* **Notes 6–8**).
5. Mayer's hematoxylin (*see* **Note 9**).

### *2.4.5. Avidin–Biotin Complex (ABC) Method*

1. Biotin-labeled lectin.
2. Lectin buffer: 60.57 g of Tris, 87.0 g of NaCl, 2.03 g of $MgCl_2$, 1.11 g of $CaCl_2$ dissolved in 1 L of distilled water. Adjust pH to 7.6 using concentrated HCl. Make up to a total volume of 10 L using distilled water (*see* **Note 5**).
3. Horseradish peroxidase avidin–biotin complex (*see* **Note 11**).
4. DAB-$H_2O_2$: 0.5 mg/mL of DAB in lectin buffer. Add $H_2O_2$ to give a concentration of 5% v/v immediately before use (*see* **Notes 6–8**).
5. Mayer's hematoxylin (*see* **Note 9**).

## *2.5. Dehydration, Clearing, and Mounting Slides for Viewing by Microscopy*

1. 70% v/v ethanol or industrial methylated spirit in distilled water.
2. 100% ethanol or industrial methylated spirit.
3. Xylene.
4. Xylene-based mounting medium, for example, Depex for enzyme-labeled preparations, or anti-fade mountant, for example, Citifluor for fluorescent-labeled preparations.
5. Coverslips.

# 3. Methods

## *3.1. Silane Treatment of Microscope Slides (see Note 1)*

1. Place up to a maximum of 1000 clean glass microscope slides in slide carriers.
2. Place the first slide carrier in a trough containing enough acetone to cover the slides; leave for 5 min.
3. Transfer, smoothly and without agitation, to a trough containing enough acetone–APES to cover the slides; leave for 5 min.
4. Transfer, smoothly and without agitation, to a trough containing enough distilled water to cover the slides; leave for 5 min. Discard the water and replace with fresh after five carriers of slides have passed through.
5. Transfer, smoothly and without agitation, to a second trough of distilled water; leave for 5 min. Discard the water and replace with fresh after five carriers of slides have passed through.

6. Transfer to absorbent paper and allow to drain.
7. Allow to dry thoroughly—either at room temperature, or more quickly in a suitable incubator, glass drying cabinet, or oven.
8. When completely dry, store in labeled boxes until required.

### 3.2. Different Types of Cell and Tissue Preparations

Cells and tissues can be prepared in a number of different ways for lectin histochemistry. Some of the more common, and more useful preparations are listed below.

#### 3.2.1. Frozen or Cryostat Sections

1. Using a sharp, clean blade, cut a solid tissue block of approx 0.5 $cm^3$ (*see* **Note 12**).
2. Place the tissue on an OCT-coated cryostat chuck (*see* **Note 13**).
3. Using long-handled tongs, pick up the chuck and immerse chuck and tissue in isopentane precooled in liquid nitrogen for approx 1–2 min (*see* **Note 14**).
4. Place the frozen chuck in the cabinet of the cryostat and leave to equilibrate for approx 30 min.
5. Using the cryostat, cut 5-µm thick sections and pick them up on silane-coated (*see* **Note 1**) clean glass microscope slides. Allow to air-dry for approx 5 min.
6. Sections may then either be stored until required by wrapping individually in aluminum foil and storing in the freezer, or may be used at once. If stored frozen, allow to thaw and equilibrate to room temperature before use.
7. Immediately before use, dip slides in acetone for 1 min and air-dry for approx 5 min.

#### 3.2.2. Fixed, Paraffin Wax-Embedded Sections

1. Cool wax-embedded tissue blocks on ice for approx 15 min.
2. Cut 5 µm thick sections by microtome (*see* **Note 15**).
3. Float sections out on a pool of 20% v/v ethanol in distilled water on a clean glass plate supported by a suitable receptacle such as a glass beaker or jar (*see* **Note 16**).
4. Carefully transfer the sections, floating on the alcohol, onto the surface of a water bath heated to 40°C—they should puff out and become flat (*see* **Note 17**).
5. Separate out individual sections very gently, using the tips of fine, bent forceps.
6. Pick up individual sections on clean, silane-coated glass microscope slides (*see* **Note 1**).
7. Allow slides to drain by up-ending them on a sheet of absorbent paper for 5–10 min.
8. Dry slides either in a 37°C incubator overnight, or on a hotplate at 60°C for 20 min. Slides may then be cooled, stacked or boxed, and stored at room temperature in a dust-tight container until required.
9. When required, soak slides in xylene for approx 15 min to remove the paraffin wax.
10. Transfer through two changes of absolute ethanol or industrial methylated spirit, then through one change of 70% v/v ethanol or industrial methylated spirit in distilled water, then distilled water, agitating the slides vigorously for 1–2 min at each stage to equilibrate (*see* **Note 18**).

#### 3.2.2.1. Carbohydrate Retrieval by Trypsinization (*see* **Note 19**).

1. Immerse slides in trypsin solution at 37°C, in an incubator or water bath, for 5–30 min (*see* **Notes 3** and **20**).
2. Wash under running tap water for 5 min.

#### 3.2.2.2. Carbohydrate Retrieval by Microwave Treatment (*see* **Note 19**)

1. Immerse slides in citrate buffer, pH 6.0, in any suitable microwavable container, such as a plastic sandwich box (*see* **Note 21**).
2. Place in a conventional microwave oven and heat on full power until the buffer boils.
3. Reduce the power to "simmer" or "defrost" for 5 min, so that the buffer boils gently. After 5 min, check the level of the buffer, and top up with hot distilled water if necessary. Heat on "simmer" or "defrost" further for 5 min.
4. Allow slides to cool at room temperature for 30 min (*see* **Note 22**).
5. Wash under running tap water for 5 min.

### 3.2.3. Cell Smears

1. Place a drop of cells in suspension approx 5 mm from one end of a clean silane-coated glass microscope slide (*see* **Note 1**).
2. Place a second clean silane-coated glass microscope slide on top of the first, allowing approx 1 cm of glass to protrude at either end, and allowing the drop to spread between the two.
3. Drag one slide over the other in a rapid, smooth movement, spreading the cells in a thin smear over the surface of both slides.
4. Air-dry the slides for approx 5 min. They may then be used at once, or wrapped individually in foil and stored in the freezer until required. If stored frozen, allow to thaw to room temperature before use.
5. When ready for use, fix by dipping in acetone for 1 min and air-dry.

### 3.2.4. Preparation of Cultured Cells

#### 3.2.4.1. Grown on Coverslips

1. Wash cultured cells, grown to 70% confluence, in fresh medium.
2. Aspirate and discard the medium.
3. Scrape cells from the flask using a rubber policeman, and resuspend in fresh medium.
4. Count cells, and subculture $1 \times 10^5$ cells into Petri dishes.
5. Place sterile coverslips in Petri dishes, and allow cells to proliferate for 24 h at 37°C under 5% $CO_2$.
6. Carefully remove coverslips using fine forceps or the edge of a scalpel blade.
7. Place coverslips, cell side up, onto a piece of dental wax or Parafilm for support, and cover each with 100 µL of cold 3% v/v paraformaldehyde in 0.1 *M* PIPES buffer, pH 6.9, for 15 min.
8. Wash thoroughly in 0.1 *M* PIPES buffer, pH 6.9.
9. Permeabilize in 0.1% v/v Triton X-100 (diluted in 0.1 *M* PIPES buffer, pH 6.9) for 10 min (*see* **Note 23**).
10. Wash thoroughly in 0.1 *M* PIPES buffer, pH 6.9.

#### 3.2.4.2. Fixed and Processed for Paraffin Wax Embedding

1. Wash cultured cells, grown to 70% confluence, in several changes of fresh medium without fetal calf serum.
2. Scrape cells from the flask using a rubber policeman.
3. Transfer to a 50-mL centrifuge tube and centrifuge at 1100 g for 5 min to pellet the cells.
4. Gently resuspend the pellet (*see* **Note 24**) in fresh medium without fetal calf serum, centrifuge at 1100 g for 5 min. Repeat.
5. Aspirate the supernatant and discard.
6. Gently resuspend the cell pellet (*see* **Note 24**) in 4% v/v formol saline for 2 h at 4°C to fix the cells.
7. Centrifuge, as previously, and aspirate and discard the supernatant.
8. Wash the cell pellet in three changes of lectin buffer—centrifuge, aspirate and discard the supernatant, add fresh buffer, resuspend the cells, as previously.
9. Gently resuspend the cell pellet in a drop of warm (37°C) liquid agarose and allow to set at room temperature (*see* **Note 25**).
10. Process the cell–agarose pellets for paraffin wax embedding and then cut sections as detailed in **Subheading 3.2.2.** Carbohydrate retrieval methods (**Subheadings 3.2.2.1.** and **3.2.2.2.**) may be appropriate.

## 3.3. Quenching Endogenous Peroxidase (see Note 26)

1. Immerse slides in methanol–hydrogen peroxide solution for 20 min.
2. Wash under running tap water for approx 5 min.

## 3.4. Examples of Some Histochemical Staining Techniques (see Note 27)

### 3.4.1. Direct Fluorescent Label Method (see **Note 28**)

1. Incubate slides with fluorescently-labeled lectin at a concentration of 10 μg/mL in lectin buffer in a humid chamber (e.g., a sandwich box lined with damp filter paper) for 1 h (*see* **Note 29**).
2. Wash in three changes of lectin buffer (*see* **Note 30**).
3. Proceed to **Subheading 3.5.**

### 3.4.2. Direct Horseradish Peroxidase Label Method (see **Note 31**)

1. Incubate slides with horseradish peroxidase labeled lectin at a concentration of 10 μg/mL in lectin buffer in a humid chamber (e.g., a sandwich box lined with damp filter paper) for 1 h (*see* **Note 29**).
2. Wash in three changes of lectin buffer (*see* **Note 30**).
3. Incubate with DAB–$H_2O_2$ for 10 min (*see* **Notes 6–8**).
4. Wash under running tap water for 5 min.
5. Counterstain in Mayer's hematoxylin for 2 min (*see* **Note 9**).
6. "Blue" under running tap water for approx 5 min (*see* **Note 32**).
7. Proceed to **Subheading 3.5.**

### *3.4.3. Biotin-Labeled Lectin: Simple Avidin–Biotin Method (see **Note 33**)*

1. Incubate slides with biotin-labeled lectin at a concentration of 10 µg/mL in lectin buffer in a humid chamber (e.g., a sandwich box lined with damp filter paper) for 1 h (*see* **Note 29**).
2. Wash in three changes of lectin buffer (*see* **Note 30**).
3. Incubate with streptavidin-labeled with peroxidase at a concentration of 5 µg/mL in lectin buffer in a humid chamber for 30 min.
4. Wash in three changes of lectin buffer (*see* **Note 30**).
5. Incubate with DAB–$H_2O_2$ for 10 min (*see* **Notes 6–8**).
6. Wash under running tap water for 5 min.
7. Counterstain in Mayer's hematoxylin for 2 min (*see* **Note 9**).
8. "Blue" under running tap water for approx 5 min (*see* **Note 32**).
9. Proceed to **Subheading 3.5.**

### *3.4.4. Indirect Antibody Method (see **Note 10**)*

1. Incubate slides with unlabeled lectin at a concentration of 10 µg/mL in lectin buffer in a humid chamber (e.g., a sandwich box lined with damp filter paper) for 1 h (*see* **Note 29**).
2. Wash in three changes of lectin buffer (*see* **Note 30**).
3. Incubate with horseradish peroxidase labeled rabbit polyclonal antibody directed against the lectin, at a dilution of 1:100 in lectin buffer in a humid chamber for 1 h.
4. Wash in three changes of lectin buffer (*see* **Note 30**).
5. Incubate with DAB–$H_2O_2$ for 10 min (*see* **Notes 6–8**).
6. Wash under running tap water for 5 min.
7. Counterstain in Mayer's hematoxylin for 2 min (*see* **Note 9**).
8. "Blue" under running tap water for approx 5 min (*see* **Note 32**).
9. Proceed to **Subheading 3.5.**

### *3.4.5. ABC Method (see **Note 34**)*

1. Incubate slides with biotin-labeled lectin at a concentration of 10 µg/mL in lectin buffer in a humid chamber (e.g., a sandwich box lined with damp filter paper) for 1 h (*see* **Note 29**).
2. Wash in three changes of lectin buffer (*see* **Note 30**).
3. Incubate with horseradish peroxidase ABC, made up according to the manufacturer's instructions (*see* **Note 11**) in a humid chamber for 30 min.
4. Wash in three changes of lectin buffer (*see* **Note 30**).
5. Incubate with DAB–$H_2O_2$ for 10 min (*see* **Notes 6–8**).
6. Wash under running tap water for 5 min.
7. Counterstain in Mayer's hematoxylin for 2 min (*see* **Note 9**).
8. "Blue" under running tap water for approx 5 min (*see* **Note 32**).
9. Proceed to **Subheading 3.5.**

### 3.5. Dehydration, Clearing, and Mounting Slides for Viewing by Microscopy (see Note 35)

1. Dehydrate by passing through 70% v/v ethanol or industrial methylated spirit in distilled water, then two changes of 100% ethanol or industrial methylated spirit, then clear in xylene. Agitate slides for 1–2 min at each stage to equilibrate.
2. Horseradish peroxidase labeled slides should be mounted in a xylene-based mounting medium, for example, Depex. Fluorescently labeled slides should be mounted in a commercially available fade-resistant mounting medium, for example, Citifluor.

### 3.6. Positive and Negative Controls and Confirmation of Specificity of Labeling

1. Positive control: A cell or tissue preparation that is known to give positive labeling from a previous experiment makes the ideal positive control. If this is not possible, kidney—either human or animal—makes an excellent positive control for labeling with many lectins owing to its diverse glycosylation patterns.
2. Negative control: Omit the lectin. Labeling should be completely abolished.
3. Confirmation of specificity: Add 0.1–0.5 *M* monosaccharide for which the lectin shows greatest affinity (e.g., for *Ulex europaeus* isolectin I add fucose; for *Helix pomatia* lectin add *N*-acetylgalactosamine) to the lectin solution, at its working dilution, about 30 min prior to incubation with the cell or tissue preparation (*see* **Note 36**).

### 3.7. Interpretation of Labeling Results

Slides should be viewed by microscope. Good labeling is indicated by a strong specific label (either deep brown for horseradish peroxidase–DAB or fluorescent) and low, or preferably nonexistent, background and nonspecific labeling (*see* **Note 37**). Omission of the lectin should abolish labeling completely. Competitive inhibition with appropriate monosaccharide should abolish or at least dramatically diminish labeling.

It is often helpful to score labeling on an arbitrary scale where the observer estimates the percentage of cells, for example, cancer cells, labeled (10%, 50%, 95%, etc.) and the intensity of labeling on a scale of – (no labeling at all), + – very weak labeling, + (weak but definite labeling) to ++++ (extremely intense labeling) to give results ranging from completely negative to 100% ++++.

## 4. Notes

1. APES alters the charge on the glass of the microscope slide so that cell and tissue preparations adhere much more firmly. We routinely APES treat *all* our microscope slides before use. It is absolutely essential if slides are to be microwave treated (*see* **Subheading 3.2.2.2.**) as the aggressiveness of the treatment will otherwise dislodge even the best cell or tissue preparation. It is desirable, but not essential, for

other applications. Alternative, commercial, brand-named preparations are also available, but tend to be more expensive. Other adhesives, for example, glycerol gelatin, hen's egg albumin, and poly-L-lysine will also improve adherence of cells/tissues to slides but are not as effective, and are certainly inadequate for preparations that are to be microwave treated. Once acetone–APES solution is prepared it should be used the same day then discarded. A 400 mL acetone–7 mL APES solution is sufficient to treat up to about 1000 microscope slides; it is therefore a good idea to treat a batch of 1000 slides, and then store them for use as required.

2. Paraffin wax-embedded tissue blocks are often the most convenient source of material for clinical studies as tumor specimens are routinely fixed in formalin and processed to paraffin wax for sectioning, staining, and histopathological diagnosis in most centers. Such blocks are often stored in hospital archives for years or decades, providing a hugely valuable resource for retrospective studies on glycosylation related to metastatic potential of tumors.
3. Crude, type II trypsin from porcine pancreas works best as the presence of impurities (e.g. chymotrypsin) assists its effect. Do not use purer (and more expensive!) products.
4. Any cells in suspension are suitable, for example, blood, cancer cells in ascites or in pleural effusions taken from patients or from animal models, cells derived from solid tissue tumors and released into suspension, or cultured cells in suspension.
5. Lectin buffer, which is basically TBS with added $CaCl_2$ and $MgCl_2$, is good for all lectin histochemistry applications. We recommend it for all dilutions and washes. Many lectins are known to require $CaCl_2$ and $MgCl_2$; the requirements of other lectins remain unknown. Never use phosphate-buffered saline (PBS) for lectin histochemistry, as the phosphate ions bind with, and sequester, the metals.
6. DAB is potentially carcinogenic and should be handled with care. Wear gloves. Avoid spillages and aerosols. Work in a fume cupboard. After use, soak all glassware etc. in a dilute solution of bleach overnight before washing. Swab down worktops with dilute bleach after use. Clean up spillages with excess water, then swab with dilute bleach.
7. DAB is available in dropper bottle kit form from major suppliers. This works out a lot more expensive option than preparing solutions from scratch, but is convenient and safer.
8. We usually make a concentrated DAB stock solution at 5 mg/mL in lectin buffer and freeze in 1-mL aliquots in 10-mL plastic screw top tubes until required. This minimizes the risk of aerosols from weighing out powder when required.
9. Mayer's hematoxylin: A number of different hematoxylin solutions are commercially available. We find Mayer's, which is a progressive stain, most convenient, although other types are equally effective.
10. Peroxidase-labeled rabbit polyclonal antibodies against lectins: Only a very limited range of polyclonal antisera against lectins are available commercially, so lack of availability may limit the applications of this method.
11. ABC is available from major suppliers, often as convenient dropper bottle kits. Avidin has four binding sites for biotin. Avidin and labeled biotin are mixed

together in such a ratio that three of avidin's possible four binding sites are saturated with labeled biotin, leaving one free to combine with the biotin label attached to the lectin. Follow manufacturer's instructions. These usually require that the avidin and labeled biotin solutions be mixed together in the appropriate ratio at least 30 min prior to use.

12. When cutting tissue, use a very sharp blade, for example, a new razor blade or scalpel, and use single, firm, swift, downward stokes. Do not hack at the tissue and avoid crushing it as this will result in poor morphology. Cut the tissue into a geometrical shape with straight edges, preferably a square, as this will make sectioning easier.
13. OCT acts as a support for the tissue block during sectioning. Be generous with it and embed the tissue block in a copious pool of it. Align the tissue block, in the OCT, with one of its straight edges parallel with the cutting edge of the chuck, as this will make sectioning easier.
14. It is possible to freeze the tissue block in other ways—for example, by immersing simply in liquid nitrogen, by using a commercially available freezing spray, by blasting with $CO_2$ gas, or even simply placing it in the chamber of the cryostat until frozen. However, to obtain optimum morphological integrity, the tissue should be frozen as rapidly as possible, avoiding the formation of morphology-destroying ice crystals, and the best way is to immerse in isopentane precooled in liquid nitrogen. Hexane precooled in liquid nitrogen is also very good. Isopentane and hexane are far better conductors of heat than liquid nitrogen alone and therefore facilitate extremely rapid and effective freezing.
15. Most tissues cut most effectively when very cold. Ice (e.g., an ice cube) should be applied to the surface of the tissue block every few minutes during cutting.
16. The idea of this step is to begin to flatten out small creases or wrinkles in the sections. Gentle manipulation using, for example, a soft paintbrush and/or forceps can aid this.
17. If sections remain wrinkled, it may indicate that the water is slightly too cold. If the wax begins to melt, it is too hot.
18. When slides are transferred from one solvent to the next, they initially appear smeary as the two solvents begin to mix. Vigorous agitation, that is, "sloshing them up and down," ensures that the slides equilibrate effectively with the new solvent. When they are equilibrated, the surface smearing will disappear. If white flecks or patches become visible around the sections, this indicates that the wax has not been adequately removed—return the sections to xylene for a further 10–15 min. A common cause of poor histochemistry results is inadequate removal of paraffin wax.
19. Carbohydrate retrieval methods: In some cases, if lectin histochemistry results on fixed, paraffin wax embedded tissues are disappointing, it may be because carbohydrate moieties have been sequestered during tissue fixation and processing to paraffin wax. Results can often be improved dramatically by carbohydrate retrieval techniques including trypsinization or microwave treatment. It is not possible to predict if carbohydrate retrieval methods will be effective, or which to choose; try both and select which conditions work best for your application. **Note:** These methods are *not* appropriate for fresh cell/tissue preparations such as frozen sections or cell smears. Other retrieval methods, such as autoclaving and pressure cooking, also exist.

20. Prewarm all glassware, solutions etc. to 37°C before use. Make trypsin solution up fresh immediately before use, as it self-digests and loses activity over time. Initially try a range of trypsinization times, for example, 0, 5, 10, 20, 30 min. Digestion times of >30 min are not recommended, as visible damage to tissue morphology becomes apparent.
21. Slides can conveniently be placed in commercially available slide carriers, which typically hold up to about 25 slides, or upright in Coplin jars. Do not overcrowd slides as this results in "hot spots" and uneven carbohydrate retrieval.
22. This cooling down period is part of the retrieval method and should not be skipped.
23. If cell surface carbohydrates only are of interest, this and subsequent steps may be omitted.
24. To resuspend cells, either vortex-mix in short sharp bursts, or tap the tube sharply. Do not suck the cells up and down in a pipet as this will damage them.
25. To suspend cells in liquid agarose at 37°C, prewarm a glass pipet to 37°C and transfer cells to a small drop (approx 0.5 mL) agarose in a small plastic tube, such as a 1.5-mL Eppendorf tube. Tap the tube several times sharply to mix. Leave agarose to set at room temperature for approx 30 min. The agarose/cell pellet can be removed by either cutting the plastic tube away, or winkling the pellet out using a syringe needle or pin.
26. If horseradish peroxidase is used to visualize lectin binding, endogenous enzyme within the cell or tissue preparation may contribute to nonspecific and confusing labeling. This may be very effectively overcome by quenching endogenous enzyme. Endogenous peroxidase is usually quenched immediately prior to incubation with the lectin. This step is not required if fluorescent labels are used. Other enzyme labels, for example, alkaline phosphatase, are also commonly used, and although they do not require this step, may require their own blocking procedure.
27. A large number of different histochemical techniques can be used to detect lectin binding to cell and tissue preparations. Choice of technique is dependent on the type of label preferred by the experimenter—fluorescent or enzyme (colored)—and the balance between quick, simple techniques which generally lack sensitivity, and the more sensitive techniques which tend to be more time consuming, contain more steps, and require more reagents (and are therefore slightly more expensive). All methods given here should give good results. Other methods also exist.
28. Fluorescent labels can give exceptionally beautiful results, but the fluorescence is ephemeral and preparations should be viewed (and if a permanent record is required, photographed) as soon as possible after labeling. Slides can be stored in the refrigerator, in the dark, for up to about 6 mo, but some preparations will deteriorate significantly during that time. This method is especially good for cells in suspension and cell smears. It may also be used for tissue sections, and use of a commercially available fluorescent nuclear counterstain is then recommended to aid interpretation of tissue morphology.
29. Lectins are usually purchased in powder form. We routinely dissolve the powder in lectin buffer to give a stock solution of 1 mg/mL, which is stable in the refrigerator for several months. Lectin stock solution can then be diluted to the optimum work-

ing dilution as required. It is generally a good idea to prepare working solutions immediately prior to use, and discard unused solution at the end of the experiment. A working concentration of 10 µg/mL is a good, optimal concentration for most lectins in histochemistry. It may be a good idea to try a range of concentrations, 2.5, 5, 10, 20, 40 µg/mL to assess which gives the best results.

30. We recommend vigorous washing in three changes of lectin buffer. Each wash should consist of vigorous "sloshing up and down" of slides in buffer for about 30 s to 1 min, then allowing slides to stand in the buffer for about 4 min. Insufficient washing, in particular omitting one or more changes of buffer, can result in unacceptably high levels of dirty background staining.
31. This method has the advantage of being quick, simple, and straightforward. It lacks the sensitivity of some of the longer, multistep and avidin–biotin methods. We have found that horseradish peroxidase labeled lectins can sometimes give subtly different binding results to native (unlabeled) or biotin-labeled lectins, possibly owing to strearic hindrance of lectin's binding site by the relatively large peroxidase molecule.
32. Mayer's hematoxylin is a deep port-wine red in color and cell/tissue preparations will stain deep red after immersion in it. The hematoxylin changes to a deep navy blue when exposed to mildly alkaline conditions (known as "bluing"), such as after washing under the tap water of most areas. If bluing in is unsuccessful owing to unusually acidic tap water, soak slides instead in tap or distilled water to which a few drops of ammonia or sodium hydroxide have been added.
33. This is a really useful method and is highly recommended for most applications; quick, straightforward, sensitive, generally giving good results. The biotin label appears to be small enough not to interfere with the combining site of the lectin and therefore does not appear, in our hands, to alter its binding characteristics (*see* **Note 31**).
34. This is a very sensitive method that gives good, clean results. It should be employed when the simpler methods, for example, the simple avidin–biotin method, give weak labeling.
35. After labeling, slides are in an aqueous medium. Before they can be mounted in a xylene-based mounting medium they must therefore be dehydrated through graded alcohols and cleared in xylene. Aqueous mounting media are also commercially available which, historically, tended to give inferior results to xylene-based mounting media. Some of the modern ones are, however, very good and may be used instead without the need for dehydration and clearing.
36. For a few lectins this may not be possible, as they are inhibited only by complex sugars, not monosaccharides (consult manufacturer's literature). If this is the case, inhibition with a heavily glycosylated glycoprotein, for example, fetuin or asialofetuin, should give effective inhibition, as should incubation of the lectin in the presence of 5% normal (human or animal) serum.
37. If dirty, nonspecific background staining is observed, this may indicate inadequate washing between stages in the staining protocol (*see* **Note 30**). If careful attention to washing does not satisfactorily limit background staining, incorporation of a blocking agent into lectin, antibody, and streptavidin horseradish peroxidase solutions may be indicated. The simplest and most effective

blocking agent is to include 3% w/v bovine serum albumin. Normal serum, often used as a blocking agent in immunocytochemistry, is not appropriate as the many heavily glycosylated molecules present will effectively inhibit specific staining.

## References


1. Goldstein, I. J., Hughes, R. C., Monsigny, M., Osawa, T., and Sharon, N. (1980) What should be called a lectin? *Nature* **285,** 66.
2. Maramatsu, T. (1993) Carbohydrate signals in metastasis and prognosis of human carcinomas. *Glycobiology* **3,** 294–296.
3. Kanangi, R. (1997) Carbohydrate mediated cell adhesion involved in hematogenous metastasis of cancer. *Glycoconjugate J.* **14,** 577–584.
4. Renkonen, R., Mattila, P., Marja-Leena, M., Rabina, J., Toppila, S., Renkonen, J., et al. (1997) *In vitro* experimental studies of sialyl Lewis x and sialyl Lewis a on endothelial and carcinoma cells: crucial glycans on selectin ligands. *Glycoconjugate J.* **14,** 593–600.
5. Nakano, T., Matsui, T., and Ota, T. (1996) Benzyl-alpha-GalNAc inhibits sialylation of O-glycosidic sugar chains and enhances experimental metastatic capacity in B16BL6 melanoma cells. *Anticancer Res.* **16,** 3577–3584.
6. Yogeeswaren, G. and Salk, P. L. (1981) Metastatic potential is positively correlated with cell surface sialylation of cultured murine tumor cells. *Science* **212,** 1514–1516.
7. Dennis, J. W., Laferte, S., Waghorne, C., Breitman, M. L., and Kerbel, R. S. (1987) Beta 1-6 branching Asn-linked oligosaccharides is directly associated with metastasis. *Science* **236,** 582–585.
8. Fernandes, B., Sagman, U., Auger, M., Demetrio, M., and Dennis, J. W. (1991) Beta 1-6 branched oligosaccharides as a marker of tumor progression in human breast and colon neoplasia. *Cancer Res.* **51,** 718–723.
9. Korczak, B., Goss, P., Fernandes, B., Baker, M., and Dennis, J. W. (1984) Branching N-linked oligosaccharides in breast cancer. *Adv. Exp. Med. Biol.* **353,** 95–104.
10. Brooks, S. A. and Leathem, A. J. C. (1991) Prediction of lymph node involvement in breast cancer by detection of altered glycosylation in the primary tumor. *Lancet* **338,** 71–74.
11. Schumacher, U., Higgs, D., Loizidou, M., Pickering, R., Leathem, A., and Taylor, I. (1994) *Helix pomatia* agglutinin binding is a useful prognostic indicator in colorectal carcinoma. *Cancer* **74,** 3104–3107.
12. Kakeji, Y., Maehara, Y., Tsujitani, S., Baba, H., Ohno, S., Watanabe, A., and Sugimachi, K. (1994) *Helix pomatia* agglutinin binding activity and lymph node metastasis in patients with gastric cancer. *Semin. Surg. Oncol.* **10,** 130–134.
13. Yoshida, Y., Okamura, T., and Shirakusa, T. (1994) An immunohistochemical study of *Helix pomatia* agglutinin binding on carcinomas of the esophagus. *Surg. Gynecol. Obstet.* **177,** 299–302.


## Suggested Further Reading


Brooks, S. A., Leathem, A. J. C., and Schumacher, U. (1997) *Lectin Histochemistry—A Concise Practical Handbook.* BIOS Scientific Publishers, Oxford, UK.

# 5

# Immunocytochemical Detection and Characterization of Individual Micrometastatic Tumor Cells

**Stephan Braun and Klaus Pantel**

## 1. Introduction

Early hematogenous spread of cancer cells must be regarded as major cause for the later development of metastatic disease in patients with completely resected solid tumors, which account for the majority of cancer-related deaths in industrialized nations. Because current procedures for tumor staging usually fail to detect such micrometastases, immunocytochemical assays with monoclonal antibodies directed against epithelial differentiation antigens have been designed for the detection of individual micrometastatic carcinoma cells in secondary organs. The recent development of monoclonal antibodies directed to epithelial differentiation proteins has opened a diagnostic window to detect such disseminated carcinoma cells. Methods for immunocytochemistry are described in detail in Chapter 2 by Brooks. Several studies have shown that bone marrow is an important, clinically relevant indicator organ for early disseminated cancer cells derived from various epithelial organs, including breast, lung, colon, rectum, prostate, kidney, ovary, esophagus, and pancreas (for review, *see* **ref.** ***1***). The use of various detection antibodies and marker antigens has contributed to an enormous heterogeneity in the methodology applied.

The critical methodological evaluation of these antibodies in respect to specificity and sensitivity revealed considerable discrepancies in their potential to build a solid basis for a reliable and reproducible assay. Most of the studies were performed with monoclonal antibodies either against cytokeratin ***(2–5)***, a major constituent of the cytoskeleton in epithelial cells, or against membrane-bound mucins, such as epithelial membrane antigen, human milk fat globule, and tumor-associated glycoprotein-12 ***(6–8)***. Analyzing bone marrow samples from noncarcinoma patients,

From: *Methods in Molecular Medicine, vol. 57: Metastasis Research Protocols, Vol. 1: Analysis of Cells and Tissues*
Edited by: S. A. Brooks and U. Schumacher © Humana Press Inc., Totowa, NJ

however, we and others ***(9–13)*** have demonstrated that—in contrast to antibodies directed against cytokeratin polypeptides—the use of antibodies directed against such epithelial mucins resulted in a considerable rate of false positive findings due to crossreactivity with autochthonous bone marrow cells. Thus, cytokeratin emerged as potential standard marker for the detection of disseminated epithelial tumor cells in bone marrow ***(13)***.

For the time, the development of standardized protocols therefore deserves the highest priority, while meta-analysis of extremely heterogeneous and thus incomparable sets of data appears to be rather premature at present ***(14)***. In this chapter, we provide a standardized protocol for the detection of minimal amounts of residual tumor cells present in bone marrow aspirations from cancer patients, at frequencies of $10^{-5}$–$10^{-6}$ nucleated cells. Epithelial tumor cells are identified with the broad-spectrum monoclonal antibody A45-B/B3 directed against the epithelial differentiation marker cytokeratin, using the sensitive alkaline phosphatase antialkaline phosphatase staining technique. Moreover, to investigate the malignant nature of such cells, we designed a double-marker assay for simultaneous detection and characterization of individual micrometastatic carcinoma cells. Here, we describe a protocol for the combined immunogold–alkaline phosphatase double-labeling of two antigens that was even applicable for the localization of two antigens within the same cellular compartment. The proposed protocols may help to improve current tumor staging (e.g., ***4,15***) and identify relevant therapeutic targets with potential consequences for adjuvant anticancer therapy ***(13)***. The reader is also referred to Chapter 19 by Haack et al., which describes the detection of metastatic tumor cells in the blood by reverse-transcriptase polymerase chain reaction (RT-PCR).

## 2. Materials

### 2.1. Bone Marrow Preparation (see Note 1)

1. 10X phosphate-buffered saline (PBS) stock solution: 1.5 *M* NaCl, 80 m*M* $Na_2HPO_4$, 20 m*M* $KH_2PO_4$, adjust pH to 7.4.
2. 1X Hank's salt solution: Available ready for use from Biochrom (Germany); store at 4°C.
3. Percoll available from Pharmacia (Germany); prepare 100% Percoll stock solution (100 mL of Percoll plus 9 mL of 1X Hank's salt solution) and dilute with sterile 0.9% NaCl to 50% Percoll (1.065 g/mL); store at 4°C. *Alternatively:* Use Ficoll (1.077 g/mL): available ready for use from Pharmacia (Germany); store at 4°C.
4. 1X Erythrocyte lysis buffer: 10 m*M* $KHCO_3$, 150 m*M* $NH_4Cl$, 0.1 m*M* EDTA; adjust pH to 7.4.

### 2.2. Immunocytochemistry

1. PBS-AB serum for dilution of all immunoglobulin stock solutions: Add 10% v/v of AB serum (Biotest, Germany) to 1X PBS.

2. Bovine serum albumin (BSA)-C serum for blocking of nonspecific binding sites: Add 0.1% v/v of acetylated bovine serum albumin (Aurion-BSA-C, Biotrend, Köln, Germany) to PBS-AB serum.
3. For single labeling, dilute monoclonal antibody A45-B/B3 (Baxter, Germany) with PBS-AB serum to a final concentration of 1.0–2.0 μg/mL (*see* **Note 2**).
4. For double labeling, dilute alkaline phosphatase-conjugated monoclonal antibody A45-B/B3 $F_{ab}$ fragment (Baxter, Germany) with PBS-AB serum to a final concentration of 1.0–2.0 μg/mL (*see* **Note 2**).
5. Bridging immunoglobulin for single labeling: Dilute rabbit antimouse antibody (e.g., Z259, Dakopatts, Germany) 1:20 in PBS-AB serum.
6. For single labeling, dilute alkaline phosphatase antialkaline phosphatase complex (e.g., D651, Dakopatts, Germany) 1:100 in PBS-AB serum.
7. For double labeling, dilute immunogold-conjugated, terminal $F_c$ fragment-specific goat antimouse antibody (Amersham, Germany) 1:40 in PBS-AB serum.
8. Immunogold postfixation buffer: 2% v/v glutaraldehyde (Sigma, Germany) in 1X PBS.
9. New fuchsin development solutions (example sufficient for four Hellendahl jars with a final volume of 253.3 mL):
   a. *Solution 1:* Levamisole (Sigma, Germany) diluted in Tris-HCl (Merck, Germany); for example, add 62.5 mL of 0.2 *M* Tris-HCl solution to 187.5 mL of $H_2O$ and 90 mg of levamisole.
   b. *Solution 2:* $NaNO_2$ (Merck, Germany) diluted in $H_2O$ and new fuchsin (Merck, Germany); for example, dissolve 50 mg of $NaNO_2$ in 1.25 mL of $H_2O$ and add 0.5 mL of new fuchsin prepared as 5% w/v solution in 2 *M* HCl; shake well until air bubbles appear.
   c. *Solution 3:* Naphthol phosphate substrate solution; for example, dissolve 125 mg of naphtol AS-BI phosphate sodium salt (Sigma, Germany) in 1.5 mL *N,N*-dimethylformamide (Sigma, Germany) (*see* **Note 2**).
10. The freshly prepared solutions 1 and 2 are mixed and stirred well before solution 3 is added; prior to use the mixture is filtered into Hellendahl jars.
11. Silver enhancement solution: Silver Intens Kit (Amersham, Germany) used as recommended by the manufacturer.
12. Cover slips are sealed with Kaiser's glycerine gelatine melted at 50°C.

It is good practice to use sterile disposable plasticware in the preparation of reagents, especially if bone marrow mononucleated cells (MNCs) are to be used for cell culture or RNA/DNA analyses. For dilution of antibodies, remove immunoglobulin solutions with sterile pipet tips. Particular care should be taken handling naphthol and *N,N*-dimethyl formamide; a fume cupboard should be used where necessary. For preparation of new fuchsin development solutions clean glassware should be used. Particular care should be taken of naphthol-containing waste. We recommend to use a laminar air-flow working bench and to wear gloves handling bone marrow aspirates.

## 3. Methods

### 3.1. Bone Marrow Preparation (see Note 1)

1. Place a maximum of 10 mL of the bone marrow aspirate in 50 mL tubes; add Hank's salt solution (4°C) to a final volume of 20 mL.
2. Centrifuge for 10 min at 180*g* at room temperature; meanwhile prepare 8 mL of 50% Percoll (4°C) in a fresh 15 mL tube. Alternatively, prepare 20 mL of Ficoll in a fresh 50 mL tube.
3. Remove serous supernatant without stirring the Hank's/bone marrow pellet; add 1 mL of 10% Percoll to 5 mL of pellet and place solution carefully onto Percoll. Alternatively, add 1X PBS to Hank's/bone marrow pellet to a final volume of 10 mL and place solution carefully onto Ficoll. Avoid mixture of cells and Percoll/Ficoll prior to centrifugation; centrifuge for 20 min at 1000*g* at room temperature and 30 min at 1230*g* at room temperature, respectively.
4. Remove the MNCs containing interface and place cellular suspension in a fresh 50-mL tube, add 1X PBS (4°C) to a final volume of 50 mL, and centrifuge for 10 min at 540*g* at 4°C.
5. Remove the washing solution and resuspend the pellet for lysis of erythrocytes—required only in case of significant contamination with erythrocytes. Add 2–10 mL of lysis buffer (4°C) depending on the degree of erythrocyte contamination, estimated by the various size of the red shaped pellet, and incubate cells for 5 min at room temperature.
6. Add 1X PBS to a final volume of 50 mL, and centrifuge for 10 min at 540*g* at 4°C.
7. Remove washing solution, resuspend cells thoroughly in a final volume of 2–5 mL of 1X PBS (4°C), count cells, and adjust cell concentration to $1 \times 10^6$ MNC/mL.
8. Prepare slides for cytocentrifugation and add a total of 0.5 mL of the cellular suspension ($= 0.5 \times 10^6$ MNC) per spot of the cell chamber (cat. no. 1666, Hettich, Germany).
9. Centrifuge with Hettich Universal 30F for 3 min at 120*g* at room temperature. Subsequently, let slides dry for 12 h and either perform immunostaining immediately or store slides at –20°C (up to 4–8 wk) or –80°C (up to 1–2 yr) for later use.

### 3.2. Immunocytochemistry

1. Thaw frozen slides of bone marrow preparation and appropriate positive and negative cell line preparation for control purpose at 37°C for at least 15 min.
2. Prepare appropriate dilutions of antibody stock solution in PBS-AB serum. Place thawed slides in a humid chamber for the subsequent incubation steps; slides must not dry out during the entire process of immunostaining.

#### 3.2.1. Single Labeling—APAAP Immunostaining

1. Blocking step: Incubate slides with PBS-AB serum for 20 min; drain off blocking solution (do not wash slides).
2. Detection antibody—A45-B/B3: Incubate antibody for 45 min, and wash thoroughly in 1X PBS for 10 min with three changes of buffer (*see* **Note 2**).

3. Bridging antibody—e.g., Z259: Incubate antibody for 30 min and wash thoroughly in 1X PBS for 10 min with three changes of buffer.
4. APAAP complex—e.g., D651: Incubate antibody complexes for 30 min and wash thoroughly in 1X PBS for 10 min with three changes of buffer.
5. New fuchsin development of alkaline phosphatase: Incubate slides with freshly prepared, mixed, and filtered alkaline phosphatase development solutions 1–3 (*see* above, *see* **Note 2**) for 20 min. Finally, wash slides in 1X PBS for 10 min with 3 changes of buffer, and seal cell preparations with cover slips and gelatine.

#### *3.2.2. Double Labeling*

1. Optionally, immerse slides in PBS supplemeted in 0.1% Triton-X-100 diluted in 1X PBS for permeabilization for 10 min at 4°C and followed by incubation in PBS with 1.0% paraformaldehyde diluted in 1X PBS for 20 min at 4°C. Fixed slides are washed in 1X PBS for 10 min with three changes of buffer.
2. Blocking step: Incubate slides with 1X PBS supplemented with 10% AB serum and 0.1% of BSA-C for 20 min; drain off blocking solution (do not wash slides).
3. Coincubation of detection antibodies: For coincubation, mix both primary antibodies—e.g., alkaline phosphatase-conjugated monoclonal antibody A45-B/B3 $F_{ab}$ fragment (for cytokeratin labeling) and a second unconjugated whole immunoglobulin (for labeling of target antigen)—at respective dilutions within the same tube. Incubate antibody mixture for 45 min, and wash thoroughly in 1X PBS for 10 min with three changes of buffer.
4. Immunogold labeling: Incubate $F_c$-terminus specific, gold-conjugated goat antimouse antibody for 45 min, and wash thoroughly in 1X PBS for 10 min with three changes of buffer. Proceed with another three washes in $H_2O$.
5. Postfixation of immunogold colloids: Immerse slides in 2% glutaraldehyde for exactly 5 min, and wash thoroughly in three changes of $H_2O$ for 10 min each.
6. Silver development of labeled immunogold: Prepare fresh silver enhancement solution according to the manufacturer's recommendations, and drop development solution onto cells as cold (4°C) as possible. Monitor silver development at regular intervalls under the microscope and stop development reaction by placing slides into $H_2O$ as soon as brown to black silver granules become visible; wash thoroughly in three changes of $H_2O$ for 10 min each. Immerse slides in PBS for 5 min (*see* **Note 3**).
7. New fuchsin development of alkaline phosphatase: Incubate slides with freshly prepared, mixed, and filtered alkaline phosphatase development solutions 1–3 for 20 min. Finally, wash slides in PBS for 10 min with three changes of buffer, and seal cell preparations with coverslips and gelatin.

It is good practice to incorporate respective positive controls—for example, tumor cell lines known to express the investigated antigens—in order to be able to judge that negative immunostaining is due to lack of antigen expression rather than methodological failure. Inversely, we recommend to use a set of slides for a negative control staining applying an antibody with nonhuman specificity. For negative control stainings we screen the identical number of cells—for example, $2.0 \times 10^6$ per patient—as in the specific immunostaining (*see* **Note 2**). To set up the described assay as a routine assay for the detection of tumor cells in bone marrow of cancer patients, we further recommend incorporation of blinded bone marrow aspirates of noncarcinoma patients in order to substantiate the specificity of the approach (*see* **Note 2**).

## 4. Notes

1. Bone marrow preparation: Besides imprecise handling, low numbers of MNC recovered from Percoll/Ficoll interface could be due to a poor quality of the bone marrow aspirate—for example, significant contamination with peripheral blood. This may also lead to an underestimation of the actual frequency of tumor cells detectable by the following immunocytochemical screening ***(1)***. Good quality aspirates of 3–5 mL of bone marrow contain $>10^7$ MNC.
2. Single labeling: Failure to produce a positive signal of the cytokeratin staining on the positive control specimens could be caused by the use of damaged control specimens—for example, repeated thawing and freezing of cytospin preparations; inadequately high dilutions of antibodies, especially of the primary antibody A45-B/B3; or problems with the preparation of the developing solutions, for example, precipitation of naphthol AS-BI phosphate in *N,N*-dimethyl formamide due to wet glassware. Since nonspecific labeling of bone marrow cells by the antibody A45-B/B3 occurs in less than 2% of noncarcinoma control patients ***(13)***, nonspecific staining signals on negative control specimens may be predominantly caused by excessively high concentrations of antibodies. So far it remained unclear why in few cases irrelevant antibodies with nonhuman—for example, murine—specificty react with human bone marrow cells on isotype control specimens of patients' bone marrow samples. In order to preserve the specificity of interpretation of staining signals, we strongly recommend to exclude these rare cases (usually below 5% in our series) of indeterminate results.
3. Double labeling: For unambiguous interpretation of silver lactate precipitates at specifically bound immunogold conjugates, we strongly recommend to use an epipolarization filter unit ***(13)***. Only the use of this device may allow recognition of brown-to-black granulations others than silver precipitates; interpretations based on light microscopy alone may lead to the registration of nonspecific signals and/or underestimation of the actual staining signals too weak to be seen by nonpolarized light microscopy. Special care must be taken to control the time- and temperature-dependent silver precipitation; we recommend continuous monitoring of the positive control slides under the microscope as soon as the first slightly brown signals become visible. Development exceeding 30 min may lead to weak new fuchsin signals owing to chemical interaction with alkaline phosphatase.

## References

1. Pantel, K., Braun, S., Passlick, B., Schlimok, G. (1996) Minimal residual epithelial cancer: diagnostic approaches and prognostic relevance. *Histochem. Cytochem.* **30,** 1–60.

2. Schlimok, G., Funke, I., Holzmann, B., et al. (1987) Micrometastatic cancer cells in bone marrow: in vitro detection with anti-cytokeratin and in vivo labeling with anti-17-1A monoclonal antibodies. *Proc. Natl. Acad. Sci. USA* **84,** 8672–8676.
3. Cote, R. J., Rosen, P. P., Lesser, M. L., et al. (1991) Prediction of early relapse in patients with operable breast cancer by detection of occult bone marrow micrometastases. *J. Clin. Oncol.* **9,** 1749–1756.
4. Lindemann, F., Schlimok, G., Dirschedl, P., et al. (1992) Prognostic significance of micrometastatic tumor cells in bone marrow of colorectal cancer patients. *Lancet* **340,** 685–689.
5. Pantel, K., Schlimok, G., Braun, S., et al. (1993) Differential expression of proliferation-associated molecules in individual micrometastatic carcinoma cells. *J. Natl. Cancer Inst.* **85,** 1419–1424.
6. Mathieu, M. C., Friedman, S., Bosq, J., et al. (1990) Immunohistochemical staining of bone marrow biopsies for detection of occult metastasis in breast cancer. *Breast Cancer Res. Treat.* **15,** 21–26.
7. Diel, I. J., Kaufmann, M., Costa, S. D., et al. (1996) Micrometastatic breast cancer cells in bone marrow at primary surgery: prognostic value in comparison with nodal status. *J. Natl. Cancer Inst.* **88,** 1652–1664.
8. Harbeck, N., Untch, M., Pache, L., et al. (1994) Tumor cell detection in the bone marrow of breast cancer patients at primary therapy: results of a 3-year median follow-up. *Br. J. Cancer* **69,** 566–571.
9. Taylor-Papadimitriou, J., Burchell, J., Moss, F., et al. (1985) Monoclonal antibodies to epithelial membrane antigen and human milk fat globule mucin define epitopes expressed on other molecules. *Lancet* **1,** 458.
10. Delsol, G., Gatter, K. C., Stein, H., et al. (1984) Human lymphoid cells express epithelial membrane antigen. *Lancet* **2,** 1124–1128.
11. Heyderman, E., Macartney, J. C. (1985) Epithelial membrane antigen and lymphoid cells. *Lancet* **1,** 109.
12. Pantel, K., Schlimok, G., Angstwurm, M., et al. (1994) Methodological analysis of immunocytochemical screening for disseminated epithelial tumor cells in bone marrow. *J. Hematother.* **3,** 165–173.
13. Braun, S., Müller, M., Hepp, F., et al. (1998) Re: Micrometastatic breast cancer cells in bone marrow at primary surgery: prognostic value in comparison with nodal status. *J. Natl. Cancer Inst.* **90,** 1099–1100.
14. Funke, I., Schraut, W. (1997) Meta-analysis of studies on bone marrow micrometastases: An independent prognostic impact remains to be substantiated. *J. Clin. Oncol.* **16,** 557–566.
15. Pantel, K., Izbicki, J. R., Passlick, B., et al. (1996) Frequency and prognostic significance of isolated tumor cells in bone marrow of patients with non-small cell lung cancer without overt metastases. *Lancet* **347,** 649–653.

# 6

# Understanding Metastasis Through Integrin Expression

**Lawrence John and Massimo Pignatelli**

## 1. Introduction

Integrins form a major family of heterodimeric cell surface receptors. Individual family members each comprise noncovalently linked, dissimilar, α- and β-subunits. Each subunit is the product of a different gene *(1)*, and α-subunits appear to have evolved separately from β-subunits *(2)*. Seventeen α-, and eight β-subunits are currently described. It is the combination of α- and β-subunits that gives rise to the individual family members. From this follows ligand-binding specificity, which may be relative (fibronectin has at least seven different binding sites), or absolute, (e.g., αEβ7 to E-cadherin *[3]*). High-affinity ligand recognition usually requires both subunits *(4)*. The redundancy in this system, with many integrins binding the same ligand, suggests an important role in cellular communication in addition to heterotypic cell–cell (e.g., the β2 integrins and α4β1) or cell–extracellular matrix (most integrins) adhesion. In addition, some integrins bind microorganisms and certain plasma proteins. Moreover, integrins are involved in fundamental cellular events such as proliferation, differentiation, apoptosis, and motility. There is overwhelming evidence for their fundamental role in the control of tumor differentiation, proliferation, invasion and metastasis.

To better understand the biological significance of differential integrin expression in tumors it is first helpful to review the structure of integrin molecules, their diversity, and functional specialization. We will then discuss both the scientific methods commonly used and the experimental evidence that they have so far provided toward the understanding of the molecular control of tumor cell behavior.

From: *Methods in Molecular Medicine, vol. 57:*
*Metastasis Research Protocols, Vol. 1: Analysis of Cells and Tissues*
Edited by: S. A. Brooks and U. Schumacher 

### 1.1. Integrin Structure

Integrins are transmembrane adhesion molecules with a short cytoplasmic domain. Electron microscopic studies suggest that these receptors have a globular head, approx 10 nm in diameter *(5)*, and two extended cytoplasmic tails. The cytoplasmic domain of the β-subunits is composed of 40–60 amino acids. The exception is the β4 subunit, which is composed of more than 1000 amino acids. The β-subunits are associated with a number of cytoskeletal proteins (e.g., α-actinin, talin, vinculin, and tensin), and they interact with the actin cytoskeleton, thus conferring mechanical stability to the cell. Again, the β4 subunit is exceptional, and instead associates with the intermediate filament cytoskeleton, forming a significant component of the hemidesmosome.

Recently, a combination of epitope mapping, site-directed mutagenesis, and domain exchange experiments have suggested that repeats 2–5 of the seven amino (N)-terminal repeats of integrin α-subunits are actually involved in the ligand binding *(6)*. Springer *(7)* has proposed that the seven N-terminal repeats of integrin α-subunits each fold into four-stranded β-sheets arranged in a torus around an axis of pseudosymmetry, the "β-propeller" model. Here the ligand binding site is predicted to be on the upper surface of the propeller. It is conceived that the lower surface, containing the cation binding repeats, interfaces with the predicted I domain-like structure present in the β-subunit.

Integrins are conformationally labile *(8)*, and are subject to disulfide bond exchange *(9)*. These characteristics, and no doubt others, form the structural basis for integrin signaling. There are two main mechanisms, "inside–out," and "outside–in."

Integrins can exert an extracellular effector response (e.g., cell anchorage) by binding ligands. This process can be regulated by cellular signaling mechanisms that modulate the binding affinity and kinetics of interaction between adhesive ligands and cell surface receptors. This is known as "integrin activation" or "inside–out signaling."

Alternatively, "post-occupancy" events such as lateral diffusion of receptors *(10)*, and interactions with, and reorganization of, the cytoskeleton *(11)*, can regulate adhesion between ligand and receptor. This is the basis of "outside–in" signaling (*see* **Fig. 1**).

Integrins bind two major classes of ligand; proteins of the extracellular matrix (e.g., fibronectin, fibrinogen, laminin, collagen, and vitronectin) and proteins of the immunoglobulin superfamily. Interestingly, most of the latter group contains immunoglobulin domains consisting of a sandwich of antiparallel β-sheets *(12)*. Examples include the intercellular adhesion molecule-1, (ICAM-1), and vascular adhesion molecule-1, (VCAM-1).

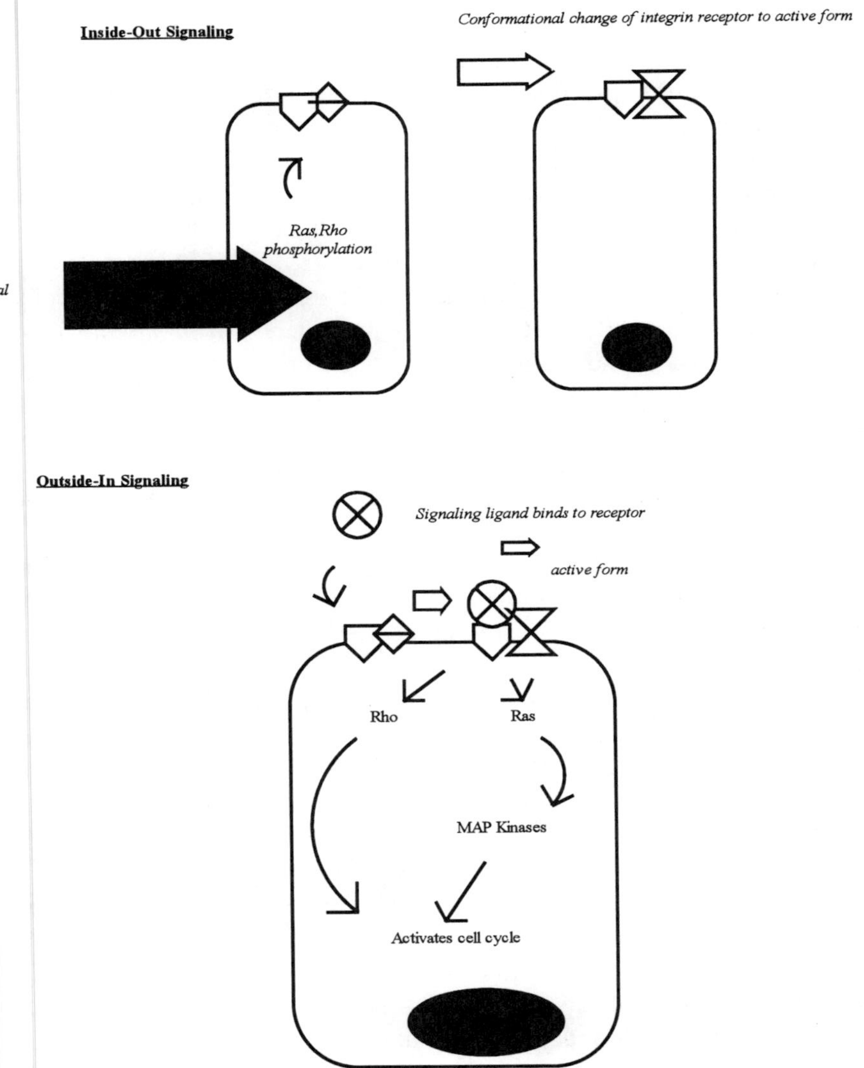


Fig. 1. Inside–out and outside–in integrin signaling.

### 1.2. Integrins in Proliferation, Apoptosis, Differentiation, Cell Migration, and Invasion

Cellular proliferation has been show to be dependent on integrin-mediated adhesion to the extracellular matrix. Specifically, the induction of cyclin D1 mRNA and protein, and the down-regulation of the cyclin-dependent kinase inhibitory proteins, both necessary for mitogen-induced progression through the $G_1$ phase, is minimal in the absence of cell attachment to a substratum ***(13)***. Due to the central role integrins play in mediating cell anchorage to the extracellular matrix, the implications of experimentally manipulating their expression in tumor growth has received much attention. Recently, α6β4 has been shown to be involved in the reversion of the malignant phenotype of human breast carcinoma cells in three-dimensional culture ***(14)***. In addition, a number of studies demonstrate that the reexpression of α2β1 or α5β1 in solid tumor cells has a negative effect on anchorage-dependent growth of these tumor cells ***(15)***.

Integrins α1β1, α5β1, and αvβ3 may protect cells from apoptosis by activating tyrosine phosphorylation of the adapter protein Shc, and activating the *Ras* extracellular signal-related protein. Certain integrins may exert control of apoptosis by their activation of the focal adhesion kinase or mitogen-activated protein kinase pathways. The latter is likely to act via *Ras* activation, which can have a direct effect on gene expression ***(16)***. In fact a new term, "anoikis," has been suggested for this specific type of apoptosis that occurs in the absence of integrin-mediated adhesion to the extracellular matrix. In α5-transfected tumor cells (Chinese hamster ovary, HT29 colon carcinoma, and MG-63 osteosarcoma cells), ligation of integrin α5β1 to fibronectin consistently suppresses anoikis ***(17)***. Furthermore, ligation of integrin αvβ3 rescues melanoma cells from anoikis in three-dimensional dermal collagen ***(18)***. The likely mechanism here involves up-regulation of the protooncogene *Bcl-2* and the down-regulation of the cyclin-dependent kinase inhibitor p21 ***(19,20)***. In addition, it has been shown that disruption of cell–matrix interactions activates the p53–p21 proapoptotic pathway ***(21)***.

Interestingly, focal adhesion kinase (FAK) is consistently overexpressed in invasive human carcinomas ***(22)***. Activation of FAK in epithelial cell lines has been shown to confer anchorage-independent (but serum-dependent) growth in soft agar. Moreover, activation of the small GTPase Rho in NIH 3T3 cells and overexpression of integrin-linked kinase (ILK) in IEC18 epithelial cells also confer anchorage-independent (but serum-dependent) growth ***(23)***. It is therefore likely that the enhanced proliferation of tumor cells, has, at least in part, a molecular basis, in the overexpression or activation of integrin signal transducers such as FAK, Rho, and ILK.

Integrins also have a critical role in differentiation. For example, gland formation by colorectal epithelial cells is β1 integrin dependent ***(24)***, and decreased expression of integrins α2β1 and α5β1 is consistently seen in poorly differentiated breast carcinomas ***(25)***.

Because integrins connect the extracellular matrix to the cytoskeleton at focal adhesion plaques, they provide the traction necessary for cell migration. Interestingly, a number of studies show changes in cell migration with alterations in integrin expression. For example, overexpression of fibronectin receptors α4β1 and α5β1 in transfected sarcoma cells enhances their motility on fibronectin in vitro, and changes their migratory properties in vivo when grafted into the neuroepithelium of chick embryos *(26)*. However, the relationship between integrin and motility is not always straightforward. For example, αvβ3 regulates α5β1-mediated cell migration toward fibronectin *(27)*. In addition, FAK also has a role. Overexpression of FAK is associated with increased cell motility *(28)*. It is therefore clear that tumor cells may utilize different integrin-dependent pathways to achieve the same goal.

Tumor invasion requuires not only cell migration, but also degradation of the surrounding extracellular structures. The matrix metalloproteinases have an important role in the latter. Significantly, matrix metalloproteinase-2 (MMP-2; a gelatinase involved in the degradation of the extracellular matrix), colocalizes with integrin αvβ3 in melanoma tumor cells in vivo and binds directly to αvβ3 in vitro *(29)*. It is possible that these proteins cooperate in such a way as to favor dissemination.

Tumors <1 $mm^3$ in size can gain sufficient nutrients by passive diffusion. Greater sizes require tumor angiogenesis. Integrins also have a role in angiogenesis. For example, it has been demonstrated that a monoclonal antibody directed against integrin αvβ5 inhibits rabbit corneal and chick cell adhesion molecule (CAM) angiogenesis induced by vascular endothelial cell growth factor *(30)*. Moreover, integrin αvβ5 is upregulated on tumor vessels, whereas it is minimally expressed on quiescent blood vessels *(31)*. It appears that the change from resting to angiogenic endothelium is mediated by the *Hox D3* homeobox gene, which increases the expression of integrin αvβ3 on human endothelial cells *(32)*.

### 1.3. Integrins in Abnormal Processes

The diffuse and yet specific expression of integrins in normal tissues (*see* **Table 1**) suggests an analogous, yet specifically altered expression in neoplastic disease. As a dynamic process, insights into the variations, or even discrete changes, in the diseased state can aid our understanding of normal function, as much as normal function can give us insight into disease. We shall discuss the role of integrins in solid tumors and their metastases, as well as in disseminated hematological malignancies.

### 1.4. Solid Tumors

Research into the expression of integrins in solid tumors involves many diverse experimental techniques. Electron microscopy, adhesion assays, cell cultures, gene transfer techniques to modulate gene expression, neutralizing monoclonal antibodies, and knockout mice have all been used to define the molecular changes in cancer cells. However, most current work is heavily dependent on the techniques of immunohistochemistry (*see* **Subheadings 2.** and **3.**), a technique described in detail in Chapter 2 by Brooks.

**Table 1**
**Integrins, Their Ligands, and Their Distribution**

| Integrin | Ligands | Size (amino acid residues) | Relative molecular mass, ×1000 | Distribution |
|---|---|---|---|---|
| α1β1 | Laminin, collagen, tenascin | | | Natural killer (NK) cells, B and activated T cells |
| α2β1 | Laminin, collagen | 1150 + 778 | 170 + 130 | NK cells, B and activated T cells, platelets, endothelial cells, fibroblasts, epithelial cells, astrocytes, Schwann cells |
| α3β1 | Laminin, collagen, fibronectin, epiligrin | 1119 + 778 | 1155 + 130 | Activated T cells, thymocytes, endothelium, fibroblasts, epithelial cells, astrocytes |
| α4β1 | α4β7, fibronectin, VCAM-1, MAdCAM-1 | 999 + 778 | 150 + 130 | NK, B and T cells, eosinophils, endothelium, muscle, fibroblasts |
| α5β1 | Fibronectin | 1008 + 778 | 160 + 130 | Activated B and T cells, memory T cells, thymocytes, fibroblasts, epithelial cells, platelets, endothelium, astrocytes |
| α6β1 | Laminin | 1050 + 778 | 150 + 130 | Leucocytes, thymocytes, epithelial cells, endothelium, glial cells |
| α7β1 | Laminin | | | Skeletal and cardiac muscle, melanoma cells |
| α8β1 | Fibronectin, vitronectin | | | Epithelial cells, neurons, oligodendroglia |
| α9β1 | | | | Respiratory epithelial cells, muscle |
| αvβ1 | Vitronectin, fibronectin | | | Oligodendroglia |

| | | | | |
|---|---|---|---|---|
| αLβ2 | ICAM–1, ICAM–2, ICAM–3 | 1145 + 747 | 180 + 95 | Leucocytes, thymocytes, macrophages, T cells, microglia |
| αMβ2 | ICAM–1, factor X, iC3b, fibrinogen, β-glucans | 1137 + 747 | 170 + 95 | Myeloid, activated B cells, NK cells, macrophages, microglia |
| αXβ2 | iC3b, fibrinogen | 1124 + 747 | 150 + 95 | Myeloid, activated B cells, NK cells, macrophages, microglia, dendritic cells |
| αIibβ3 | Fibronectin, vitronectin, von Willebrand's factor, thrombospondin | 1008 + 762 | 142 + 110 | Platelets |
| αvβ3 | Fibronectin, vitronectin, von Willebrand's factor, thrombospondin, PE CAM-1 | 1018 + 762 | 150 + 110 | Activated B and T cells, endothelium, tumors, glia |
| α6β4 | Laminin-5 | | | Schwann cells, endothelium, epithelial cells, fibroblasts |
| αvβ5 | Vitronectin, fibronectin | | | Fibroblasts, macrophages, epithelial cells, tumors |
| αvb6 | Fibronectin | | | |
| αvβ8 | Fibronectin | | | Oligodendroglia, Schwann cells |
| α4β7 | Fibronectin, VCAM-1, MAdCAM-1 | 999 + 770 | 150 + 120 | NK, B and T cells |
| αEβ7 | E-cadherin | 1160 + 779 | 175 + 120 | Lymphocyte subset |

Many solid tumors exhibit altered integrin expression, whether in their growth, invasive or metastatic phases of their life cycle (*see* **Table 2** and **Fig. 2**). There is currently much investigation into the specifics of these processes and the differential integrin expression upon which they depend.

### 1.5. Localization of Integrin Ligands Is Specific to Their Functional Specialization

Disseminated hematological malignancies, such as mucosal associated lymphoid tissue (MALT) lymphoma, depend on many aspects of integrin function, and give considerable insight into how solid tumors may metastasize. Both metastasizing tumor cells and (activated) lymphocytes demonstrate "invasive" behavior, migration involving reversible adhesive contacts, accumulation and expansion in draining lymphoid tissue, release into the circulation, and extravasation.

Under normal physiological conditions lymphocytes continuously recirculate throughout the body, homing specifically, and localizing to certain tissues and lymphoid organs. Lymphocytes expressing α4β7 home specifically to the intestine through recognition of the mucosal addressin cell adhesion molecule-1 (MAdCAM-1), expressed in Peyer's patch endothelial venules. When extravasated, the lymphocytes also expressing αEβ7 are retained. Ninety percent of intraepithelial lymphocytes (IELs), and 40% of *lamina propria* lymphocytes express αEβ7 (compared to fewer than 2% of circulating lymphocytes). The αEβ7 mediates binding to E-cadherin. It is this complex that is involved in retaining lymphocytes in association with the epithelial cells of the gut. Thus, α4β7 is involved in extravasation of lymphocytes (stage 1), then αEβ7 is involved in retaining lymphocytes (stage 2), where they can perform their important role in mucosal immunity. In fact, different functionally specific integrins are active in different anatomical sites. In inflamed endothelium (other than in Peyer's patches), the marginalized lymphocyte is tethered and rolls along the endothelium (dependent on the interaction between α4β1 integrin and the ligand E-selectin), before flattening and undergoing diapedesis (under the influence of β2 integrins). In the Peyer's patches β2 integrin is also important for flattening and diapedesis. In other words, what appears to be the same process is in fact biologically distinct, as evidenced by the integrin expression. This has important consequences for understanding the mechanisms in various diseased states, particularly the specific locations to which specific tumor types have a propensity to metastasize.

Interestingly, it has been shown that α4β7, αEβ7, and L-selectin, which mediate the tissue-specific positioning of normal lymphocytes in the skin, mucosa, epithelium, and lymph nodes, respectively, are selectively expressed in lymphomas localized at these sites. In addition, low-grade B-cell MALT lymphomas and intestinal T-cell lymphomas express α4β7. This is not

**Table 2**
**Integrins and Tumor Associations**

| Tumor | Integrin | Association |
|---|---|---|
| Invasive melanoma | $\alpha 2\beta 1$ | Increased expression in invasive compared with *in situ* melanoma ***(33)*** |
| | $\alpha v\beta 3$ | Strong correlation with acquisition and both vertical growth phase of melanoma and metastasis ***(34)*** |
| | | Promotes M21 melanoma growth by regulating tumor cell survival ***(35)*** |
| | $\beta 3$ | Transfection of $\beta 3$ with melanoma cells promotes cell migration and metastasis ***(36,37)*** |
| | $\alpha v\beta 3$ | Modulates release of MMP-2 and subsequent invasive behavior in melanoma cell lines ***(37)*** |
| | $\alpha 6\beta 1$ | Facilitates extravasation of B16F1 melanoma cells during haematogenous metastasis to the liver ***(38)*** |
| Colon cancer | $\alpha 5\beta 1$ | Expression in HT29 colon cancer cells significantly inhibits proliferation ***(39)*** |
| | $\alpha v\beta 6$ | $\alpha v\beta 6$ Mediated signaling controls release of MMP-9, a gelatinase enzyme ***(40)*** |
| | $\alpha 2\beta 1$ | Mediates the migration of colon carcinoma cells on collagen ***(41)*** |
| Glioma | $\alpha 3\beta 1$ | Migration and invasion of glioma cell lines specifically inhibited by antibodies against $\alpha 3\beta 1$ ***(42)*** |
| Breast cancer | $\alpha 6\beta 4$ | Upregulation and redistribution of this integrin to discrete adhesive structures is associated with a significant reduction in invasive behavior in breast cancer cell lines ***(43)*** |
| | $\alpha v\beta 3 + \alpha v\beta 5$ | Differentially regulated and expressed in breast cancer cells by overexpression of protein kinase C-$\alpha$ ***(44)*** |
| | $\alpha 1 + \alpha 2$ | Mediate invasive activity of mouse mammary carcinoma cells through regulation of stromelysin-1 expression ***(45)*** |
| Oral squamous cell carcinomas | $\beta 1, \alpha 2, \alpha 6$ | Significant reduction in expression ***(46)*** |
| Ovarian cancer | $\alpha 2\beta 1$ | Promotes metastatic dissemination of human ovarian epithelial carcinoma by its interaction with type 1 collagen ***(47)*** |
| | $\beta 1$ | Regulates MMP-2 and membrane type 1 matrix metalloproteinase in ovarian carcinoma cell lines ***(48)*** |

*(continued)*

**Table 2**
**Integrins and Tumor Associations (cont.)**

| Tumor | Integrin | Association |
|---|---|---|
| Urinary bladder carcinomas | $\beta 4$ | Reduction in association with intraepithelial spreading of bladder carcinoma *(49)* |
| Hepatocellular carcinoma | $\alpha 5\beta 1$ | Expressed in well-differentiated carcinoma, decreased or absent in moderately or poorly differentiated human tumors *(50)* |
| | $\beta 1$ | Involved in invasion in hepatocellular carcinoma cell lines *(51)* |
| Thyroid carcinoma | $\alpha 5\beta 1$ | Involved in the attachment of metastatic follicular carcinoma cell lines to bone *(52)* |
| | $\alpha 2\beta 1$ | Upregulated in anaplastic thyroid carcinoma *(53)* |
| | $\beta 1$ | Increased expression in human papillary thyroid carcinoma *(54)* |
| Prostate carcinoma | $\alpha 2\beta 3$ | Participates in the metastatic progression of prostatic adenocarcinoma *(55)* |
| | $\beta 4$ | Downregulated in prostatic carcinoma and prostatic intraepithelial neoplasia *(56)* |
| Gastric carcinoma | $\alpha 2\beta 1$ | Associated with lymph node and liver metastasis |
| | $\alpha 3\beta 1$ | Associated with liver and peritoneal metastasis *(57)* |

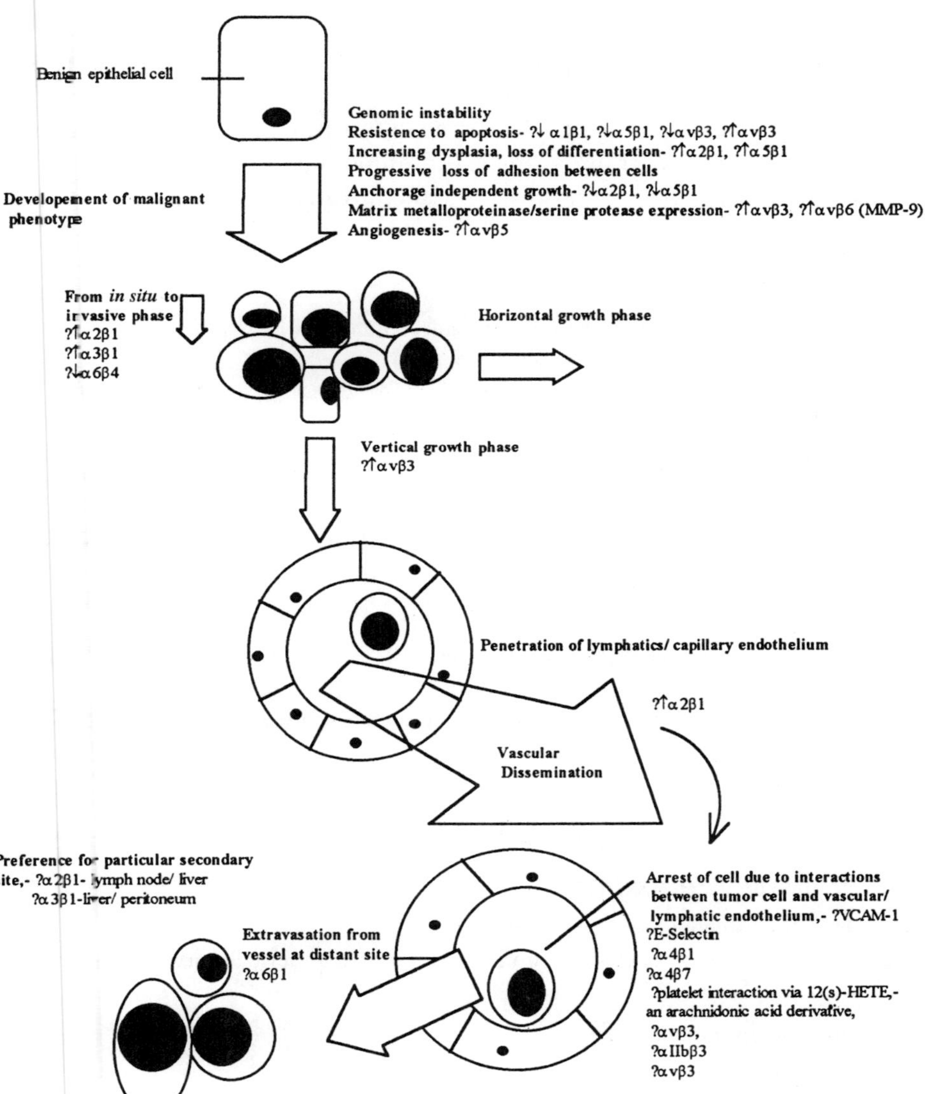


Fig. 2. A generalized scheme for tumor differentiation, growth, and metastasis, illustrating putative site/function-specific integrin expression (for specific references and tumor type involved, *see* text).

commonly expressed in primary lymphomas from nonmucosal sites *(58)*. Moreover, αEβ7 expression has been shown to parallel the degree of epidermotropism of T cells, in skin, in both neoplastic (cutaneous T-cell lymphoma) and inflammatory disorders of the skin *(59)*.

Clearly, many common malignancies are associated with differential integrin expression at different stages of their life cycle. As work progresses, the significance of various associations will become clearer. However, there are currently many exciting developments in this field. For example, integrins are being investigated for use as tracers for tumor targeting ***(60)*** and for use in synthetic gene delivery systems for gene therapy ***(61)***.

## 2. Materials

### *2.1. Dewaxing*

1. Paraffin sections of interest on aminoalkylsilane (APES) slides.
2. 1X Metal rack to hold the slides in solutions.
3. 2X 250 mL (enough to cover slides) of 99.0% xylene in a glass wash bath.
4. 2X 250 mL of 99.0% alcohol in a glass wash bath.

### *2.2. Hydrogen Peroxide Treatment*

1. 250 mL of 0.6% hydrogen peroxide in water in a glass wash bath.
2. 1X wax pen (this is a wax pen for defining the area around the section for reagent contact).

### *2.3. Trypsin Pretreatment*

1. 0.1% Trypsin: Make fresh, for example, 250 mg of powder in 250 mL of calcium chloride for a 0.1% solution.
2. 250 mL (for 0.1% solution) of 0.1% calcium chloride in water (kept at 37°C in the oven).
3. Hotplate stirrer to mix.
4. NaOH to adjust the pH to 7.8.
5. One 500-mL plastic beaker to hold the solution.

### *2.4. Protease Pretreatment*

1. One protease enzyme pretreatment kit.
2. One humid chamber (to hold up to 20 slides, the practical maximum if doing it by hand).

### *2.5. Microwave Pretreatment*

1. One 750 W microwave oven.
2. Plastic slide rack to hold up to 20 slides.
3. 2-L Plastic beaker to hold the rack plus buffer.
4. Citrate buffer solution, (10.5 g of citric acid to 5 L of distilled water): Approx 800 mL, to cover the slides during the full duration of pretreatment.

### *2.6. Pressure Cooking*

1. One pressure cooker.
2. One hot plate.
3. 800 mL of either citrate buffer, (*see* **Subheading 2.5., step 4** for quantities), or an EDTA buffer (mix 1.85 g of EDTA with 5 L of distilled water).

### 2.7. Primary Antisera

1. Concentrated antibody of interest, stored in refrigerator, according to manufacturer's instructions.
2. 1% Phosphate-buffered saline (PBS) solution, (about 1 L per 20 slides depending on the volume of Coplin jars used).
3. Pipets to dilute the antibody to test concentrations.
4. Two disposable plastic pipets to apply the antibody.
5. Plastic test tubes to house various dilutions of antibody.
6. One permanent marker.
7. One humid chamber to house the slides when incubating.
8. One refrigerator at 4°C if leaving the antibody overnight.

### 2.8. Secondary and Tertiary Antibodies

1. Concentrated secondary and tertiary antibodies are supplied in kit form, often to be diluted at 1:60 (DAKO).
2. PBS for washes.
3. Two Coplin jars for washes (depending on the number of slides used).
4. Two disposable pipets for application.
5. One timer.

### 2.9. DAB Chromogen

1. DAB chromogen is supplied in kit form (e.g., DAKO), one drop of DAB to 1 mL of buffer solution.
2. One pair of gloves, as DAB is mutagenic (*see* **Note 9** for appropriate disposal of DAB solutions).

### 2.10. Counterstaining and Mounting

1. 250 mL of Mayer's hematoxylin in a glass wash bath.
2. 250 mL of Scott's acid alcohol in a glass wash bath.
3. Xylene and alcohol solutions (use the same as used in the pretreatment, *see* **Subheading 2.1.**).
4. Coverslips and DPX (mountant), glue.

## 3. Method

Immunohistochemistry is a sensitive and specific technique, involving the application of a primary monoclonal antibody of interest to a tissue of interest. To this complex, a biotinylated secondary antibody is applied, followed by a third antibody with a streptavidin–peroxidase tail. Finally, a chromogen is added. This adheres to the biotin–avidin–peroxidase complex and renders it visible to microscopic evaluation (*see* **Fig. 3** for an illustration of a generalized method for immunohistochemistry, in combination with the text that follow).

### 3.1. Sample Preparation

Cut the required paraffin section and mount on APES slides (*see* **Note 1**).

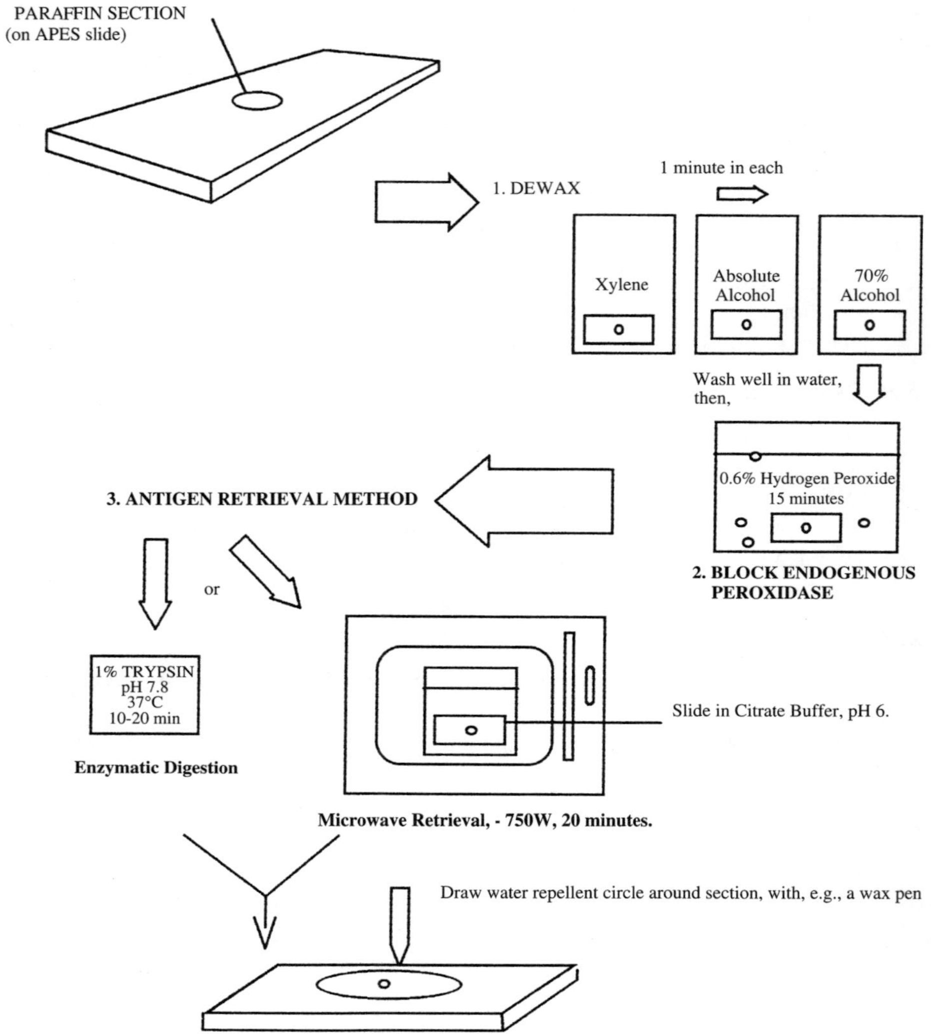


Fig. 3. Generalized immunohistochemical method.

### 3.2. Dewax Slides

Put the slides in a metal slide rack and place in a wash bath containing 99% xylene for approx 1 min. Next, move the rack into another wash bath containing fresh 99% xylene, again for approx 1 min (*see* **Note 2**). Now transfer the sections to a wash bath containing alcohol, again for approx 1 min, and then a fresh change of alcohol, for approx 1 min, to completely remove the xylene from the sections. Next wash the slides well in water.

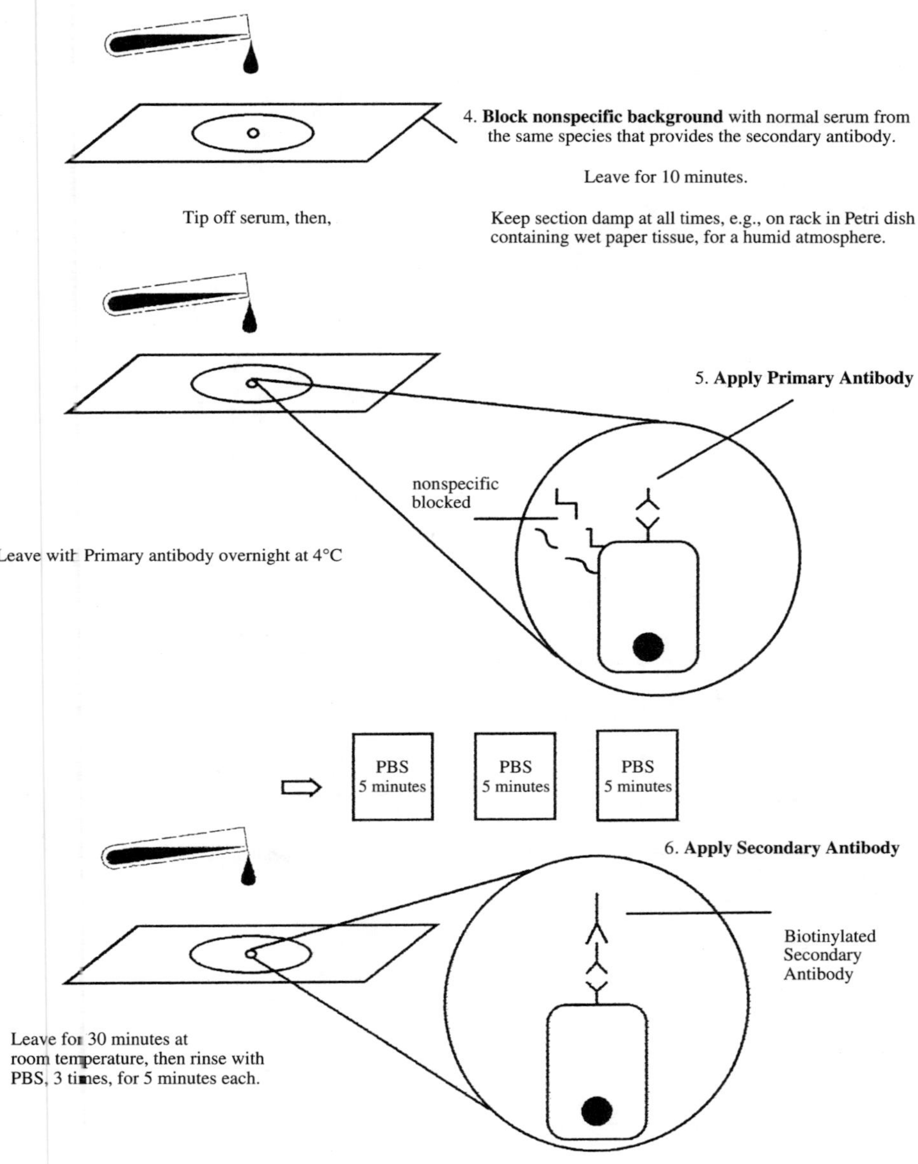


Fig. 3.

### 3.3. Block Endogenous Peroxidase

Place the slides in 0.6% hydrogen peroxide for about 15 min at room temperature (*see* **Note 3**).

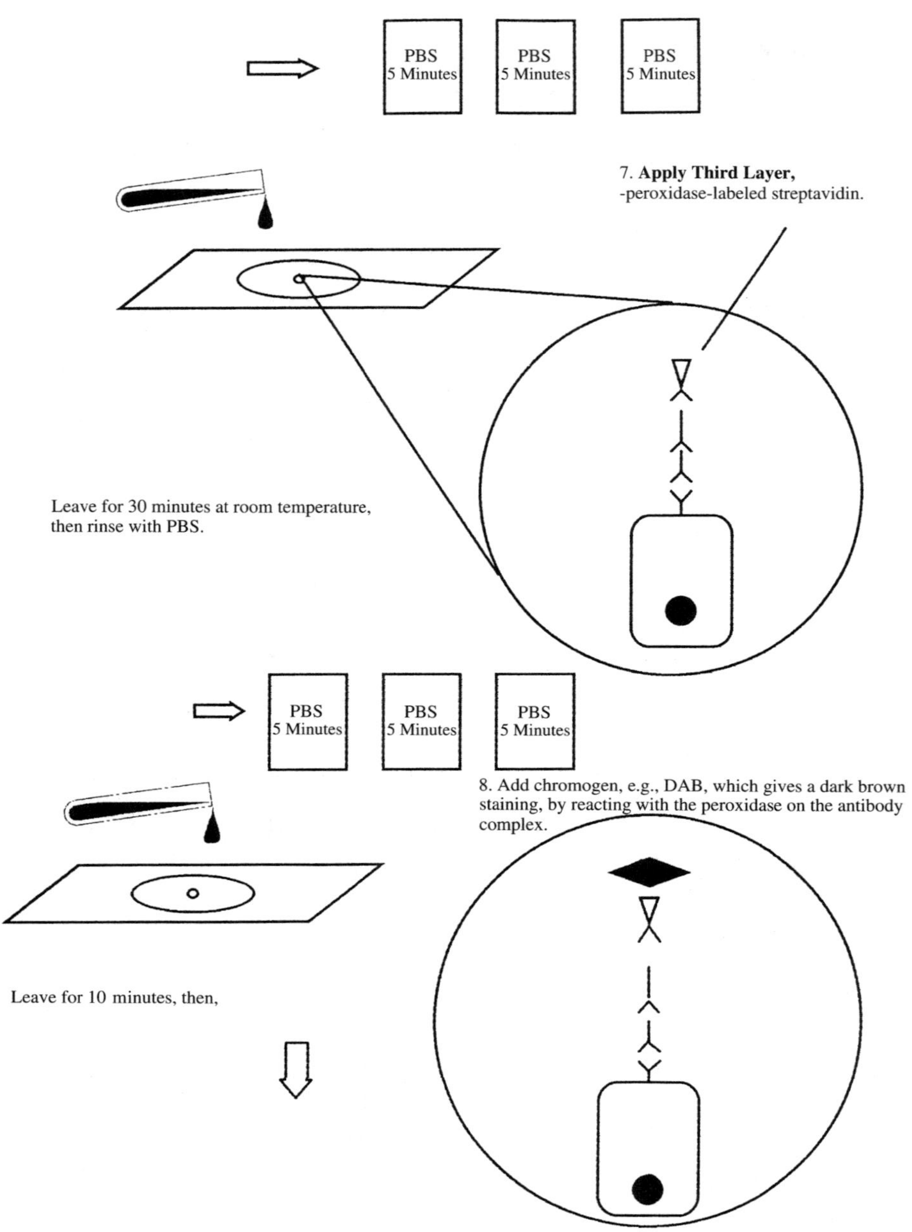
PBS
5 Minutes
PBS
5 Minutes
PBS
5 Minutes
7. Apply Third Layer,
-peroxidase-labeled streptavidin.
Leave for 30 minutes at room temperature, then rinse with PBS.
PBS
5 Minutes
PBS
5 Minutes
PBS
5 Minutes
8. Add chromogen, e.g., DAB, which gives a dark brown staining, by reacting with the peroxidase on the antibody complex.
Leave for 10 minutes, then,

Fig. 3.

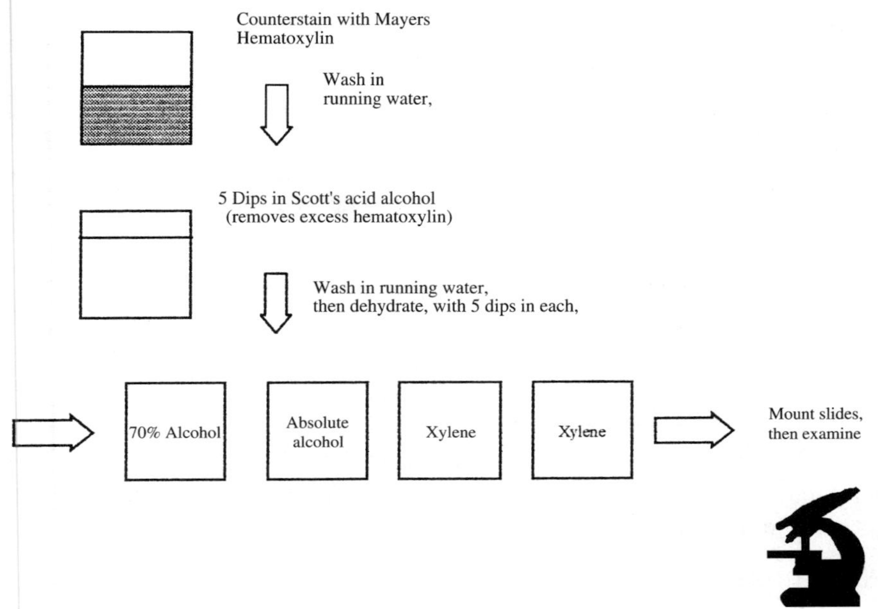


Fig. 3.

### 3.4. Retrieve Antigen

Use any one of the following (*see* **Note 4**):

#### 3.4.1. Enzymatic Digestion

Leave the slides in a solution of 0.1% trypsin, pH 7.8, at 37°C (make fresh; 250 mL of 0.1% calcium chloride to 250 mg of trypsin, then add a few drops of sodium hydroxide until the pH is 7.8), for 10–20 min, after washing off the hydrogen peroxide in a wash bath of tap water. Then proceed to **Subheading 3.5.** Alternatively, use a commercially available protease kit.

#### 3.4.2. Pressure Cooking

After removing from the hydrogen peroxide, wash the slides well in a bath of tap water. Place in a metal rack in a pressure cooker and cover with citrate buffer. Heat until the buffer boils, then fasten the lid. When maximum pressure is reached, leave for a further 2 min. Proceed to **Subheading 3.5.**

#### 3.4.3. Microwave Retrieval

Again, wash off the hydrogen peroxide, then place the slides in a plastic rack, in a plastic beaker, topped up well with citrate buffer solution. This is heated in a 750 W microwave oven for 20 min. Proceed to **Subheading 3.5.**

### 3.5. Identify Section

Pick up slides individually, dry around the edges with a piece of tissue, and draw a water repellant circle around each section with the wax pen. The slide should then be placed back in the rack in a beaker containing water, to prevent the section from drying out. When all the slides are marked in this way, they should be placed on a damp slide rack, section side up, so that the reagents can be applied (*see* **Note 5**).

### 3.6. Block Nonspecific Background

Leave the serum in contact with the section for 15 min, then drain off (*see* **Note 6**).

### 3.7. Apply Primary Antibody

Make up fresh primary antibody, diluted with PBS, according to the manufacturer's instructions. Apply to the sections and leave in the refrigerator overnight at 4°C (*see* **Note 7**).

### 3.8. Washes

Leave the slides to soak in wash jars containing PBS, for three changes of 5 min each (*see* **Note 8**).

### 3.9. Apply the Secondary Antibody

Make up the biotinylated secondary antibody according to the manufacturer's instructions. Apply to the sections, and leave at room temperature for 30 min. Next wash the slides, as before, in three washes of PBS, for 5 min each.

### 3.10. Apply the Third Layer

Make up the third layer labeled with peroxide according to the manufacturer's instructions, and apply to the sections. Leave for 30 min at room temperature. Wash as before, 3× in PBS, for 5 min each.

### 3.11. Apply the Chromogen

Make up the DAB chromogen (*see* **Note 9**), according to the manufacturer's instructions, for example, one drop per milliliter of buffer (DAKO), and drip onto the sections. Leave until the color is developed (5–30 min at room temperature). Drain off the DAB and wash the slides briefly under running tap water, and place in a metal slide rack. Immerse the slides in Mayer's hematoxylin for 30 s, and then wash under running tap water, followed by 5–10 dips in Scott's acid alcohol. Next, leave the sections in two changes of absolute alcohol, for approx 1 min in each, to dehydrate the slides, and then take through three changes of 99% xylene, for approx 1 min in each. The slides are now ready for mounting (for troubleshooting, *see* **Note 10**).

## 4. Notes

1. Three-micron sections are probably optimal, as they give good optical results, and are not too thin to fall off the slides regularly. The sections should be mounted on APES slides, which are coated with an adhesive, to further prevent sections from falling off. This procedure is described in detail in Chapter 2 by Brooks.
2. Make sure that the slides are completely covered by the xylene during this stage. The times are only approximate; thicker sections will require longer immersion in the xylene.
3. The hydrogen peroxide blocks the endogenous (background) peroxidase activity, which is a potential artifact. The hydrogen peroxide solution has reduced oxidizing power with repeated use, and so needs to be changed regularly for optimum effect.
4. Antigen retrieval is a way of exposing "natural" antigens, which may have been "lost" (e.g., by protein folding), during routine processing (e.g., because of formalin fixation). Either one of the three types of antigen retrieval may be optimal for any given antibody to an integrin, and it is difficult to predict which one that is. In addition, for any given method, the exposure time is crucial. However, for any given method there is usually a working range, where the optimal time is likely to reside; for example, for protease digestion, it is most often between 2 and 10 min at room temperature.

   It is often more convenient to use a commercially available protease kit than to make up fresh trypsin. However, if the trypsin method is used, the trypsin must be prepared fresh, as it deteriorates rapidly at room temperature.
5. If a heat-based retrieval is used, care must be taken to cool the slides before this stage. Setting the beaker down under a stream of running cold tap water until cool is quicker than leaving the fluid to air-cool. The wax circle around the section is important, as it limits the spreading of the droplet of applied reagent, and consequently maximizes the contact with the section.

   The slide chamber can be kept humid by placing tissue in the base, sprayed with distilled water. It is important to have a well-fitting lid for this chamber, to prevent the sections from drying out, both when the reagents are applied and between applying reagents.
6. Although the primary antibodies are manufactured to be (relatively) specific to the desired antigen, they will necessarily possess some affinity for other, "nondesired," antigenic sites. This is a potential confounder, so these sites should now be blocked with normal serum from the same species that provides the secondary antibody. This will contain antibodies with a higher affinity for background sites than the primary antibody, and as binding is competitive, and these antibodies do not contain a tail to bind with the secondary antibody, should render these sites invisible. A cheap alternative to commercially available normal serum is to use a teaspoon of skim milk powder, made up with 50 mL of distilled water.
7. Make sure no air bubbles are present in the droplet. This can produce artifactually poor staining areas. Contact time with the primary antibody is an important variable. If it is too long, there may be too much background positivity, and if too short, there may be inadequate staining of the desired area. There are reports that

microwaving the slide with the primary antibody is effective at lowering the required contact time by both increasing the temperature and by independently increasing molecular motion, and therefore increasing the chance of antibody–antigen contact. At room temperature, an hour is usually sufficient, and in the refridgerator at 4°C, they can be left overnight.

8. Washing is important for clean background, but needs to be fairly passive so as not to destroy the sections.
9. It is important to dispose of DAB appropriately: DAB is carcinogenic, and it is possible that neutralizing DAB with bleach yields carcinogenic products. It is best to follow applicable local regulations. Pharmingen Canada's technical protocols suggests, for each 10 mL of DAB solution, adding 5 mL of 0.2 *M* potassium permanganate and 5 mL of 2.0 *M* sulfuric acid. This mixture is left to stand for 10 h, during which time the DAB should become nonmutagenic. Next, the mixture can be decolored with ascorbic acid, then neutralized with sodium bicarbonate, and then, if compatible with local regulations, poured down the sink.
10. However carefully one adheres to the above guidelines, there are still some potential problems to be aware of. These include:
    a. Positive staining of control sections: This can occur if the staining dishes are not kept completely clear of eosin, and fluorescence is used rather than DAB, as eosin can give a fluorescence signal. In addition, some wax pens can be dissolved by Vectastain mountant and leach a background green fluorescent signal across the whole section.
    b. Problems with endogenous biotin after pressure cooker retrieval: This can be particularly troublesome with kidney and liver. Many commercial kits are available specifically for this problem and are often effective. Using a citrate buffer of pH 6.0 rather than Tris/EDTA of pH 9.0 for antigen retrieval also reduces endogenous biotin staining.
    c. Adherence of the primary antibody to the blocking agent: This can occur infrequently, giving negative results, and is remedied by using a different blocker/serum for the nonspecific background.
    d. Selecting an appropriate negative control: A good negative control is a tissue in which the signal is eliminated by preincubating the antibody with an excess of the peptide or protein to which it was raised. A Western blot can suggest specificity of the interaction if only one band is seen.
    e. Sections falling off the slide: This is unfortunately common, even when APES slides are used, owing to the mechanical trauma imposed by the method. It may be beneficial to postfix slides for a few minutes in the original fixative, after blocking the endogenous peroxidase. This is useful even if the sections do not have a tendency to completely detach, as antibody can become trapped under the section, when just the edges detach, and therefore give rise to a high background staining.
    f. High background staining: This can be because of many factors, such as the above, poor specificity of the primary antibody, incubating for too long a period of time, inadequate blocking of background, etc. However, one simple measure is to cool the DAB solution and check the staining under the microscope at 2, 4, and 6 min, to find the optimum time.

For success, it is crucial to try a number of different techniques/incubation times/antigen retrieval methods, etc. It is well worth getting all the manufacturers' technical information sheets and finding out what other researchers use with any particular antibody. Such information often appears on the Internet well before it appears in print, and can save many wasted hours.

## References

1. Hynes, R. O. (1992) Integrins: versatility, modulation, and signaling in cell adhesion. *Cell* **69,** 11–25.
2. Takada, Y. and Hemler, M. E. (1989) The primary structure of the VLA-2/collagen receptor a2 subunit (platelet GPlα): homology to other integrins and the presence of a possible collagen binding domain. *J. Cell Biol.* **109,** 397–407.
3. Higgins, J. M., Mandelbrot, D. A., Shaw, S. K., et al. (1998) Direct and regulated interaction of integrin αEβ7 with E-cadherin. *J. Cell Biol.* **140,** 197–210.
4. Fitzgerald, L. A. and Phillips, D. R. (1985) Calcium regulation of the platelet membrane glycoprotein Iib–IIIa complex. *J. Biol. Chem.* **260,** 11,366–11,374.
5. Nermut, M. V., Green, N. M., Eason, P., et al. (1998) Electron microscopy and structural model of human fibronectin receptor. *EMBO J.* **7,** 4093–4099.
6. Lofus, J. C., Halloran, C. E., Ginsberg, M. H., et al. (1996) The amino-terminal third of aliβ defines the ligand recognition specificity of integrin αIIbβ3. *J. Biol. Chem.* **271,** 2033–2039.
7. Springer, T. A. (1997) Folding of the N-terminal, ligand binding region of integrin α-subunits into a β propeller domain. *Proc. Natl. Acad. Sci. USA* **94,** 65–72.
8. Parise, L. V., Helgerson, S. L., and Steiner, B. (1987) Synthetic peptides derived from fibrinogen and fibronectin change the conformation of purified platelet glycoprotein Iib–IIIa. *J. Biol. Chem.* **262,** 12,597–12,602.
9. Phillips, D. R. and Agin, P. P. (1977) Platelet plasma membrane glycoproteins. Evidence for the presence of non-equivalent disulphide bonds using non-reduced-reduced two dimensional gel electrophoresis. *J. Biol. Chem.* **252,** 2121–2126.
10. Chan, P., Lawrence, M. B., Dustin, M. L., et al. (1991) Influence of receptor lateral mobility on adhesion strengthening between membranes containing LFA-3 and CD2. *J. Cell Biol.* **115,** 245–255.
11. Lotz, M. M., Bundsal, C. A., Erickson, H. P., et al. (1989) Cell adhesion to fibronectin and tenacin: quantitative measurements of initial binding and subsequent strengthening response. *J. Cell Biol.* **109,** 1795–1805.
12. Leahy, D. J. (1997) Implications of atomic resolution structures for cell adhesion. *Annu. Rev. Dev. Biol.* **13,** 363–369.
13. Bohmer, R. M., Scharf, E., and Assoian, R. K. (1996) Cytoskeletal integrity is required throughout the mitogen stimulation phase of the cell cycle and mediates the anchorage dependant expression of cyclin D1. *Mol. Biol Cell.* **7,** 101–111.
14. Weaver V., Peterson O., Wang F., Larabell C., Briand P., and Damsky C. (1997) Reversion of the malignant phenotype of human breast cells in three dimensional culture and in vivo by integrin blocking antibodies. *J. Cell Biol.*, **137,** 231–245.
15. Varner, J. and Cheresh, D. (1996) Integrins and cancer. *Curr. Opin. Cell. Biol.* **8,** 724–730.

16. Zhu, X. and Assoian, R. K. (1995) Integrin-dependant activation of MAP Kinase: a link to shape-dependant cell proliferation. *Mol. Biol. Cell.* **6,** 273–282.
17. Frisch, S. and Ruoshlati, E. (1997) Integrins and anoikis. *Curr. Opin. Cell Biol.* **9,** 701–706.
18. Montgomery, A., Reisfeld, R., and Cheresh, D. (1994) Integrin αvβ3 rescues melanoma cells from apoptosis in three dimensional dermal collagen. *Proc. Natl. Acad. Sci. USA* **91,** 8856–8860.
19. Stromblad, S., Becker, J., Yebra, M., Brooks, P., and Cheresh, D. (1996) Suppression of p53 activity and p21 expression by vascular cell integrin αvβ3 during angiogenesis. *J. Clin Invest.* **98,** 426–433.
20. Wu, R. and Schonthal, A. H. (1997) Activation of p53-p21 pathway in response to disruption of cell–matrix interactions. *J. Biol. Chem.* **272,** 29,091–29,098.
21. Owens, L., Xu, L., Craven, R., Dent, G., Weiner, T., and Kornberg, L. (1995) Overexpression of the focal adhesion kinase in invasive human tumors. *Cancer Res.* **55,** 2752–2755.
22. Schwartz, M., Toksoz, D., and Khosravi-Far, R. (1996) Transformation by Rho exchange factor oncogenes is mediated by activation of an integrin-dependent pathway. *EMBO J.* **15,** 6525–6530.
23. Radeva, G., Petrocelli, T., Behrend, E., Leung-Hagesteijn, C., Filmus, J., and Slingerland, J. (1997) Overexpression of the integrin-linked kinase promotes anchorage-independent cell cycle progression. *J. Biol. Chem.* **272,** 13,937–13,944.
24. Pignatelli, M. and Bodmer, W. F. (1989) Integrin-receptor-mediated differentiation and growth inhibition are enhanced by transforming growth factor-beta in colorectal tumor cells grown in collagen gel. *Int. J. Cancer* **44,** 518–523.
25. Zutter, M., Mazoujian, G., and Santoro, S. (1990) Decreased expression of integrin adhesive protein receptors in adenocarcinoma of the breast. *Am. J. Pathol.* **137,** 863–870.
26. Beauvais, A., Erickson, C., Goins, T., et al. (1995) Changes in the fibronectin-specific integrin expression pattern modify the migratory behaviour of sarcoma S180 cells in vitro and in the embryonic environment. *J. Cell Biol.* **128,** 699–713.
27. Simon, K., Nutt, E., Abraham, D., et al. (1997) The αvβ3 integrin regulates α5β1-mediated cell migration toward fibronectin. *J. Biol. Chem.* **272,** 29,380–29,389.
28. Cary, L., Chang, J., and Guan, J. (1996) Stimulation of cell migration by overexpression of focal adhesion kinase and its association with src and Fyn. *J. Cell Sci.* **109,** 1787–1794.
29. Brookes, P., Stromblad, S., Sanders, L., et al. (1996) Localization of matrixmetalloproteinase MMP-2 to the surface of invasive cells by interaction with integrin αvβ3. *Cell* **85,** 683–693.
30. Friedlander, M., Brooke, P., Shaffer, R., et al. (1995) Definition of two angiogenic pathways by distinct αv integrins. *Science* **270,** 1500–1502.
31. Brookes, P., Stromblad, S., Klemke, R., et al. (1995) Antiintegrin αvβ3 blocks human breast cancer growth and angiogenesis in human skin. *J. Clin. Invest.* **96,** 1815–1822.
32. Boudeau, N., Andrews, C., Srebrow, A., et al. (1997) Induction of the angiogenic phenotype by Hox D3. *J. Cell Biol.* **139,** 257–264.
33. van Duinen, C. M., van den Broek, L. J., Vermeer, B. J., et al. (1994) The distribution of cellular adhesion molecules in pigmented skin lesions. *Cancer* **73,** 2131–2139.

34. Nesbit, M. and Herlyn, M. (1994) Adhesion receptors in human melanoma progression. *Invas. Metastas.* **14,** 131–146.
35. Petitclerc, E., Stromblad, S., von Schalscha, T. L., et al. (1999) Integrin αvβ3 promotes M21 melanoma growth in human skin by regulating tumor cell survival. *Cancer Res.* **59,** 2724–2730.
36. Filardo, E. J., Brooks, P. C., Deming, S. L., et al. (1995) Requirement of the NPXY motif in the integrin β3 cytoplasmic tail for melanoma cell migration in vitro and in vivo. *J. Cell Biol.* **130,** 441–450.
37. Seftor, R. E., Seftor, E. A., Stetler Stevenson, W. G., et al. (1993) The 72 kDa type IV collagenase is modulated via expression of αvβ3 and α5β1 integrins during human melanoma cell invasion. *Cancer Res.* **53,** 3411–3415.
38. Hangan, D., Morris, V., Boeters, L., et al. (1997) An epitope on VLA-6 (α6β1), integrin involved in migration but not adhesion is required for extravasation of murine melanoma B16F1 cells in liver. *Cancer Res.* **57,** 3812–3817.
39. Varner, J. A., Emerson, D. A., and Juliano, R. L. (1995) Integrin α5β3 expression negatively regulates cell growth: reversal by attachment to fibronectin. *Mol. Biol. Cell.* **6,** 725–740.
40. Niu, J., Gu, X., and Turton, J., et al. (1998) Integrin-mediated signaling of gelatinase B secretion in colon cancer cells. *Biochem. Biophys. Res. Commun.* **249,** 287–291.
41. Pignatelli, M., Karayiannakis, A. J., Noda, M., Efstathiou, J., and Kmiot, W. A. (1997) E-cadherin/catenin complex in the gastrointestinal tract, in *The Gut as a Model in Cell and Molecular Biology*. (Halter, F., Winton, D., Wright, N. A., eds.) Kluwer, Lancaster, pp. 194–203.
42. Fukushima, Y., Ohnishi, T., Arita, N., et al. (1998) Integrin α3β1-mediated interaction with laminin 5 stimulates adhesion, migration and invasion of malignant glioma cells. *Int. J. Cancer* **76,** 63–72.
43. Jones, J. L., Royall, J. E., Critchley, D. R., et al. (1997) Modulation of myoepithelial associated α6β4 integrin in a breast cancer cell line alters invasive potential. *Exp. Cell Res.* **235,** 325–333.
44. Carey, I., Williams, C. L., Ways, D. K., and Noti, J. D. (1999) Overexpression of protein kinase C-alpha in MCF-7 breast cancer cells results in differential regulation and expression of αvβ3 and αvβ5. *Int. J. Oncol.* **15,** 127–136.
45. Lochter, A., Navre, M., Werb, Z., and Bissell, M. J. (1999) a1 and a2 integrins mediate invasive activity of mouse mammary carcinoma cells through regulation of stromelysin-1 expression. *Mol. Biol. Cell.* **10,** 271–282.
46. Maragou, P., Bazopoulou-Kyrkanidou, E., Panotopoulou, E., et al. (1999) Alteration of integrin expression in oral squamous cell carcinomas. *Oral Dis.* **5,** 20–26.
47. Fishman, D. A., Kearns, A., Chilukuri, K., et al. (1998) Metastatic dissemination of human ovarian epithelial carcinoma is promoted by α2β1-integrin-mediated interaction with type 1 collagen. *Invas. Metastas.* **18,** 15–26.
48. Ellerbroek, S. M., Fishman, D. A., Kearns, A. S., et al. (1999) Ovarian carcinoma regulation of matrix metalloproteinase-2 and membrane type 1 matrix metalloproteinase through β1 integrin. *Cancer Res.* **59,** 1635–1641.
49. Harabayashi, T., Kanai, Y., Yamada, T., et al. (1999) Reduction of integrin β4 and enhanced migration on laminin in association with intraepithelial spreading of urinary bladder carcinomas. *J. Urol.* **161,** 1364–1371.

50. Zhao, J., Zhai, W., and Zhang, Y. (1997) Expression and detection of integrin α5β1, fibronectin and its fragment in patients with hepatocellular carcinoma. *Chung Hua Ping Li Hsueh Tsa Chih* **26,** 65–69.
51. Masumoto, A., Arao, S., and Otsuki, M. (1999) Role of β1 integrins in adhesion and invasion of hepatocellular carcinoma cells. *Hepatology* **29,** 68–74.
52. Smit, J. W., van der Pluijm, G., Vloedgraven, H. J., et al. (1998) Role of integrins in the attachment of metastatic follicular thyroid carcinoma cell lines to bone. *Thyroid* **8,** 29–36.
53. Dahlam, T., Grimelius, L., Wallin, G., et al. (1998) Integrins in thyroid tissue: upregulation of α2β1 in anaplastic thyroid carcinoma. *Eur. J. Endocrinol.* **138,** 104–112.
54. Ensinger, C., Obrist, P., Bacher-Stier, C., et al. (1998) β1-integrin expression in papillary thyroid carcinoma. *Anticancer Res.* **18,** 33–40.
55. Trikha, M., Raso, E., Cai, Y., Fazakas, Z., et al. (1998) Role of α2(b)β3 integrins in prostatic cancer metastasis. *Prostate* **35,** 185–192.
56. Allen, M. V., Smith, G. J., Juliano, R., et al. (1998) Downregulation of the β4 integrin subunit in prostatic carcinoma and prostatic intraepithelial neoplasia. *Hum. Pathol.* **29,** 311–318.
57. Ura, H., Denno, R., Hirata, K., Yamaguchi, K., et al. (1998) Separate functions of α2β1 and α3β1 integrins in the metastatic process of human gastric carcinoma. *Surg. Today* **28,** 1001–1006.
58. Pals, S. T., Drillenburg, P., Radaszkiewicz, T., Manten-Horst, E., et al. (1997) Adhesion molecules in the dissemination of non-Hodgkins lymphomas. *Acta Haematol.* **97,** 73–80.
59. Dietz, S. B., Whitaker-Menezes, D., and Lessin, S. R. (1996) The role of αEβ7 integrin and E-cadherin in epidermotropism in cutaneous T-cell lymphoma. *J. Cutan. Pathol.* **23,** 312–318.
60. Haubner, R., Wester, H. J., Reuning, T., et al. (1999) Radiolabelled αvβ3 integrin antagonists: a new class of tracers for tumor targeting. *J. Nucl. Med.* **40,** 1061–1071.
61. Hart, S. (1999) Use of adhesion molecules for gene therapy. *Exp. Nephrol.* **7,** 193–199.

7

# Surrogate Markers of Angiogenesis and Metastasis

Girolamo Ranieri and Giampietro Gasparini

## 1. Introduction

Angiogenesis, the formation of new blood vessels from a preexisting vascular network, is a complex multistep process under the control of positive and negative factors *(1,2)*. Growth, progression, and metastasis of malignant tumors are angiogenesis-dependent processes *(3–5)*. There is compelling evidence that development of angiogenic phenotype is a common pathway in the biology of tumor progression and metastasis, and the angiogenesis is correlated with other molecular mechanisms (e.g., the control of wild-type p53 on vascular endothelial growth factor or thrombospondin-1) *(6,7)*. Angiogenesis can be directly determined, and so it may be useful as a biomarker *(8–10)*. Tumor microvessel density *(11–13)* or tissue angiogenic factors expression may predict which patients are at different risk of metastasis *(14–16)*, and serum levels of angiogenic factors may serve as tumor markers *(17,18)*.

At present the most widely used method to assess angiogenesis is quantification of intratumoral microvessel density (IMD) of the primary tumor by using specific markers for endothelial cells such as factor VIII-related antigen (FVIII-RA), anti-CD31 (platelet endothelial cell adhesion molecule, PECAM), or anti-CD34 antibodies *(19–22)*, using a standard immunocytochemistry immunoperoxidase technique to stain microvessels *(23,24)*. The basic principles of immunocytochemistry are described in Chapter 2 by Brooks and use of immunocytochemistry employing antibodies against factor VIII-related antigen and CD31 is described in Chapter 8 by Turner and Harris. The above markers are unable to discriminate between quiescent and activated proliferating endothelium. The LM-609 antibody to integrin $\alpha v\beta_3$ seems to be specific for activated/proliferating endothelial cells and for the smaller intratumoral microvessels, but the related antigen is largely lost during fixation and/or paraffin embedding

From: *Methods in Molecular Medicine, vol. 57: Metastasis Research Protocols, Vol. 1: Analysis of Cells and Tissues*
Edited by: S. A. Brooks and U. Schumacher © Humana Press Inc., Totowa, NJ

procedures; thus its use is possible only on fresh or frozen samples ***(25)***. Using Weidner's method, areas representative of the invasive component of the tumor are selected from paraffin block sections stained with hematoxylin and eosin ***(26)***. The areas of highest neovascularization are found by scanning the tumor sections with light microscopy at low magnification (40× and 100×) and identifying the microvessel stained using an endothelial marker. Individual microvessels are then counted on 200× and 400× fields. ***(26)***. Once neovasculature is highlighted it may be evaluated also using other techniques: (1) Chalkley counting ***(27)*** is a manual method to be performed by scanning the tumor section at low magnification to identify the fields of intense vascularization. At higher magnification (200×–250×), an eyepiece graticule containing 25 randomly positioned dots is rotated so that the maximum number of points are within the vessels of the vascular "hot spots." Individual microvessels on the overlaying dots are counted, and (2) with a multiparametric computerized image analysis system (CIAS) other morphometric parameters can be detected: vessel lumen area (MVA), vessel lumen parameter (MVP), and the percentage of immunostained area for microscopic field ***(28,29)***. This semi-automated technique has been suggested to be a more objective method of assessing IMD ***(30,31)***. Microvessel counting techniques are also described and evaluated in Chapter 8 by Turner and Harris. The expression of endothelial cell growth factors—vascular endothelial growth factor (VEGF), thymidine phosphorylase (TP), and basic fibroblast growth factor (bFGF)—can also be determined in tumor tissue sections ***(32–36)*** using immunoperoxidase techniques. The determination of levels of the above growth factors in the cytosolic extract of tumors ***(37–39)***, in serum ***(40–42)***, in whole blood, in peripheral blood mononuclear cells (PBMNCs), or in platelets ***(43–45)*** may be performed using enzyme-linked immunosorbent assays (ELISA) ***(46)***. Most authors who correlated the markers of angiogenic activity with clinical outcome in patients with early-stage breast cancer or with other types of solid tumors found that the patients with primary tumors with high angiogenic indexes have poorer prognosis as compared to those with low angiogenic tumors ***(47–52)*** (*see* **Note 1**).

## 2. Materials

### *2.1. Materials for Immunohistochemical Assays*

1. 10% Formalin.
2. Paraffin wax.
3. Xylene.
4. Ethanol.
5. Poly-L-lysine coated slides.
6. A wax pen for drawing hydrophobic ring around the tissue section of the slides.
7. 0.5% Bovine serum albumin (BSA).
8. Trypsin or pronase solution.

9. Phosphate-buffered saline (PBS), pH 7.2–7.4: 0.2 g of $KH_2PO_4$, 28 g of $Na_2HPO_4 \cdot 12H_2O$, and 0.2 g of KCl. Make up to 1 L with distilled water.
10. Microvave oven with variable temperature.
11. 100% Cold acetone (for frozen tissue).
12. 0.05% v/v hydrogen peroxide in methanol or hydrogen peroxide 0.3%–0.1% $NaN_3$ in distilled water for fresh/frozen tissue sections.
13. Citrate buffer, pH 6.0: Dissolve 21.0 g of citric acid in 10 L distilled water and add 265 mL of 1 *M* NaOH.
14. 0.05 *M* Tris–HCl buffer, pH 7.6.
15. Humid chamber.
16. Primary monoclonal (or polyclonal) antibody using commerical kits. The primary antibody can be prediluted or concentrated.
17. Secondary antibody. Use an appropriate biotinylated antibody raised against the species used to produce the primary antibody. Secondary is obtained using commercial kits.
18. Avidine peroxidase complex (HRP) using a commercial kit.
19. Diaminobenzidine (DAB) solution prediluted in a commercial kit (or 3-amino, 9-etyl carbazole [AEC]).
20. Buffered gylcerin: 9 mL of glycerin, 1 mL of 9% NaCl, 100 mg of Tris, or DPX from a commercial manufacturer, as a substrate for coverslips.
21. Mayer's hematoxylin or Gill's hematoxylin for nuclear counterstain.

### 2.2. ELISA in Tissues

1. 96-Well microtiter plates.
2. Liquid nitrogen.
3. Rabbit or mouse antiangiogenic factors obtained using commercial kits.
4. Peroxidase-conjugated Fab' of the antiangiogenic factor antibody, using commercial kits.
5. 0.1 *M* NaCl.
6. 0.15 *M* NaCl.
7. 0.2 *M* Carbonate buffer, pH 9.5.
8. 0.2 *M* Carbonate buffer, pH 9.0.
9. 0.025 *M* Carbonate buffer, pH 9.0.
10. $H_2SO_4$.
11. PBS, pH 7.4.
12. Skim milk in PBS.
13. 1.5. m*M* Ethylenediaminetetraacetic acid (EDTA).
14. 10 m*M* Sodium molybdate.
15. 0.05% Tween-20.
16. 3 m*M* $NaN_3$.
17. 10% Glycerol.
18. 0.1% $NaN_3$.
19. BSA 1%.
20. Substrate solution with *o*–phenylenediaminidine or TMB substrate system or K-Blue substrate from a commercial manufactuer.

### 2.3. ELISA of Plasma, Whole Blood, Peripheral Blood Mononuclear Cells, or Platelets

1. Venoject blood system.
2. Sterile test tubes containing sodium citrate as anticoagulant.
3. Sterile $H_2O$.
4. Sterile test tubes anticoagulated with acid-citrate dextrose and containing 0.18 μ*M* prostaglandin $E_1$.
5. Ficoll hypague.
6. Sodium metrizoate from a commercial manufacturer.
7. 10% Fetal bovine serum (FBS).
8. Sterile PBS.
9. Commercial kit for VEGF immunoassay.

## 3. Methods

### 3.1. Immunohistochemical Assay to Determine Microvessel Count Using Pan-Endothelial Markers on Paraffined Tissue

1. Surgically removed tumor specimens are fixed in 10% formalin for 24 h, and then embedded in paraffin wax.
2. Four sections of 4 μm thickness are cut from each paraffin wax block.
3. Tissue sections are mounted on ploy-L-lysine-coated slides.
4. The sections are dewaxed in xylene and then in ethanol and water.
5. Before immunostaining, an antigen retrieval system is applied; the slides are placed in a plastic container filled with 0.01 *M* sodium citrate buffer, pH 6.0, which is heated in a microwave oven for 3–5 min.
6. The sections are allowed to cool for 30 min.
7. A ring is drawn around the stained tissue section with a hydrophobic pen.
8. Enzymatic pretreatment is performed with 0.05% trypsin and 0.1% $CaCl_2$ in PBS for 30 min at 37°C.
9. The sections are washed in water and rinsed in PBS.
10. Endogenous peroxidase in the histological section is quenched by incubation with 0.05% v/v hydrogen peroxide in methanol.
11. The sections are washed in water.
12. Incubation with 60 μL of primary antibody (mouse monoclonal anti-CD31 or anti-CD34 antibodies). Dilutions suggested: 1:50 and 1:100, respectively, in PBS, 0.5% BSA is performed overnight in a humid chamber at room temperature. Incubation with 60 μL of rabbit polyclonal antibody anti F-VIII-RA (dilution proposed 1:200) is performed for 2 h in a humid chamber at room temperature.
13. Sections are washed twice in PBS.
14. 60 μL of secondary antibodies are incubated for 1 h using biotinylated sheep antimouse (can be obtained for Amersham International PLC, Bucks, UK) for a primary monoclonal antibody (at dilution 1:200 in PBS, 0.5% BSA) or biotinylated donkey antirabbit (Amersham International PLC, Bucks, UK) for a primary polyclonal antibody, both incubated for 30 min.

15. Sections are washed twice in PBS.
16. Incubation with diluted avidin–biotin peroxidase complex/HRP 60 µL (DAKO Ltd., Bucks, UK) for 1 h at room temperature in a humid chamber.
17. Sections are washed twice in PBS.
18. A fresh solution is prepared of 0.05% DAB in 0.05 *M* Tris-HCl, pH 7.6 plus 33 µL of 30% vol hydrogen peroxide/100 mL.
19. The slide is covered with the above solution or commercial DAB solution for 5 min at room temperature.
20. Sections are counterstained with Mayer's hematoxylin for 5 min.
21. Sections are washed in tap water.
22. Coverslips are mounted with DPX.
23. Negative control; no primary antibody is added.
24. A known vascular lesion is prepared as a positive control.
25. Areas representative of the invasive component of the tumor are selected with light microscopy at low magnification from paraffin sections stained with hematoxylin and eosin.
26. In several adjacent sections immunostained with one of the above pan-endothelial markers, the invasive tumor component containing the most intense vascularization ("hot spot") is identified by scanning at low power (40× and 100×) (*see* **Notes 2–4**).
27. Individual microvessels are counted on 200× (i.e., 20× objective lens and 10× ocular lens; 0.74 $mm^2$ per field and 400× fields (i.e., 40× objective lens and 10× ocular lens; 0.19 $mm^2$ per field) (*see* **Notes 5** and **6**).
28. Each count is expressed as the highest number of microvessels identified within any 200× or 400× field **(Fig. 1)** (*see* **Notes 7–9**).
29. An alternative technique for assessment of tumor angiogenesis is Chalkley counting. At higher magnification (200–250×), an eyepiece graticule containing 25 randomly positioned dots is rotated so that the maximum number of points are on or within the microvessels of the "vascular hot spots" (*see* **Note 10**).
30. The overlaying dots are counted.
31. Another method for evaluation of vascularization utilizes a computerized image analysis system (CIAS). The sections are evaluated at 200× magnification (0.73 $mm^2$ area). Each frame image is automatically detected by the system, allowing for human intervention to correct some details (elimination of background or nonspecific staining). The parameters automatically detected include number of microvessels, the overall area (MVA), the perimeter (MVP) of microvessels, and the percentage of immunostained area in each single microscopic field.

### 3.2. Immunohistochemical Assay to Determine Microvessel Count Using the Antiintegrin $\alpha v\beta 3$ LM609 Monoclonal Antibody on Fresh or Frozen Tissue

1. A representative part of each surgically removed tumor specimen is snap frozen in liquid nitrogen and stored at –80°C until determination.
2. Sections of 4–6 µm thickness are cut.
3. Tissue sections are mounted on poly-L-lysine-coated slides.

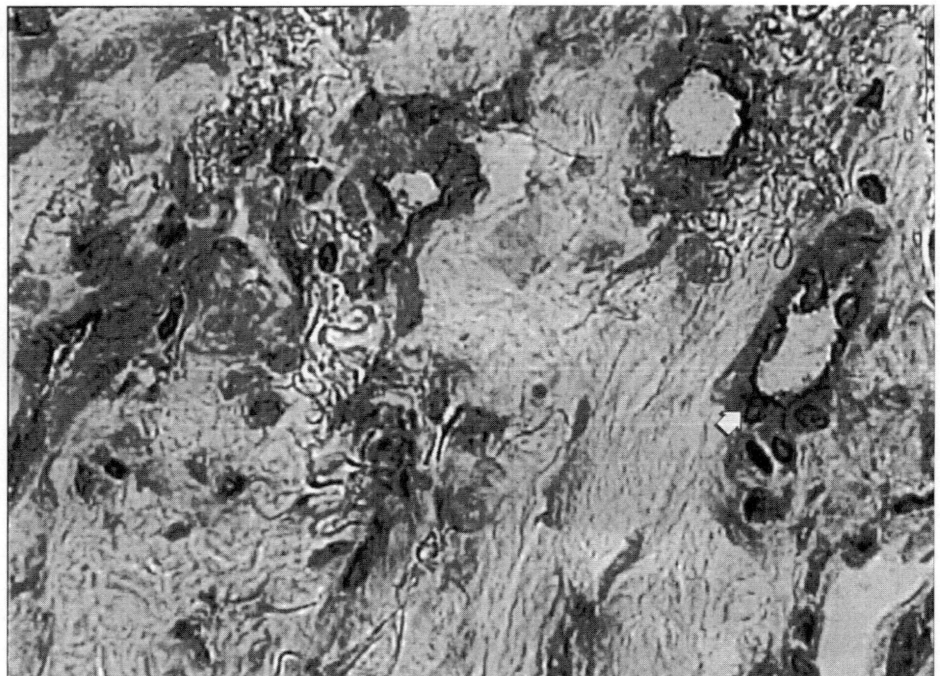

Fig. 1. Immunoperoxidase reaction with the panendothelial marker anti-CD34 antibody. "Hot spot" field ×400 (i.e., 40× objective and 10× ocular lens).

4. The sections are fixed in cold acetone at 4°C for 10 min.
5. Sections are washed with PBS.
6. Endogenous peroxidase is quenched by incubation with hydrogen peroxide/0.1–0.3% $NaN_3$ in distilled water.
7. Sections are washed with PBS.
8. Incubation with 60 µL of the antiintegrin avb3 LM609 monoclonal antibody is performed for 1 h in a humid chamber at room temperature.
9. The sections are washed twice in PBS.
10. 60 µL of secondary antimouse polyclonal biotinylated antibody is incubated for 30 min in a humid chamber at room temperature.
11. Sections are washed in PBS.
12. Incubation with 60 µL of avidin–biotin–peroxidase complex–HRP (DAKO) for 30 min at room temperature in a humid chamber.
13. Sections are washed with PBS.

Next steps are similiar to **steps 18–23** in **Subheading 3.1.**

14. An adjacent serial tumor section is immunostained with the pan-endothelial marker anti-CD31 as an internal positive control.
15. Microvessel counts is evaluated on five "hot-spot" fields (*see* **Note 11**).

### 3.3. VEGF Expression Using Immunohistochemical Methods

1. Paraffin-embedded cancer tissue sections are used.
2. The primary antibody is a rabbit anti-VEGF polyclonal antibody produced by Santa Cruz Biotechnology.
3. The secondary antibody is a goat antirabbit biotin-labeled reagent.
4. VEGF immunoreactivity is detected in the cytoplasm of neoplastic cells and the staining intensity is graded: (–) no staining; (+) weak; (++) moderate; or (+++) strong staining.
5. The proportion of stained cells can be evaluated in accordance with the percentage of immunoreactive cells counting at least 1000 tumor cells.

### 3.4. Platelet-Derived Endothelial Cell Growth Factor/Thymidine Phosphorylase Expression Using Immunohistochemical Method

1. Paraffin-embedded tissue sections are used.
2. The primary antibody is an antihuman plasma monoclonal TP antibody (from Nippon Roche Research Center, Kanagawa, Japan).
3. The secondary antibody is a polyclonal biotinylated reagent.
4. TP immunoreactivity is observed in cancer and stromal cells, such as macrophages and fibroblasts. Endothelial cells staining is also evaluable. Immunostaining is mainly cytoplasmic.
5. Intensity of staining is: (–) no staining cells; (+) weak; (++) moderate; (+++) strong staining.
6. The proportion of cells staining is graded as 0–24%; 25–74%; 75–100%.
7. A tumor is classified TP-positive if more than 25% of tumor cells show moderate staining.

### 3.5. VEGF Assay in the Cytosol: Colorimetric Enzymatic Immunoassays

1. A representative part of a tumor specimen is snap frozen in liquid nitrogen and stored at –80°C until determination.
2. For the preparation of cytosolic extracts, the frozen tissue samples are homogenized using a microdismembrator.
3. The samples are homogenized with 10 vol of cold (i.e., 4°C) 10 m*M* PBS, pH 7.4, containing 1.5. m*M* EDTA, 10 m*M* sodium molybdate, 3 m*M* $NaN_3$, and 10% glycerol.
4. Low salt extract (cytosol) is obtained by centrifugation at 100,000*g* for 1 h at 4°C.
5. The supernatant fraction is appropriately diluted according to the protein concentration.
6. 96-well microtiter plates are coated with 10 ng/mL of the purified anti-$VEGF_{121}$ polyclonal antibody (Toagosei Ibaraki Japan No 091895) in 0.1 *M* NaCl and 0.025 *M* carbonate buffer, pH 9.0. For determination of $VEGF_{165}$ levels, the commercial Kit Quantikine R & D System, Minneapolis, MN, can be used. Subsequently the well microtiter plates are coated with the mouse antihuman VEGF monoclonal antibody.
7. The reaction is blocked with 1% BSA, 0.2 *M* carbonate buffer, pH 9.5, 0.1 *M* NaCl, and 0.1% $NaN_3$.

8. Aliquots 100 µL of tumor cytosolic extracts serially diluted with human recombinant VEGF (standard) are added to the live wells and incubated for 1 h at 22°C.
9. Six washes in BSA are performed.
10. 100 µL of peroxidase-conjugated Fab' of the anti-VEGF antibody is added. (For $VEGF_{165}$ the secondary is a VEGF specific enzyme-linked polyclonal goat antibody.)
11. The cytosol is incubated for 1 h at 22°C.
12. The enzyme reaction is performed at 22°C for 30 min with *o*-phenylenediaminidine (Sigma, St. Louis, MO) as a substrate.
13. The 490 nm absorption is measured by a microplate reader.
14. Cytosolic concentrations of VEGF are estimated from a standard curve determined from the serially diluted VEGF.
15. The minimal detectable dose level is 5 pg/mg of protein for $VEGF_{121}$ and 0.1 pg/µg of protein for $VEGF_{165}$.

### 3.6. TP Assay in the Cytosol Using Colorimetric Sandwich Enzymatic Immunoassays

1. A representative part of tumor specimen is snap frozen in liquid nitrogen and stored at –80°C.
2. For the preparation of cytosolic extracts, the frozen tissue samples are homogenized using a microdismembrator.
3. The samples are homogenized with 10 vol of cold (i.e., 4°C) 10 m*M* PBS, pH 7.4. containing 1.5 m*M* EDTA, 10 m*M* sodium molybdate 3 m*M* $NaN_3$, and 10% gylcerol.
4. Low-salt extract (cytosol) is obtained by centrifugation at 100,000*g* for 1 h at 4°C.
5. The supernatant fraction is appropriately diluted according to the protein concentration.
6. 96-well microtiter plates are coated with 10 mg/mL of the purified anti-TP antibody 104B (Nippon Roche Research Centre, Kamakura, Japan) in PBS.
7. The reaction is blocked with 30 mg/mL of skim milk in PBS (Difco, Detroit, MI).
8. Aliquots of 50 µL of tumor cytosolic extracts and serial dilutions of TP standard recombinant human TP proteins, with a known enzymatic activity of TP in duplicate, are added to the wells and then incubated for 2 h at 37°C.
9. Four washes with PBS containing 0.05% Tween-20 (KPL, Gaithersburg, MD) are performed.
10. 50 µL of 1 µg/mL concentration is added of the antihuman TP232-2 monoclonal antibody (Nippon Roche Research Centre, Kamakura, Japan) in dilution buffer, to each well.
11. The cytosol is incubated for 2 h at room temperature.
12. Four washes with PBS are performed.
13. 50 µL is added of antimouse IgG antiserum conjugated with peroxidase (KPL) (1:10,000 in dilution buffer).
14. Cytosol is incubated for 1 h at room temperature.
15. The enzyme reaction is carried out at room temperature for 10 min with a prepared substrate solution (TMB substrate system).

16. Microplate reader and TP concentrations are estimated from the standard curve determined from the serially diluted human TP at the 492 nm absorption.
17. Calculations are performed using a microplate manager computer software.
18. The minimal detectable dose is 2 U/mg.
19. One unit of TP is equivalent to the enzymatic activity that generates 1 mg of 5-fluorouracil from 5'-deoxy-5-flourouridine/h.

### 3.7. bFGF in the Cytosol

1. A colorimetric sandwich enzymatic immunoassay may be used to determine the concentrations of bFGF in cytosolic extracts using the Quantikine human bFGF immunoassay (R & D System, Inc., Minneapolis, MN).
2. The primary antibody is a murine monoclonal antibody against bFGF.
3. The enzyme-linked polyclonal antibody specific for bFGF is added to "sandwich" the bFGF immobilized during the first incubation.
4. The enzyme reaction is performed with *o*-phenylenediaminidine as a substrate. The absorbance is detected at 450 nm.
5. The concentrations of bFGF are determined with a standard curve of known bFGF. A substrate solution for HRP is added and with *o*-phenylenediaminidine.
6. Concentrations of bFGF are given as picograms per milligram of cytosolic protein.

### 3.8. Hepatocyte Growth Factor (Scatter Factor, HGF/SF) in the Cytosol

1. The assay is performed using α murine a chain HGF-specific monoclonal antibody (Genentech, Inc., San Francisco, CA).
2. The polyclonal sheep anti-HGF/SF antibody (Genentech, Inc., San Francisco, CA) is used.
3. A biotinylated donkey antisheep antibody (ICN Biomedicals, Thomas, UK) is added.
4. A substrate solution for HRP is added containing *o*-phenylenediaminidine. The absorbance is detected at 492 nm.
5. The concentrations of HGF/SF are calculated from a standard curve of known HGF.
6. The minimal detection limit is 0.25 ng/mL.
7. The concentrations of HGF are given as ng/100 mg of total protein.

### 3.9. VEGF ELISA in Serum

1. Blood samples are drawn into a test tube.
2. Blood is centrifuged at 2000 rpm for 20 min (or at 2500 rpm for 15 min).
3. Serum is stored in aliquots at –20°C until determination.
4. 96-well microtiter plates are coated with 10 ng/mL of rabbit antihuman VEGF polyclonal antibody (Cabru, Peregallo di Lesmo, Italy) in 0.1 *M* NaCl and 0.025 *M* carbonate buffer, pH 9.0 (For $VEGF_{165}$ the commercial kits from Quantikine R & D System, Minneapolis MN can be utilized and the primary antibody is an antihuman VEGF monoclonal antibody, incubation of 2 h).
5. Plates are incubated overnight at 4°C.
6. Plated are coated with 1% BSA, 0.2 *M* carbonate buffer, pH 9.0, and 0.1 *M* NaCl for 3 h at 25°C.

7. 100 µL of serum sample and serially diluted human recombinant VEGF (Sigma), as a standard, are added to the wells and incubated overnight at 4°C.
8. Three washes with PBS, 0.15 *M* NaCl, and 100 µL, of murine antihuman VEGF monoclonal antibody (Sigma, 1 µg/mL in PBS, 0.15 *M* NaCl, and 0.1% BSA) are added to each to each well (for $VEGF_{165}$ the secondary antibody is an enzyme-linked polyclonal goat antibody).
9. Plates are incubated for 1 h at 25°C.
10. Sections are washed several times with PBS.
11. Wells are incubated with 100 µL of antimouse IgG (fc-specific) peroxidase conjugate (Sigma).
12. The reaction is developed with K-Blue substrate (Oxford Biomedical Research, Inc., Oxford) and then it is blocked with 1 *M* $H_2SO_4$ after 30 min at room temperature.
13. The absorption is measured by a microplate reader at 490 nm.
14. Results are calculated from a known VEGF standard curve.

### 3.10. VEGF ELISA in Plasma, Whole Blood, Peripheral Blood Mononuclear Cells, or Platelets

1. Peripheral venous blood samples are collected using a venoject blood system.
2. All samples are taken between 8:00 and 8:45 A.M.
3. Plasma and whole blood samples are collected in sterile test tubes, containing sodium citrate as anticoagulant.
4. The samples are incubated at +4°C for 60–240 min.
5. The plasma samples are centrifuged at 2000*g* for 10 min at +4°C, and then stored in aliquots at –70°C until determination.
6. The cells of the whole blood sample are lysed by adding 2 vol of sterile $H_2O$, and subsequently, the sample is freeze–thawed twice.
7. Peripheral venous blood samples are collected in sterile test tubes anticoagulated with acid-citrate dextrose and containing 0.18 µ*M* prostaglandin $E_1$ (Sigma Chemical, St. Louis, MO).
8. Platelet-rich plasma is obtained by centrifugation (120*g* for 20 min, +37°C for 180*g* for 10 min).
9. Platelet-rich plasma is collected and transferred to new tubes, and the cell counts are determined using a differential cell counter.
10. Platelets from a cancer patient and healthy control are obtained.
11. Prior to VEGF immunoassay, the platelets are lysed by adding 2 vol of sterile $H_2O$, and subsequently the sample is freeze-thawed twice.
12. Peripheral venous blood samples are collected in sterile test tubes containing a sodium citrate as anticoagulant.
13. Peripheral blood mononuclear cells (PBMNCs) suspension is prepared using a density gradient centrifugation on a mixture of Ficoll and sodium metrizoate.
14. Blood anticoagulated with sodium citrate is diluted with an equal volume of PBS and layered on top of Ficoll Hypaque (Pharmacia Biotech, Uppsala, Sweden) and centrifuged at 400*g* at +18°C for 30 min.
15. The PBMNC layer is then collected, washed 3× with PBS containing 10% fetal bovine serum (Sigma), and resuspended in sterile PBS.

16. The identity of cells is then confirmed, and cell counts are assessed using a differential cell counter.
17. The cells are lysed.
18. VEGF concentration is determined in different blood compartments using a sandwich enzyme immunoassay technique (Quantikine Human VEGF Immunoassay R & D).
19. Next steps: *See* **steps 4–14, Subheading 3.9.**
20. Optical densities are determined using a microtiter plate reader at 450 nm.
21. The minimum detectable dose is 9.0 pg/mL.

## 4. Notes

1. Comparative studies are needed to identify the method of choice, the most appropriate sample to be tested (entire tumor tissue, tumor cytosol, serum, whole blood, peripheral blood mononuclear cells, and platelets) and the criteria of evaluation to classify a tumor as positive or negative for expression of a certain angiogenic marker.
2. Although neoplasms are assumed not to elicit the formation of a new lymphatic system, evaluation of preexisting lymphatic vessels may induce false-positive microvessel counts using the anti-FVIII-RA and anti-CD34 antibodies.
3. FVIII-RA is highly specific for vasculature, but it may not be reactive with part of capillary endothelium.
4. The occasional anti-CD31 monoclonal antibody signal on inflammatory cells can be easily differentiated from endothelial cells on the basis of the morphological differences of the cell types.
5. The anti-CD34 monoclonal antibody is one of the most sensitve pan-endothelial cell markers, but it highlights a wide variety of stromal elements.
6. Pan-endothelial markers are not available to discriminate between quiescent and activated/proliferating endothelium, and IMD assessed by using these markers cannot accurately represent the dynamic angiogenic activity of a tumor.
7. The vascular "hot-spot" is usually recognized at the tumor margins.
8. The major potential methodological pitfall of determination of IMD using Weidner's method is related to the choice of the "hot-spot;" consequently the training and experience of the investigator is an absolute requisite for its identification.
9. According to Weidner's method, any highlighted endothelial cell or cell cluster clearly separate from adjacent microvessels, tumor cells, or other connective tissue elements should be regarded as a distinct countable microvessel (*see* **Fig. 2**). Presence of vascular lumen is not required, nor is presence of red blood cells. A cutoff vessel caliber size is not required. Sclerotic hypocellular areas within tumor and the immediately adjacent benign tissues are not considered for vessel count.
10. CIAS is not completely automated and still requires a high degree of operator interaction. Some weakness of the method are: high costs, the time needed to select the "hot-spot," and the lack of valid software for automated identification of the vascular "hot-spot."
11. The monoclonal antibody LM609 to integrin $\alpha v\beta 3$ is specific for activated/proliferating endothelial cells and for the smaller microvessels, but it can be used only on fresh or frozen samples.

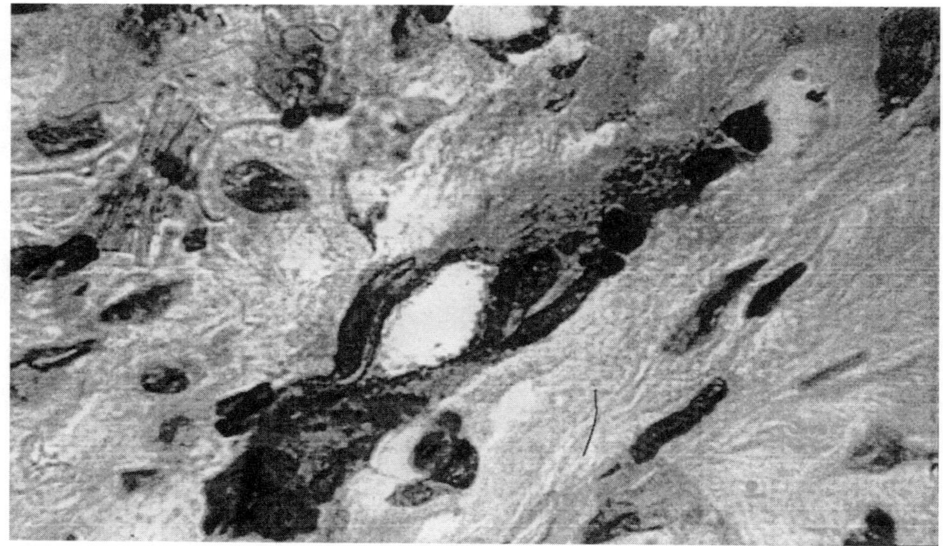

Fig. 2. Immunoperoxidase reaction with the lm-609 antibody to integrin αvβ3 at high magnification ×1000.

## References


1. Folkman, J. (1992) The role of angiogenesis in tumour growth. *Semin. Cancer Biol.* **3,** 65–72.
2. Folkman, J. (1993) Tumor angiogenesis, in *Cancer Medicine*, 4th ed. (Holland, J., Frei, E., and Bast, R. C., eds.), Lea and Febiger, Melbourne, pp. 153–171.
3. Folkman, J. (1990) What is the evidence that tumors are angiogenesis dependent? *J. Natl. Cancer Inst.* **82,** 4–6.
4. D'Amore, P. A. (1992) Capillary growth: a two-cell system. *Cancer Biol.* **3,** 49–56.
5. D'Amore, P. A. and Smith, S. R. (1993) Growth factors effects on cells of the vascular wall: a survey. *Growth Factors* **8,** 61–75.
6. Dameron, K. M., Volpert, O. V., Tainsky, M. A., and Bouck, N. (1994) Control of angiogenesis in fibroblasts by p53 regulation of thrombospondin-1. *Science* **265,** 1582–1585.
7. Kieser, A., Wiech, H. A., Bradner, G., Marmé, D., and Kolch, W. (1994) Mutant p53 potentiates protein kinase C induction of vascular endothelial growth factor expression. *Oncogene* **9,** 963–969.
8. Gasparini, G. (1996) Clinical significance of the determination of angiogenesis in human breast cancer: update of the biological background and overview of the Vicenza studies. *Eur. J. Cancer* **32A,** 2485–2493.
9. Gasparini, G. and Harris, A. L. (1999) Prognostic significance of tumour vascularity, in *Antiangiogenic Agents in Cancer Therapy* (Teicher, B. A., ed.), Humana Press, Totowa, NJ, pp. 317–319.

10. Gasparini, G. and Harris, A. L. (1995) Clinical importance of the determination of tumor angiogenesis in breast carcinoma: much more than a new prognostic tool. *J. Clin. Oncol.* **13,** 765–782.
11. Gasparini, G. (1994) Quantification of intratumoral vascularization predicts metastasis in human invasive solid tumors. *Oncol. Rep.* **1,** 7–12.
12. Gasparini, G. (1997) Prognostic and predictive value of intra-tumoral microvessel density in human solid tumors, in *Tumor Angiogenesis* (Bicknell, R., Lewis, C. E., Ferrara, M., eds.), Oxford University Press, Oxford, pp. 29–44.
13. De Jong, J. S., Van Diest, P. J., and Baak, J. P.A. (1995) Methods in laboratory investigation. Heterogeneity and reproducibility of microvessel counts in breast cancer. *Lab. Invest.* **73,** 922–926.
14. Ambs, S., Bennett, W. P., Merriam, W. G., Ogunfusika, M. O., Oser, S. M., Khan, M. A., and Harris, C. C. (1998) Vascular endothelial growth factor and nitric oxide synthase expression in human lung cancer and the relation to p53. *Br. J. Cancer* **78,** 233–239.
15. Gasparini, G., Fox, S. B., Verderio, P., Bonoldi, E., Boracchi, P., et al. (1996) Determination of angiogenesis adds information to estrogen receptor status in predicting the effacy of adjuvant tamoxifen in node-positive breast cancer patients. *Clin. Cancer Res.* **2,** 1191–1198.
16. Gasparini, G., Bonoldi, E., Viale, G., Verderio, P., Boracchi, P., Panizzoni, G. A., et al. (1996) Prognostic and predictive value of tumor angiogenesis in ovarian carcinomas. *Int. J. Cancer (Pred. Oncol.)* **69,** 205–211.
17. Vermeulen, P. B., Gasparini, G., Fox, S. B., Toi, M., Martin, L., McCulloch, P., et al. (1996) Quantification of angiogenesis in solid human tumours: an international consensus on the methodology and criteria of evaluation. *Eur. J. Cancer* **32,** 2478–2484.
18. Obermair, A., Tempfer, C., Hefler, L., Preyer, O., Kaider, A., Zeillinger, R., et al. (1998) Concentration of vascular endothelial growth factor (VEGF) in the serum of patients with suspected ovarian cancer. *Br. J. Cancer* **77,** 1870–1874.
19. Parums, D. V., Cordell, J. L., Micklem, K., Heryet, A. R., Gatte, K. C., and Mason, D. Y. (1990) JC70: a new monoclonal antibody that detects vascular endothelium associated antigen on routinely processed tissue sections. *J. Clin. Pathol.* **43,** 752–757.
20. Horak, E., Leek, R., Klenk, N., Lejeune, S., Smith, K., Stuart, N., et al. (1992) Angiogenesis, assessed by platelet/endothelial cell adhesion molecule antibodies, as an indicator of node metastasis and survival in breast cancer. *Lancet* **340,** 1120–1124.
21. Ruiter, D. J., Schlingemann, R. O., Rietveid, F. J. R., and de Waal, R. M. W. (1989) Monoclonal antibody-defined human endothelial antigens as a vascular marker. *Invest Dermatol.* **93,** 25–32.
22. Traweek, S. T., Kandalaft, P. L., Mehta, P., and Battifora, H. (1991) The human hematopoietic progenitor cell antigen (CD34) in vascular neoplasia. *Am. J. Clin. Pathol.* **96,** 25–31.
23. Sternberger, S. S. and De Lellis, R. A., eds. (1982) *Diagnostic Immunohistochemistry*, Masson, New York.
24. Bullock, G. R. and Petrusz, P., eds. (1982 and 1983) *Techniques in Immunochemistry*, vols. 1 and 2. Academic, London.

25. Gasparini, G., Brooks, P. C., Biganzoli, E., Vermeulen, P. B., and Dirix, L. Y., and Ranieri, G., et al. (1998) Vascular integrin $\alpha v\beta 3$: a new prognostic indicator in breast cancer. *Clin. Cancer Res.* **4,** 2625–1264.
26. Weidner, N., Semple, J. P., Welch, W. R., and Folkman, J. (1991) Tumor angiogenesis and metastasis-correlation in invasive breast carcinoma. *N. Engl. J. Med.* **324,** 1–8.
27. Fox, S. B., Leek, R. D., Weekes, M. P., Withehouse, R. M., Gatter, K. C., and Harris, A. L. (1995) Quantitation and prognostic value of breast cancer angiogenesis: comparison of microvessel density, Chalkey count and computer image analysis. *J. Pathol.* **77,** 277–283.
28. Visscher, D., Smilanetz, S., Drozdowicz, S., Wilkes, S. (1993) Prognostic significance of image morphometric microvessel enumeration in breast carcinoma. *Anal. Quant. Cytol.* **15,** 88–92.
29. Simpson, J. F., Ahn, C., Battifora, M., and Esteban, J. M. (1996) Endothelial area as a prognostic indicator for invasive breast carcinoma. *Cancer* **77,** 2077–2085.
30. Barbareschi, M., Weidner, N., Gasparini, G., Morelli, L., Forti, S., Eccher, C., et al. (1995) Microvessel density quantification in breast carcinomas. Assessment by light microscopy vs computers aided image analysis system. *Appl. Immunohis*tochem. **3,** 75–84.
31. Barth, P. J., Weingartner, K., Kohler, H. H., and Bittinger, A. (1996) Assessment of the vascularization in prostatic carcinoma: a morphometric investigation. *Hum. Pathol.* **27,** 1306–1310.
32. Toi, M., Ueno, T., Matsumoto, H., Saji, H., Funata, N., Koike, M., and Tominaga, T. (1999) Significance of thymidine phosphorylase as a marker of tumor monocytes in breast cancer. *Clin. Cancer Res.* **5,** 1131–1137.
33. Toi, M., Hoshina, S., Taniguchi, T., Yamamoto, Y., Ishitsuka, H., and Tominaga, T. (1995) Expression of platelet–derived endothelial cell growth factor/thymidine phoshorylase in human breast cancer. *Int. J. Cancer (Pred. Oncol.)* **64,** 79–82.
34. Fox, S. B., Engels, K., Comley, M., Witehouse, R. M., Turley, H., Gatter, K. C., and Harris, A. L. (1997) Relationship of elevated tumour thymidine phosphorylase in node–positive breast carcinomas to the effects of adjuvant CMF. *Ann. Oncol.* **8,** 271–275.
35. Fox, S. B., Westwood, M., Moghaddam, A., Comley, M., Turley, H., Witehouse, R. M., et al. (1996) The angiogenic factor platelet-derived endothelial cell growth factor/hymidine phosphorylase is up-regulated in breast cancer epithelium and endothelium. *Br. J. Cancer* **73,** 275–280.
36. Polak, J. M. and van Noorden, S., eds. (1983) I*mmunocytochemistry*. John Wright and Sons, Bristol, UK.
37. Nagy, J., Curry, G. W., Hillan, K. J., Purushotham, A. D., and George, W. D. (1996) Hepatocyte growth factor/scatter factor, angiogenesis and tumour cell proliferation in primary breast cancer. *Breast* **5,** 105–109.
38. Colomer, R., Aparicio, J., Montero, S., Guzmán, C., Larrodera, L., and Cortés-Funes, H. (1997) Low levels of basic fibroblast growth factor (bFGF) are associated with a poor prognosis in human breast carcinoma. *Br. J. Cancer* **76,** 1215–1220.
39. Gasparini, G., Toi, M., Miceli, R., Vermeulen, P. B., Dittadi, R., Biganzoli, E., et al. (1999) Clinical relevance of vascular endothelial growth factor and thymidine phosphorylase in patients with node-positive breast cancer treated with either adjuvant chemotherapy or hormone therapy. *Cancer J. Sci. Am.* **5,** 101–111.

40. Hyodo, I., Doi, T., Endo, H., Hosokawa, Y., Nishikawa, Y., Tanimizu, M., et al. (1998) Clinical significance of plasma vascular endothelial growth factor in gastrointestinal cancers. *Eur. J. Cancer* **34,** 2041–2045.
41. Toi, M., Taniguchi, T., Yamamoto, Y., Kurisaki, T., Suzuki, H., and Tominaga, T. (1996) Clinical significance of the determination of angiogenic factors. *Eur. J. Cancer* **32A,** 2513–2519.
42. Morelli, D., Lazzerini, D., Cazzaniga, S., Squicciarini, P., Bignami, P., Maier, J. A., et al. (1998) Evaluation of the balance between angiogenic and antiangiogenic circulating factors in patients with breast and gastrointestinal cancers. *Clin. Cancer Res.* **4,** 1221–1225.
43. Salven, P., Orpana, A., and Joensuu, H. (1999) Leukocytes and platelets of patients with cancer contain high levels of vascular endothelial growth factor. *Clin. Cancer Res.* **5,** 487–491.
44. Banks, R. E., Forbes, M. A., Kinsey, S. E., Stanley, A., Ingham, E., Walters, C., and Selby, P. J. (1998) Release of the angiogenic cytokine vascular endothelial growth factor (VEGF) from platelets: significance for VEGF measurements and cancer biology. *Br. J. Cancer* **77,** 956–964.
45. Wynendaele, W., Derua, R., Hoylaerts, M. F., Pawinski, A., Waelkens, E., Bruijn, E. A., et al. (1999) Vascular endothelial growth factor measured in platelet poor plasma allows optimal separation between cancer patients and volunteers: a key to study an angiogenic marker in vivo? *Ann. Oncol.* **10,** 965–971.
46. Waterborg, J. H. and Matthews, H. R. (1984) The Lowry method for protein quantitation, in *Methods in Molecular Biology*, vol. 1 (Walker, J., ed.), Humana Press, Totowa, NJ, pp. 1–3.
47. Fontanini,G., Vignati, S., Lucchi, M., Mussi, A., Calcinai, A., Boldrini, L., et al. (1997) Neoangiogenesis and p53 protein in lung cancer: their prognostic role and their relation with vascular endothelial growth factor (VEGF) expression. *Br. J. Cancer* **75,** 1295–1301.
48. DePlacido, S., Carlomagno, C., Ciardiello, F., DeLaurentiis, M., Pepe, S., Ruggiero, A., et al. (1999) Measurement of neovascularization is an independent prognosticator of survival in node-negative breast cancer patients with long-term follow-up. *Clin. Cancer Res.* **5,** 2854–2859.
49. Dickinson, A. J.,Fox, S. B., Persad, R. A., Hollyer, J., Sibley, G. N. A., and Harris, A. L. (1994) Quantitation of angiogenesis as an independent predictor of prognosis in invasive bladder carcinomas. *Br. J. Urol.* **74,** 762–766.
50. Gasparini, G., Toi, M., Gion, M., Verderio, P., Dittadi, R., Hanatani, M., et al. (1997) Prognostic significance of vascular endothelial growth factor protein in node-negative breast carcinoma. *J. Natl. Cancer Inst.* **89,** 139–147.
51. Toi, M., Gion, M., Biganzoli, E., Dittadi, R., Boracchi, P., Miceli, R., et al. (1997) Co-determination of the angiogenic factors thymidine phosphorylase and vascular endothelial growth factor in node-negative breast cancer: prognostic implications. *Angiogenesis* **1,** 71–83.
52. Linderholm, B., Tavelin, B., Grankvist, K., and Henriksson, R. (1998) Vascular endothelial growth factor is of high prognostic value in node-negative breast carcinoma. *J. Clin. Oncol.* **16,** 3121–3128.

# 8

# Measurement of Microvessel Density in Primary Tumors

**Helen E. Turner and Adrian L. Harris**

## 1. Introduction

Angiogenesis describes the development of new vessels from existing blood vessels and is a crucial physiological process involved in embryo development, wound healing, and the female reproductive cycle. However, angiogenesis has also been shown to be required for tumor growth and metastasis *(1)*. On the basis of experiments showing that tumors implanted into isolated perfused organs failed to develop whereas the same tumors implanted within 6 mm of blood vessels induced angiogenesis, grew, and metastasized *(2,3)*, Folkman proposed that solid tumors are dependent on angiogenesis for growth beyond a few millimeters in size, and that increase in tumor diameter required a corresponding increase in vascularization *(4)*. It is becoming clear that there is a complex balance between pro-angiogenic growth factors and endogenous inhibitors that determine the final angiogenic phenotype *(5)*.

### *1.1. Why Measure Angiogenesis?*

Angiogenesis (measured as tumor microvessel density) has been shown to be correlated with tumor behavior. In many human tumors including breast, bladder, and stomach, angiogenesis has been shown to be correlated with development of metastasis *(6)*, prognosis *(7,8)*, and survival *(9,10)*.

### *1.2. Is Microvessel Density an Adequate Reflection of Angiogenesis?*

This has been answered to some extent by different means—microvessel density has been shown to correlate with the concentration and expression of pro-angiogenic growth factors (fibroblast growth factor [bFGF]) and vascular

From: *Methods in Molecular Medicine, vol. 57:*
*Metastasis Research Protocols, Vol. 1: Analysis of Cells and Tissues*
Edited by: S. A. Brooks and U. Schumacher © Humana Press Inc., Totowa, NJ

endothelial growth factor [VEGF]) *(11,12)* and also to be associated with enzymes involved in the early stages of angiogenesis as well as tumor growth and occurrence of distant metastasis *(13)*.

### 1.3. How Is Vascular Density Measured?

Vascular density of different tumors has been assessed by counting vessels labeled using immunohistochemistry with antibodies to different endothelial markers, on both frozen and paraffin-embedded sections. The antibodies that are most commonly used are to factor VIII related antigen (FVIIIAg), CD31 (platelet endothelial cell adhesion molecule), CD34, and the lectin *Ulex europaeus* agglutinin 1 (UEA-1). Methods for immunocytochemistry are described in Chapter 2 by Brooks, and lectin histochemistry in Chapter 4 by Brooks and Hall. It has been shown that these markers have different sensitivities for the detection of endothelium *(8,14,15)* (*see* **Table 1**). CD31 and CD34 are the most sensitive and specific markers currently available for use on paraffin wax-embedded sections, although other markers may be easier to use *(13,16)* Chapter 7 by Ranieri and Gasparini also describes related approaches to determining markers of angiogenesis by immunocytochemistry, and also by ELISA.

The use of these markers reflects total vascular density, and new markers are being developed that are more specific for activated endothelium (e.g., TEC-11, E-9, 4A11, and H4/18) but their disadvantage is that at present the majority, need to be used on frozen tissue. They may, however, be useful to assess specific targets for therapy or degree of remodeling of the vasculature.

### 1.4. Quantification of Microvessel Density

There are several different ways of assessing microvessel counts (*see* **Table 2**). The original technique described selection of the most highly vascular areas (hot spots) and counting of individual microvessels *(6)*. Other techniques that have been described are an overall semiquantitative grading technique, Chalkley counting using a graticule containing 25 randomly positioned dots *(17)*, and computerized image analysis *(13,18)*. These approaches are also described in Chapter 7 by Ranieri and Gasparini.

The accuracy of any of these techniques depends mainly on the correct localization of the hot spot in addition to whether a representative block of tissue is chosen. Despite these potential problems, microvessel counts from all the techniques have been shown to correlate with tumor prognosis and behavior.

## 2. Materials

1. Tris-buffered saline (TBS): Make up Tris stock (121 g of Tris in 1000 mL of distilled water) and 2 *M* NaCl stock (116.9 g of sodium chloride in 1 L of distilled water). To make up TBS dilute 50 mL of Tris stock and 75 mL of 2 *M* sodium chloride to 1 L with distilled water, and adjust the pH to 7.6 using HCl or NaOH.

**Table 1**
**Advantages and Disadvantages of Endothelial Markers**

| Antibody | Endothelial marker | Advantages | Disadvantages |
|---|---|---|---|
| UEA-1 | Fucose residues on endothelial cells. | Easy to use. Stains all microvessels. | Stains Golgi in some neoplastic cells. |
| Factor VIII-related antigen | Factor VIII-related antigen. | Stains large vessels. | Does not stain all vascular endothelium, e.g., poor staining of smaller microvessels. Stains lymphatics. |
| CD31 | Platelet/endothelial cell adhesion molecule. | Sensitive marker. Stains all microvessels. | Staining can be variable. Stains plasma cells. |
| CD34 | Human hemopoietic progenitor cell antigen. | Sensitive marker. Stains all microvessels. | Variable perivascular stromal staining. |

**Table 2**
**Advantages and Disadvantages of Microvessel Counting Techniques**

| Technique | Advantages | Disadvantages |
|---|---|---|
| Hot spot vessel count | Vessels with variable vascular staining still assessable. | Requires subjective assessment of hot spot. Time consuming. |
| Semiquantitative grading | Time efficient. | Highly subjective. Less reproducible. |
| Chalkley counting | Objective. Relatively time efficient. | Requires subjective assessment of hot spot. |
| Computerized image analysis | Additional parameters assessable including microvessel perimeter and microvessel area. Objective. | Heterogeneous staining patterns and intensity within tumors. |

2. Phosphate-buffered saline (PBS): To make a 10X concentrated solution of PBS add 80 g of sodium chloride, 2 g of potassium chloride, 11.5 g of disodium hydrogen orthophosphate, and 2 g of potassium dihydrogen orthophosphate; make up to 1 L of distilled water, and adjust pH to 7.4. Dilute 100 mL of 10X stock with 900 mL of distilled water for PBS ready to use.
3. 0.01 *M* Sodium citrate buffer, pH 6: 2.49 g of trisodium citrate in 1 L of distilled water; adjust the pH to 6.
4. 0.1 *M* Trypsin: Add 1 g of calcium chloride to 1 L of TBS, and adjust the pH to 7.8 with sodium hydroxide. Add 1 g of trypsin to the solution and check that the pH is 7.8. Avoid strong acids/alkalis to change pH at this stage, as this may denature the enzyme.
5. 3% v/v Hydrogen peroxide: Dilute 30% hydrogen peroxide 1:10 with distilled water.
6. Antibodies: Primary antibodies are used at dilutions depending on optimization (*see* **Note 1**). For example, CD31 may be tried at 1:20 and 1:40, and biotinylated UEA1 at 1:200 and 1:100. Secondary antibodies (biotinylated anti-immunoglobulin—mouse, rabbit, rat—depending on primary) are used at 1:200 to 1:400 diluted with fetal calf serum (1:20) with TBS/PBS for 30 min. Streptavidin–horseradish peroxidase 1:400 is diluted with fetal calf serum diluted 1:20 with TBS or PBS for 30 min. Biotinylated UEA1 and secondary antibodies are required.
7. Diaminobenzidine (DAB): Made up 1:10 with peroxide substrate buffer.

## 3. Methods

### 3.1. Deparaffinization

1. Use 4-μm sections of paraffin wax-embedded tissue on APES (3-aminopropyl triethoxy silane) (Sigma)-coated slides.
2. Place sections sequentially for 5 min each in three containers with xylene (*see* **Note 2**), followed by two containers with 100% ethanol, followed by 70% ethanol and then distilled water.
3. Block endogenous peroxidase by placing the sections in 3% hydrogen peroxide for 5 min followed by distilled water.

### 3.2. Pretreatment

1. No pretreatment is required for UEA1 staining
2. CD31, CD34, and F8 require trypsin pretreatment. Warm trypsin (*see* **Note 3**) to 37°C in water bath and add slides for 15 min. Place slides immediately in distilled water at end of the treatment.
3. CD31 sections require further microwave pretreatment. Sections are microwaved in sodium citrate buffer, pH 6.0 (*see* **Note 4**) until the solution is seen to boil (approx 3–4 min depending on microwave power). The sections are left in the hot buffer for 20 min prior to washing in distilled water for 5 min (*see* **Notes 5** and **6**).

### 3.3. Immunostaining

1. Block nonspecific antibody binding by incubating the sections with fetal calf serum diluted 1:20 with TBS or PBS (*see* **Notes 7** and **8**) for 20 min at room temperature.

2. Add primary antibody at dilution, temperature, and for a duration determined by prior optimization. An idea of optimal dilutions will be available with the antibody (*see* **Note 1**). Wash thoroughly in PBS–TBS (*see* **Notes 7** and **8**).
3. Add biotinylated secondary antibody—anti-immunoglobulin (mouse, rabbit, rat—depending on primary) at a dilution of 1:200 for 30 min. This step is not required for UEA1 as it is already biotinylated. Then wash thoroughly in PBS–TBS.
4. Add streptavidin–horseradish peroxidase 1:400 for 30 min (*see* **Note 9**), then wash thoroughly in PBS–TBS.
5. Wash in distilled water.
6. Use DAB and peroxide substrate buffer (made up 1:10) applied to slides for up to 15 min to allow color development (*see* **Note 10**).
7. Counterstain lightly with hematoxylin (*see* **Note 11**), for example 10–30 s, then place sections in warm water (approx 30°C to allow color development).
8. Dehydrate in graded alcohols (70%, 100%, 100%) followed by xylene, prior to mounting in a xylene-based mountant, e.g., Depex.

### 3.4. Counting

1. Sections are examined under low power and the area containing the maximum number of discrete microvessels ("hot spot") (*see* **Note 12**) is identified. It is helpful to have two observers to agree on this (*see* **Note 13**).
2. Discrete microvessels are identified as any highlighted endothelial cell separate from adjacent microvessels, even if the vessel gives the impression of being part of a vessel transected by the plane of the tissue section more than once. A vessel lumen, presence of red blood cells, and a certain caliber of vessel are not required.
3. Vessel density may then be graded semiquantitatively into low, medium, and high grade.
4. An alternative is to use the Chalkley graticule; after localization of the hot spot under low power, an eyepiece graticule containing 25 randomly positioned dots is rotated so that the maximum number of points are on or within the vessels of the vascular hot spot and the number of overlying dots are counted at ×250.
5. Three hot spots should be measured and a mean of the counts calculated.
6. Quantification by counting microvessels on a ×400 field can be performed manually using a counting grid or an alternative is to set up a program on a computerized image analysis system.

## 4. Notes

1. The optimal concentration for each antibody should be determined by using a control tumor specimen with a serial dilution of the primary antibody over different time periods. In general, it is worth comparing 1 h at room temperature at dilutions similar to those quoted by the supplier, with overnight at 4°C at similar dilutions.
2. Xylene must be kept in a fume cupboard. An alternative to xylene is to use histoclear (National Diagnostics), which can be used on the bench.
3. It is important to use trypsin at 37°C and pH 7.8. Other proteases may be used in addition to trypsin, for example, pronase or pepsin.

4. The PBS concentrate and sodium citrate buffers can be made in large quantities in advance if thiomersal (1 mL of 10% thiomersal solution to 1 L of solution) is added. Similarly concentrated Tris and saline can be kept ready for dilution as needed.
5. A proportion of paraffin-embedded tumors do not stain using CD31. This is probably related to the fixation technique, for example, fixatives containing acetic acid frequently lead to antigen loss.
6. Sections may be left overnight in buffer (avoid detergent) following pretreatment.
7. The pH of the PBS and TBS are very important for optimal staining (7.4 and 7.6, respectively).
8. Triton or Tween may be added to buffer to improve tissue permeation.
9. An alternative to using the horseradish peroxidase system is the alkaline phosphatase label, which produces a red color rather than brown, and does not require blocking of endogenous peroxidase activity, for details, *see* Chapter 2 by Brooks.
10. DAB is toxic and carcinogenic. It should be disposed of appropriately and should not come in contact with exposed skin.
11. Filter hematoxylin daily before use.
12. The potential for inaccurate counts are mainly related to the subjectivity involved in hot spot localization. Thus it is important to measure intra-observer variation and inter-observer variation, and also if possible to be trained by someone experienced in this technique.
13. If distinct staining of vessels is not achieved even after repeating the staining, the case should be excluded from study or else another marker tried, as otherwise this may lead to inaccurate evaluation of each element that should be considered as a single countable microvessel.

## References

1. Folkman, J. (1990) What is the evidence that tumors are angiogenesis dependent? *J. Natl. Cancer Inst.* **82,** 4–6.
2. Gimbrone, M. A., Jr., Leapman, S., Cotran, R. S., and Folkman, J. (1972) Tumor dormancy in vivo by prevention of neovascularisation. *J. Exp. Med.* **73,** 461–473.
3. Gimbrone, M. A., Jr., Leapman, S., Cotran, R. S., and Folkman, J. (1973) Tumor angiogenesis: iris neovascularisation at a distance from intraocular tumors. *J. Natl. Acad. Inst.* **50,** 219–228.
4. Folkman, J. (1972) Anti-angiogenesis: new concept for therapy of solid tumors. *Ann. Surg.* **175,** 409–416.
5. Hanahan, D. and Folkman, J. (1996) Patterns and emerging mechanisms of the angiogenic switch during tumorigenesis. *Cell* **86,** 353–364.
6. Weidner, N., Semple, J. P., Welch, W. R., and Folkman, J. (1991) Tumor angiogenesis and metastasis: correlation in invasive breast carcinoma. *N. Engl. J. Med.* **324,** 1–8.
7. Weidner, N., Folkman, J., Pozza, F., Bevilacqua, P., Allred, E. N., Moore, D. H., et al. (1992) Tumor angiogenesis: a new significant and independent prognostic indicator in early stage breast carcinoma. *J. Natl. Cancer Inst.* **84,** 1875–1887.

8. Horak, E. R., Leek, R., Klenk, N., LeJeune, S., Smith, K., Stuart, N., et al. (1992) Angiogenesis, assessed by platelet/endothelial cell adhesion molecule antibodies, as indicator of node metastases and survival in breast cancer. *Lancet* **340,** 1120–1124.
9. Maeda, K., Chung, Y. S., Takasuka, S., Ogawa, Y., Sawada, T., Yamashita, Y., et al. (1995) Tumor angiogenesis as a predictor of recurrence in gastric carcinoma. *J. Clin. Oncol.* **13,** 477–481.
10. Bochner, B. H., Cote, R. J., Weidner, N., Groshen, S., Chen, S. C., Skinner, D. G., et al. (1995) Angiogenesis in bladder cancer: relationship between microvessel density and tumor prognosis. *J. Natl. Cancer Inst.* **87,** 1603–1612.
11. Li, V. W., Folkerth, R. D., Watnabe, H., Yu, C., Pupnick, M., Barnes, P., et al. (1994) Microvessel count and cerebrospinal fluid basic fibroblast growth factor in children with brain tumors. *Lancet* **344,** 82–86.
12. Toi, M., Inada, K., Suzuki, H., and Tominaga, T. (1995) Tumor angiogenesis in breast cancer: its association with vascular endothelial growth factor expression. *Breast Cancer Res. Treat.* **36,** 193–204.
13. Vermeulen, P. B., Gasparini, G., Fox, S. B., Toi, M., Martin, L., McCulloch, P., et al. (1996) Quantification of angiogenesis in solid human tumors: an international consensus on the methodology and criteria of evaluation. *Eur. J. Cancer* **32A,** 2474–2484.
14. Holthofer, H., Virtanen, I., Kariniemi, A. L., Hormia, M., Linder, E., and Miettinen, A. (1982) *Ulex europaeus* 1 lectin as a marker for vascular endothelium in human tissues. *Lab. Invest.* **47,** 60–66.
15. Schlingeman, R. O., Rietveld, F. J. R., de Waal, R. M. W., Bradley, N. J., Skene, A. I., Davies, A. J. S., et al. (1990) Leukocyte antigen CD34 is expressed by a subset of cultured endothelial cells and on endothelial abluminal microprocesses in the tumor stroma. *Lab. Invest.* **62,** 690–696.
16. Martin, L., Green, B., Renshaw, C., Lowe, D., Rudland, P., Leinster, S. J., and Winstanley, J. (1997) Examining the technique of angiogenesis assessment in invasive breast cancer. *Br. J. Cancer* **76,** 1046–1054.
17. Fox, S. B., Leek, R. D., Weekes, M. P., Whitehouse, R. M., Gatter, K. C., and Harris, A. L. (1995) Quantification and prognostic value of breast cancer angiogenesis: comparison of microvessel density, Chalkley count and computer image analysis. *J. Pathol.* **177,** 275–283.
18. Barbareschi, M., Weidner, N., Gasparini, G., Morelli, L., Forti, S., Eccher, C., et al. (1995) Microvessel density quantification in breast carcinomas. *Appl. Immunohistochem.* **3,** 75–84.

# 9

# Assessment of Cellular Proliferation by Calculation of Mitotic Index and by Immunohistochemistry

**Cheryl E. Gillett**

## 1. Introduction

The ability of most carcinomas to metastasize to different organs, where they grow and gradually destroy the surrounding tissues, is the feature of this disease that both increases morbidity and significantly reduces a patient's survival time. One of the ways of contributing to the prediction of the likely course of a malignancy is by measuring the proliferative activity of the primary tumor. This allows treatment to be tailored according to the predicted aggressiveness of the disease (extrapolated from the likely growth rate).

A crude estimate of a tumor's proliferation rate is to measure the change in clinical size over a period of time. However, this is not particularly accurate, as, first, clinical size and pathological tumor size are often different, and second, an increase in clinical size could be due to either increased cell size (hypertrophy) or increased cell number (hyperplasia) or even a variation in non-neoplastic elements such as fibrotic tissue.

For many years microscopists used the number of mitotic figures present in a tissue section to estimate how rapidly the cells are proliferating and thereby gauge how aggressive the tumor was likely to be. With a few refinements, mainly associated with the method of evaluation, this technique continues to be used to measure the proliferative activity of tumors today *(1)*. For some tumors, mitotic activity is used in conjunction with other histological or clinical parameters to "grade" or "index" a tumor, as described in Chapter 1 by Roskell and Buley, and this information is used to determine the most appropriate way of treating the patient *(2–5)*.

From: *Methods in Molecular Medicine, vol. 57: Metastasis Research Protocols, Vol. 1: Analysis of Cells and Tissues*
Edited by: S. A. Brooks and U. Schumacher © Humana Press Inc., Totowa, NJ

Mitotic figures can be easily demonstrated using conventional formalin-fixed paraffin wax embedded tissues, which are used for the histological diagnosis. There has been and, to some extent, remains some controversy over the most appropriate way of "counting" the number of mitoses in the section. While some workers recommend that the number of mitoses in a sample of the section should be counted as a percentage of the number of malignant interphase cells *(6)*, others recommend that the number of mitoses be counted over 10 high-power fields *(7)*. The latter is a far less arduous task and, therefore, a more popular choice. Both methods give comparable results provided that the high-power field method is performed accurately and consistently *(8)*.

A number of alternative methods of measuring proliferative activity have also been reported. One of these is the detection of the Ki-67 antigen, which is expressed in the nuclei of all proliferating cells but is absent from those cells that are not actively involved in the cell cycle *(9)*. The Ki-67 antigen is localized primarily in the nucleolus in interphase cells and is present on all chromosomes during the mitotic phase *(10)*. The levels of this protein can be evaluated using immunohistochemistry, a method that is used routinely in diagnostic laboratories and is described in detail in Chapter 2 by Brooks. A number of antibodies have been generated against Ki-67 but the most reliable and robust one is the monoclonal antibody, MIB1, which can be used on formalin-fixed, paraffin wax-embedded material *(11)*. The technique does require an antigen recovery step, which is achieved by heating the sections in citrate buffer using either a pressure cooker *(12)* or microwave oven *(13)*. The pressure cooker method of antigen retrieval tends to give a more even staining result and causes slightly less damage to the tissue. The proliferative activity of the tumor is estimated by counting the proportion of MIB1-positive cells in a sample of the histological section.

Other methods for measuring proliferative activity have been described and can provide reliable and prognostically useful results. These include flow cytometry, which measures the amount of DNA in the cell and can be used to estimate the proportion of cells in the tumor that are in S phase of the cell cycle *(14)*, as described in Chapter 10 by Camplejohn; and the incorporation of either tritiated thymidine or bromodeoxyuridine into the DNA during synthesis, which can then be demonstrated using either autoradiography or immunohistochemistry respectively. Both methods also estimate the proportion of cells in S phase *(15,16)*. These techniques for measuring proliferative activity are not applicable to laboratories without the specialized equipment or skills and overall do not provide any other prognostic information in addition to that obtained from using either mitotic index or MIB1 immunohistochemistry.

## 2. Materials

### 2.1. Mitotic Index

1. Hematoxylin: Hematoxylins such as Gill's, Harris's, or Mayer's can be used. Although these can be made in the laboratory, it is easier to obtain them from a commercial source.
2. Eosin: 1% Eosin (yellowish) in tap water, mixed and filtered.
3. Acid alcohol: 1% Concentrated HCl in 70% alcohol.

### 2.2. MIB1 Immunohistochemistry

1. 0.005 *M* Tris-buffered saline (TBS): 50 m*M* Tris-HCl, 0.14 *M* NaCl, pH 7.6. Keep at room temperature. TBS can be made in large volumes and kept for 2–3 wk; however, check the pH weekly and discard if necessary.
2. 0.2 *M* Tris-HCl, pH 7.6: Keep stock and working solutions at 4°C until required. Allow the solution to reach room temperature before use.
3. Citrate buffer: 0.01 *M* citric acid, pH 6.0. Use 1 *N* NaOH to adjust the pH.
4. Endogenous peroxidase inhibitor: 0.5% hydrogen peroxide in methanol. Make up fresh.
5. Blocking reagent: 20% normal rabbit serum in TBS.
6. Anti-Ki-67 antibody: Monoclonal mouse MIB1 antibody (Immunotech, Masseine, France).
7. Biotinylated rabbit antimouse Ig (Dako, Glostrup, Denmark, E0354): Use diluted 1:400 in 3% normal human serum.
8. Peroxidase-conjugated streptavidin–biotin complex (Dako, Glostrup, Denmark, K0377): Make up as directed in the kit instructions. Allow 30 min for the streptavidin and biotin to complex before applying to the section.
9. 3,3'-diaminobenzidine tetrahydrochloride (DAB) 10 mg tablets (Sigma-Aldrich, Poole, UK D5905).

## 3. Methods

### 3.1. Mitotic Index

1. Cut 3-µm sections from the paraffin wax blocks, float out on warm water, and pick up on clean glass slides. Allow the water to drain from the section then place the slide in a rack and into a 58°C oven for 30–60 min. Sections are now ready for staining.
2. Dewax the sections by placing the rack in xylene (2 × 3 min).
3. Remove the xylene by placing the rack in absolute alcohol (2 × 3 min) and 70% alcohol (3 min), then wash the slides in running tap water (2 min).
4. Shake off excess water and place the rack of slides in hematoxylin (3 min).
5. Place the rack in running tap water to remove the excess dye.
6. Remove the nonspecific hematoxylin staining by placing the rack in acid alcohol (5 s), then immediately placing the rack back in running tap water. This step is known as "differentiation."

7. Leave the rack of slides in running tap water for at least 5 min. The alkalinity of the tap water changes the hematoxylin from red/purple to blue; hence, this step is known as allowing time for the sections to "blue." The "bluing" time can be shortened by using warm water. If the tap water is acidic then an alkaline solution such as Scott's tap water should be used.
8. Check microscopically if the excess dye has been removed from all structures except the nuclei. If there is still some nonspecific staining then repeat **steps 6** and **7**, reducing the time in acid alcohol accordingly (*see* **Note 1**).
9. Shake off excess water and place the rack in eosin (5 min).
10. Drain off excess eosin, then rapidly place the rack of slides in tap water followed by a series of graded alcohols. As eosin is water soluble it is rapidly removed from the section when placed under the running tap water and should therefore be put in water only for a couple of seconds. Agitating the rack allows more even removal of the dye from the sections. From tap water place the rack in 70% alcohol (5 s), then absolute alcohol (2 × 2 min). This step is known as "dehydration."
11. Place the rack in xylene (3 × 2 min). This is known as "clearing," as it removes the alcohol and allows the sections to be mounted permanently.
12. "Mount" the slides by placing a coverslip on the section using a resinous mountant such as Eukitt, DPX, or Pertex. Place a small amount of mountant either on the coverslip or the section and gently lower one onto the other, taking care not to form any bubbles between them. Allow the mountant to set (approx 30 min) before examining the section under the microscope.
13. All nuclei are stained blue while other tissue elements are a varying shade of pink. Mitotic figures are intensely stained (because of the concentration of hematoxylin in the condensed chromatids).

### 3.2. MIB1 Immunohistochemistry (see Note 2)

1. Cut 3-μm sections from the paraffin wax blocks, float out on warm water, and pick up on clean glass slides that have been coated with an adhesive such as Vectabond. Allow the water to drain from the section and place the slide into a rack and then in a 37°C oven at least overnight. On the day prior to staining, sections should be baked onto the slide by placing them in a 58°C oven overnight.
2. With each batch of test sections *always* include a positive control section that is known to contain MIB1-positive nuclei, such as tonsil. A "negative" control section should also be included for each test case, where the primary antibody is replaced by TBS (*see* **step 12**). This will demonstrate any staining not associated with binding of the MIB1 antibody.

   All incubations are at room temperature and reagents are diluted in TBS, unless stated otherwise. Allow approx 100 μL per section when making up the reagents.
3. Dewax the sections by placing the rack of slides in xylene (2 × 3 min).
4. Remove the xylene by placing the rack in absolute alcohol (2 × 3 min).
5. Inhibit endogenous peroxidase activity by transferring sections to a humid chamber and covering with freshly prepared 0.5% $H_2O_2$ in methanol (10 min).
6. Return the sections to the rack and wash in running tap water (5 min).

7. Use pressure cooker heat-mediated antigen retrieval to reveal Ki-67 antigenic sites:
   a. Fill a 5 L pressure cooker (stainless steel) with 3 L of citrate buffer, loosely place the lid on top, and heat on an electric boiling ring. When the buffer is boiling carefully place the racks, into the pressure cooker.
   b. Put the lid on and lock it.
   c. Follow the manufacturer's guidelines to set the cooker to reach pressure.
   d. Once the cooker has reached sufficient pressure to seal the valves (103 kPa/15 psi), "boil" the sections for 2 min. Timing should start immediately after full pressure is reached.
   e. After 2 min release the pressure and remove the pressure cooker from the heat source.
   f. Once the pressure has dropped slide the top and bottom handles apart. Place the cooker in a sink and flush out the buffer with running tap water (10 min).
8. Transfer the slides to a humid chamber and cover with TBS (2 × 5 min).
9. Drain sections and wipe away the excess buffer.
10. Incubate sections in freshly prepared 20% normal rabbit serum in TBS (10 min).
11. Drain sections and wipe away excess serum (*do not rinse*).
12. Cover sections with optimally diluted monoclonal MIB1 (45 min). For the negative control use TBS.
13. Rinse sections in TBS (2 × 3 min), drain off excess buffer, and wipe around section.
14. Cover sections in biotinylated rabbit antimouse Ig diluted 1:400 in 3% normal human serum (30 min).
15. Make up the horseradish peroxidase-conjugated streptavidin–biotin complex (HRP-StABC) at this stage. To 1 mL of TBS add 10 µL from bottle A (streptavidin) and 10 µL from bottle B (biotinylated HRP).
16. Rinse sections in TBS (2 × 3 min), then drain off excess buffer and wipe around section.
17. Apply the preformed HRP-StABC to the sections and incubate (30 min).
18. Rinse sections in TBS (2 × 3 min), then drain off excess buffer and wipe around section.
19. Demonstrate peroxidase by incubating in freshly prepared diaminobenzidine (DAB) solution, 0.5 mg/mL in 0.2 *M* Tris/HCl, pH 7.6; containing 0.6 µL of $H_2O_2$ (10 min). Take care when handling DAB, as it is a potential carcinogen.
20. Wash with TBS, then put slides into a rack and place in running tap water (5 min).
21. Lightly stain the nuclei with hematoxylin (1 min).
22. Rinse in tap water to remove excess hematoxylin. Place rack of sections in 1% acid alcohol (5 s) to differentiate, then immediately return them to tap water. Leave sections in tap water (at least 3 min) so that the nuclei "blue."
23. Dehydrate through 70%, then 2× absolute alcohol (2 min each), clear in xylene, and place a coverslip on the section using a resinous mountant such as Eukitt, DPX, or Pertex.
24. MIB1-positive nuclei are stained brown, while the remaining nonproliferating nuclei are blue.

### 3.3. Evaluation

#### 3.3.1. Mitotic Activity (see **Note 3**)

1. Using a low-power objective lens, scan the section to find the most cellular area of the tumor, which is used to evaluate the mitotic activity.
2. At high power (×40 objective lens) count the number of mitotic figures present in the field and make a note of this number.
3. Move to an adjacent field (it does not matter which direction provided that an area is never assessed more than once) and again count the number of mitoses.
4. Continue in this way until 10 high-power fields have been examined.
5. The mitotic activity is the mean number of mitoses present in the fields sampled and is described as the number of mitoses per high-power field.
6. Interpretation of results depends on the organ being studied, that is, what its proliferative activity is in normal tissue.

#### 3.3.2. MIB1 Immunohistochemistry

1. Using a low-power objective lens, scan the section to find the most cellular area of the tumor, which is used to evaluate the MIB1 index.
2. At high power (×40 objective) count the number of malignant MIB1-positive nuclei in the field. Then count the total number of malignant cells in the field.
3. Move to an adjacent field and repeat **step 2**. Continue in this way until a total of at least 1000 cells has been evaluated (*see* **Note 4**).
4. The MIB1 index is the proportion of positive cells in the sample.
5. Interpretation of results depends on the organ being studied, that is, what its proliferative activity is in normal tissue.

## 4. Notes

1. Mitotic index demonstration: The staining times with all three reagents may vary. Hematoxylin continues to oxidize and becomes more resistant to differentiation in acid alcohol. The solution should be replaced fortnightly to maintain crisp blue nuclear staining. Acid alcohol should be replaced when it stops rapidly removing the excess hematoxylin. Eosin gradually weakens over time as water is introduced into the solution.

   It is important that the dehydrating alcohols are replaced after approx 200 sections. If the sections look cloudy when transferred from alcohol into xylene then the sections have not been fully dehydrated. Place the sections in new absolute alcohol (two changes) before xylene.

   Sections should be quickly checked using a microscope prior to mounting to ensure that all tissue elements are well stained. Weakly stained sections can be returned to either hematoxylin or eosin. However, if they are already in xylene they must be put through graded alcohol to running tap water first.

   Mitotic figures are not always easy to identify and may, in particular, be confused with apoptotic bodies, which have a similar condensed chromatin appearance. Mitoses tend to have irregular extensions of chromatin that are visible from late prophase to telophase and lack a nuclear membrane ***(17)***. Mitotic figures can be particularly hard to identify if (a) the section has been overstained, when they appear similar to interphase cells; (b) if the tissue has been fixed in an alcohol-based fixative such as Carnoy's, which condenses all nuclei, again making it difficult to distinguish mitoses from interphase cells; and (c) the sections are thick, which leads to overstaining and poor morphology.

2. MIB1 demonstration: Small pieces of fibrous tissue tend to adhere more easily to the slide than larger fatty ones. Section adhesion can be improved by allowing the slides to dry at 37°C or room temperature for at least 24 h prior to baking them in the oven. Sections for immunohistochemistry should only be baked on the slide if they are to undergo heat-mediated antigen retrieval. Some antigenicity can be lost when the slides are baked.

   The deleterious effects of antigen retrieval can be reduced by limiting the amount of friction exerted by bubbles rising between the slides. This is done by loading the slides into every other slot in the racks, thus increasing the space between slides, and then placing these racks on top of some empty racks (without handles) in the pressure cooker.

   The optimal dilution of the MIB1 antibodies varies according to the commercial source and also according to the type of fixative used on the tissue and the length of fixation. The optimal antibody concentration should be established by testing out a series of MIB1 dilutions. Likewise, the dilution of the biotinylated antibody may also vary according to the length and temperature of incubation and should be optimised using a series of different concentrations.

   Other peroxidase-labeled detection systems can be used such as the Super sensitive multilink kit (Biogenex, San Ramon, CA, LP000-UL) or the Vectastain elite ABC kit (Vector Laboratories Inc., Burlingame, CA, PK-6200).

   The hematoxylin staining of the nuclei must be light so that it does not conceal any weakly expressed Ki-67. Thick sections may lead to more intense MIB1 staining, increased uptake of hematoxylin, and poor morphology.

3. Mitotic index evaluation: The area of a ×40 high-power field varies considerably between microscopes and this could potentially lead to differences in the size of the sample being examined. To standardize the evaluation either the same microscope should always be used to carry out the assessment or the number of mitoses per high-power field can be related to the area of the high-power field. For example a mitotic index may be quoted as 3.5 per hpf (0.18 mm$^2$). Using this format, the results can easily be compared with those of other laboratories.

   If the section is very small, such as a needle core biopsy, and there are fewer than 10 assessable high-power fields, then a slightly deeper second section should be cut from the block, stained and evaluated to achieve a sufficiently large sample size.

4. MIB1 immunohistochemistry evaluation: Not all nuclei have a homogeneous appearance when stained but may exhibit staining only around the nuclear membrane and as a "granular" pattern within it. All nuclei that express some degree of MIB1 staining should be considered as positive for the evaluation. If the section is small or very acellular, a second level should be cut from the block to be able to assess 1000 cells.

Reducing the sample size significantly increases the chances of either overestimating or underestimating the MIB1 score in heterogeneous tissue. With practice, it is possible to assess 1000 cells in 10 min.

The use of an indexed square eye piece graticule can also help with the speed and accuracy of the evaluation, ensuring that the cells are never assessed more than once.

## References


1. Baak, J. P. A. (1990) Mitosis counting in tumors. *Hum. Pathol.* **21,** 683–685.
2. Trojani, M., Contesso, G., Coindre, J. M., Rouesse, J., Bui, N. B., Mascarel, A., et al. (1984) Soft tissue sarcoma of adults; study of pathological prognostic variables and definition of a histopathological grading system. *Int. J. Cancer* **33,** 37–42.
3. Elston, C. W. and Ellis, I. O. (1991) Pathological prognostic factors in breast cancer. 1. The value of histological grade in breast cancer: experience from a large study with long term follow-up. *Histopathology* **19,** 403–410.
4. Dougherty, M. J., Compton, C., Talbert, M., and Wood, W. C. (1991) Sarcomas of the gastrointestinal tract. Separation into favorable and unfavorable prognostic groups by mitotic count. *Ann. Surg.* **214,** 569–574.
5. Galea, M. H., Blamey, R. W., Elston, C. W., and Ellis, I. O. (1992) The Nottingham Prognostic Index in primary breast cancer. *Breast Cancer Res. Treat.* **22,** 207–219.
6. Woosley, J. T. (1991) Measuring cell proliferation. *Arch. Pathol. Lab. Med.* **115,** 555–557.
7. Aaltomaa, S., Lipponen, P., Eskelinen, M., Kosma, V.-M., Marin, S., Alhava, E., and Syrjänen, K. (1992) Mitotic indexes as prognostic predictors in female breast cancer. *J. Cancer Res. Clin. Oncol.* **118,** 75–81.
8. Jannink, I., van Diest, P. J., and Baak, J. P. A. (1996) Comparison of the prognostic value of mitotic frequency and mitotic activity index in breast cancer. *The Breast* **5,** 31–36.
9. Gerdes, J., Schwab, U., Lemke, H., and Stein, H. (1983) Production of a mouse monoclonal antibody reactive with a human nuclear antigen associated with cell proliferation. *Int. J. Cancer* **31,** 13–20.
10. Kill, I. R. (1996) Localisation of the Ki-67 antigen within the nucleolus: evidence for a fibrillarin-deficient region of the dense fibrillar component. *J. Cell Science* **109,** 1253–1263.
11. Cattoretti, G., Becker, M. H. G., Key, G., Duchrow, M., Schluter, C., Galle, J., and Gerdes, J. (1992) Monoclonal antibodies against recombinant parts of the Ki67 antigen (MIB1 and MIB3) detects proliferating cells in microwave-processed formalin fixed paraffin sections. *J. Pathol.* **169,** 357–363.
12. Norton, A. J., Jordon, S., and Yeomans, P. (1994) Brief high temperature heat denaturation (pressure cooking): a simple and effective method of antigen retrieval for routinely processed tissue. *J. Pathol.* **173,** 371–379.
13. Shi, S. R., Key, M. E., and Kalra, K. L. (1991) Antigen retrieval in formalin-fixed paraffin-embedded tissues: an enhancemnet method for immunohistochemical

staining based on microwave oven heating of sections. *J. Histochem. Cytochem.* **39,** 741–748.

14. Quirke, P. (1990) Flow cytometry in the quantitation of DNA aneuploidy and cell proliferation in human disease, in *Current Topics in Pathology—Pathology of the Nucleus* (Underwood, J. C. E., ed.), Springer-Verlag, Heidelberg, pp. 215–256.
15. Meyer, J. S. and Conner, R. E. (1977) In vitro labeling of solid tissues with tritiated thymidine for autoradiographic detection of S-phase nuclei. *Stain Technol.* **52,** 185–195.
16. Sasaki, K., Ogiono, T., and Takahashi, M. (1987) In vitro BrdUrd labeling of solid tumors and immunological determination of labeling index. *Histotechnology* **10,** 47–49.
17. van Diest, P. J., Baak, J. P. A., Matze-Cok, P., Wisse-Brekelmans, E. C. M., van Galen, C. M., Kurver, P. H. J., et al. (1992) Reproducibility of mitosis counting in 2,469 breast cancer specimens: results from a multicenter morphometric mammary carcinoma project. *Hum. Pathol.* **23,** 603–607.

# 10

# Flow Cytometric Measurement of Cell Proliferation

**Richard S. Camplejohn**

## 1. Introduction

The simplest guide to cell proliferation that can be obtained by the use of flow cytometry is the S-phase fraction (SPF) calculated from DNA histograms. Measurement of such histograms was one of the earliest applications of flow cytometry, being first reported in the late 1960s. SPF is the fraction of cells in the S phase of the cell cycle and is broadly equivalent to a tritiated thymidine labeling index ([$^3$H]TdR LI). The advantages of SPF as a proliferative index include that it can be obtained rapidly from fresh, frozen, or paraffin wax-embedded tissue without the need for radioactive chemicals. The disadvantages include the need to disaggregate solid tissues, thus losing tissue morphology, and the fact that, like mitotic index and [$^3$H]TdR LI, SPF is only a crude proliferative index, which gives no details of the rate of cell proliferation. Flow cytometry offers a wide range of more sophisticated methods that allow more detailed analysis of the cell cycle and the rate of cell proliferation, and one such method involving bromodeoxyuridine (BrdUrd) labeling is described in this chapter. In the section on further reading, reference is made to sources that describe the combined measurement of DNA content and cell cycle related proteins ***(1,2)***. Also beyond the scope of this chapter are applications of flow cytometry that allow the measurement of cell death and differentiation, but references to these areas are also included ***(3)***.

The literature on DNA flow cytometric studies is enormous, and a considerable number of such studies have looked at various aspects of the relationship between proliferation and metastasis. Many of these studies have examined the power of parameters such as SPF to predict the probability of metastasis ***(4)*** or have compared the levels of SPF in primary and metastatic tumors ***(5)***. As SPF seems often to be a guide to the level of aggressiveness of individual tumors it is perhaps not surprising that many studies find a positive correlation between the level of metastasis in a patient and tumor SPF.

From: *Methods in Molecular Medicine, vol. 57:*
*Metastasis Research Protocols, Vol. 1: Analysis of Cells and Tissues*
Edited by: S. A. Brooks and U. Schumacher © Humana Press Inc., Totowa, NJ

## 2. Materials

### 2.1. Simple Mechanical Method of Tissue Disaggregation

1. A stainless steel grid (pore size 0.5 mm).
2. Petri dish.
3. 5-mL Disposable syringes with 19-, 21-, 23-, and 25-gauge needles.
4. 35-µm mesh nylon filter.
5. Standard tissue culture medium such as minimum essential Eagle medium (MEM).

### 2.2. Enzymatic Enucleation Method of Tissue Disaggregation and Staining

1. Solutions A, B, and C are quite complex and their compositions are detailed in **Table 1**. The reagents are also available as a kit from Becton–Dickinson (CycleTEST PLUS DNA reagent kit cat. no. 340242).

### 2.3. Dissociation of Paraffin Wax-Embedded Archival Material

1. Thick (50 µm) tissue sections in individual fine pore biopsy cassettes.
2. Xylene or one of the citrus-derived dewaxing agents and graded alcohols from 100% to 50% v/v. The method is most conveniently carried out using a tissue processing machine such as Histokinette, but it can be done manually with the various solutions in liter beakers.
3. 0.5% Pepsin (Sigma cat. no. 7012) in 0.9% saline with the pH adjusted to 1.5 using 2 *M* HCl.
4. Propidium iodide (PI, 50 µg/mL) and RNase (200 µg/mL, both from Sigma) dissolved in phosphate-buffered saline (PBS).
5. 35-µm mesh nylon filter.

### 2.4. BrdUrd Staining Protocol

1. Bromodeoxyuridine (Sigma) dissolved in tissue culture medium at concentrations between 2 and 10 µ*M* for in vitro incorporation.
2. 1 *M* HCl.
3. 20 µL of mouse anti-BrdUrd antibody (Becton–Dickinson) added to 80 µL of PBS containing 0.5% Tween-20 and 1% serum.
4. 10 µL of $F(ab)_2$ fluorescein isothiocyanate (FITC)-labeled rabbit antimouse antibody (Dako) added to 90 µL of PBS containing 0.5% Tween-20 and 1% serum.
5. Propidium iodide (50 µg/mL) and RNase (200 µg/mL, both from Sigma) dissolved in PBS.

## 3. Methods

### 3.1. Tissue Disaggregation

An absolute prerequisite for any cellular flow cytometric technique is the production of a high-quality single-cell suspension (i.e., a high percentage of intact, single cells with few clumps, damaged cells, or debris). Some sources

**Table 1**
**Preparation of Buffer and Staining Solutions for the Enucleation Method**

| | |
|---|---|
| *Citrate buffer* | |
| Sucrose (BDH) | 85.50 g (250 m*M*) |
| Trisodium citrate, 2 $H_2O$ (Merck) | 11.76 g (40 m*M*) |
| Dissolve in distilled water | approx 800 mL |
| Dimethyl sulfoxide (DMSO) (Merck) is added | 50 mL |
| Distilled water is added to a total volume of | 1000 mL |
| pH is adjusted to | 7.6 |
| *Stock solution* | |
| Trisodium citrate, 2 $H_2O$ (Merck) | 1000 mg (3.4 m*M*) |
| Nonidet P-40 (Shell) | 1000 µL (0.1% v/v) |
| Spermine tetrahydrochloride (Serva, 35300) | 522 mg (1.5 m*M*) |
| TRIS (Sigma 7-9, T-1378) | 61 mg (0.5 m*M*) |
| Distilled water is added to a total volume of | 1000 mL |
| *Solution A* | |
| Stock solution | 1000 mL |
| Trypsin (Sigma, T-0134) | 30 mg |
| pH is adjusted to | 7.6 |
| *Solution B* | |
| Stock solution | 1000 mL |
| Trypsin inhibitor (Sigma T-9253) | 500 mg |
| Ribonuclease A (Sigma, R-4875) | 100 mg |
| pH is adjusted to | 7.6 |
| *Solution C* | |
| Stock solution | 1000 mL |
| PI (Fluka) | 416 mg |
| Spermine tetrahydrochloride (Serva, 35300) | 1160 mg |
| pH is adjusted to | 7.6 |
| Should be light protected | |

All chemicals should be reagent grade. This information is taken from Table 1, p. 755, Vindelov and Christensen (1990) *(7)*.

of material such as peripheral blood lymphocytes and suspension culture cells need little processing to produce such a suspension and monolayer cultures can often be treated with a simple enzymatic protocol (typically trypsin–EDTA) to produce predominantly single cells. The production of a single-cell suspension from solid tissues is more problematic (*see* **Note 1**); in this chapter three examples of such methods are given. The first of these describes a simple mechanical method that is suitable for easily disaggregated tissue such as lymph node, tonsil, or many lymphoma samples, while the second method involves the use of a carefully controlled cocktail of chemicals including enzymes. The third method is applicable to paraffin-embedded tissue from any source, providing fixation was carried out appropriately in simple or buffered formalin without excessive heat.

### 3.1.1. Simple Mechanical Method of Tissue Disaggregation

1. Place tissue sample on stainless steel grid (pore size 0.5 mm) placed over a Petri dish. Cut tissue into small (1–2 $mm^3$) pieces.
2. Moisten the tissue fragments with tissue culture medium such as MEM containing 5% serum and using the rubber plunger from a 5-mL disposable syringe, gently push the tissue through the steel mesh into the Petri dish. Rinse steel sieve thoroughly with 4 mL of MEM.
3. Take up the crude suspension from the Petri dish into a 5-mL syringe and pass it sequentially twice through each of a 19-, 21-, 23-, and finally 25-gauge needle. (Needle aspirates taken directly from solid tumors can be entered into this method here.)
4. Filter the suspension through a 35-µm mesh nylon filter. Adjust the cell concentration to the desired level.

Fine-needle aspiration of solid tumors, including those that are not otherwise easily dissociated, may be an effective method to obtain a crude cell suspension, which can be further disaggregated as described from **step 3** of **Subheading 3.1.1.**, or can be entered into an enucleation technique (**Subheading 3.1.2.**). Fine-needle aspiration may be performed either *in situ* on tumors from suitable sites such as breast cancer ***(6)***, or may be taken from excised tumor biopsies ***(7)***. In both cases high success rates were reported. Whichever disaggregation method is used the single-cell suspension can either be stained fresh or fixed in 70% ethanol at 4°C. Care should be taken if fixing the cells to avoid excessive clumping by agitating during addition of fixative and not fixing excessive concentrations of cells. Cells at a concentration of approx $1 \times 10^6$/mL can be stained in PBS containing final concentrations of 50 µg/mL of PI, 200 µg of RNase, and for unfixed cells 0.2% Triton X-100. Cells fixed in 70% ethanol should, after removal of the alcohol, be washed in PBS prior to staining.

### 3.1.2. Enzymatic Enucleation Method of Tissue Disaggregation and Staining

1. Cells are obtained in citrate buffer by fine-needle aspiration either *in situ* or following surgical removal.
2. Add chicken and trout erythrocytes to the needle aspirate containing approx $10^6$ cells in 200 µL of citrate buffer.
3. Add 1.8 mL of solution A containing trypsin to the suspension in **step 2**, invert the tube to mix. Leave for 10 min at room temperature, mixing 2–3×.

4. Add 1.5 mL of solution B containing trypsin inhibitor and RNase, mix, and leave for a further 10 min at room temperature.
5. Add 1.5 mL of ice-cold solution C, containing PI and spermine hydrochloride. Mix the solutions and filter through a 25-µm nylon mesh into plastic tubes protected from the light.
6. Keep samples on ice for a minimum of 15 min (maximum 3 h).

### *3.1.3. Dissociation of Paraffin Wax-Embedded Archival Material*

1. Place one 50-µm tissue section for each sample in a separate small biopsy cassette (pore size 300 µm) with a label written in pencil.
2. Place up to 12 cassettes in a basket and run the batch of samples on a tissue processing machine such as a histokinette with the following program. The samples are agitated for 15 min in the following:

| | |
|---|---|
| Xylene | 3 changes |
| 100% Alcohol | 2× |
| 90% Alcohol | 2× |
| 70% Alcohol | 2× |
| 50% Alcohol | 2× |

   This process can be performed overnight (total time, 2 h 45 min).
3. Wash the basket of cassettes thoroughly, leaving them to soak for 10 min, in two changes of distilled water.
4. Drain the water from the cassette, open it carefully, then using fine forceps transfer the dewaxed section to a universal container with pepsin solution made up as follows: 0.5% pepsin in 0.9% saline with the pH adjusted to 1.5 using 2 *M* HCl.
5. Transfer the batch of universals to a 37°C water bath and incubate for 30 min.
6. Centrifuge at 800*g* (typically 2000 rpm) for 3 min at room temperature.
7. Carefully remove and discard the supernatant with a Pasteur pipet.
8. Resuspend the tissue fragments and released nuclei in 2 mL of PBS and agitate vigorously using a vortex mixer.
9. Filter the suspension through a 35-µm pore size nylon filter and adjust the concentration of nuclei as desired.
10. Stain the nuclei with PI (50 µg/mL) and RNase (200 µg/mL) solution for a minimum of 14 min in the dark at room temperature (*see* **Note 2**).

### 3.2. Running Samples on a Flow Cytometer

It is impossible to describe in a short chapter such as this how samples are actually run on a flow cytometer, as there are a number of different companies selling such machines and each company generally sells a number of models of cytometer. In addition, a variety of older machines, which are no longer sold, are still used in various laboratories. For each type of machine, there is a detailed handbook describing its operation. However, whatever type of flow cytometer is used to produce DNA histograms, certain general principles apply if good quality results are to be obtained, including:

1. Producing a high-quality single cell suspension.
2. Making measurements on a sufficient number of cells/nuclei (typically not fewer than $10^4$ cells).
3. Making use of all the information available such as light scatter and pulse area/width data to reduce the contamination of DNA histograms by clumps and debris *(8)*.

### 3.3. Data Analysis

As with the running of samples, the analysis of data is difficult to describe in any detail, as most laboratories now use commercial software to analyze flow cytometric data including DNA histograms (*see* **Note 3**). In my laboratory a program called *Modfit* ™ (supplied by Verity Software House) is used and this program allows DNA histograms to be fitted according to a number of mathematical models. These models use a variety of functions to fit areas under parts of the DNA histogram corresponding to $G_1$, S phase, and $G_2$/M. The precise answers for these cell cycle phase distributions produced by such models depend on the assumptions underlying the computer algorithms used to fit the curves. With *Modfit* it is possible to remove debris using a model developed for use with paraffin wax-embedded material and a different model suitable for fresh/frozen tissue. Similarly, there is an option to autodetect aggregates, fit curves with distinct or indistinct $G_2$/M peaks, detect an apoptotic peak, and so forth. In addition different models can be used to fit S phase including a rectangular one similar to that described in **Subheading 3.3.3.** Such commercial software can be very convenient to use, and if used with caution, can be a valuable aid to data analysis. However, when using such commercial packages it is salutary to try out different models and note the different answers produced from the same set of data. Ideally, all users of such programs would understand the basis of the models applied but this is often not the case. Described in the following subheadings briefly is a simple manual system of data analysis that was used in our laboratory up until 1990 and that gave clinically useful SPF data.

#### 3.3.1. Calculation of the Coefficient of Variation (CV)

The CV, usually of the DNA diploid $G_1$ peak, is a useful shorthand way of representing the quality of DNA data. The CV is calculated from the following equation, on the basis that the $G_1$ peak can be represented by a normal distribution.

$$\text{CV} = W/(M \times 2.35)$$

where $W$ = full width of the $G_1$ peak at the half-maximum height and $M$ = peak channel number of the $G_1$ peak.

High-quality DNA histograms will have a low CV (typically <5%) and the $G_1$ peak approximates very closely to a normal distribution. Many real DNA histograms are not so good and it is important in clinical DNA studies to quote both the average CV and the range of CV seen in the series of cases reported. This enables readers to judge the quality of the data presented.

### 3.3.2. Calculation of the DNA Index (DI)

The basic principles by which DNA aneuploidy is recognized and by which the DI is calculated are widely accepted. In histograms that exhibit two or more $G_1$ peaks (*see* **Fig. 1**, lower panel) the DI of the DNA aneuploid peak(s) is calculated by reference to the position of the diploid $G_1$ peak. For example, in **Fig. 1** (lower panel), the tumor $G_1$ peak has twice the amount of DNA compared with the DNA diploid $G_1$ peak and thus the sample is tetraploid.

### 3.3.3. Calculation of SPF

For DNA diploid histograms, SPF can be calculated by a very simple method first described by Baisch et al. ***(9)***. As illustrated in the upper panel of **Fig. 1**, this method involves fitting a rectangle to represent S phase and calculating the number of cells within this area. The sides of the rectangle are determined by the peak channel numbers of the $G_1$ and $G_2$ peaks, and the height of the rectangle is determined from the average number of cells/channel in the middle 10 channels of the DNA histogram. A similar method can be used to estimate the aneuploid SPF as illustrated in **Fig. 1** (lower panel). The fractions of cell in $G_1$ and $G_2$/M can be estimated by subtraction if required. This very simple method showed good agreement with BrdUrd labeling in a series of cell lines and gave good prognostic information in a large series of breast cancer cases ***(10)***.

## 3.4. Multiparametric Flow Cytometry Using BrdUrd Labeling

A real strength of flow cytometry is the ability to measure a number of parameters simultaneously for each individual cell. Thus, DNA content can be combined with surface or internal markers. One example of such a technique involves the staining of cells that have incorporated the thymidine analog bromodeoxyuridine into their DNA. Such cells can be recognized using antibodies specific to DNA containing BrdUrd. By fixing cells at various times after a pulse of BrdUrd, which can be administered either in vitro or in vivo, detailed information on the length of the cell cycle phases and the rate of cell proliferation can be obtained. A modification of this technique has been applied to the measurement of cell proliferation in clinical cancers following administration of BrdUrd to patients ***(11)***. The staining procedure necessary to measure DNA content simultaneously with BrdUrd uptake is relatively simple. Cells or tissue are normally fixed in alcohol until required and then the cells are subjected to a procedure that partially denatures DNA to allow access of the antibody to its epitope. This is usually done using either dilute acid or heat. The harshness of the denaturation step required, does vary between different cells, but the aim is to achieve enough denaturation to allow access of antibody without denaturing so much DNA that measurement of total DNA content is disrupted. A protocol that works with most cells is described in the following subheading.

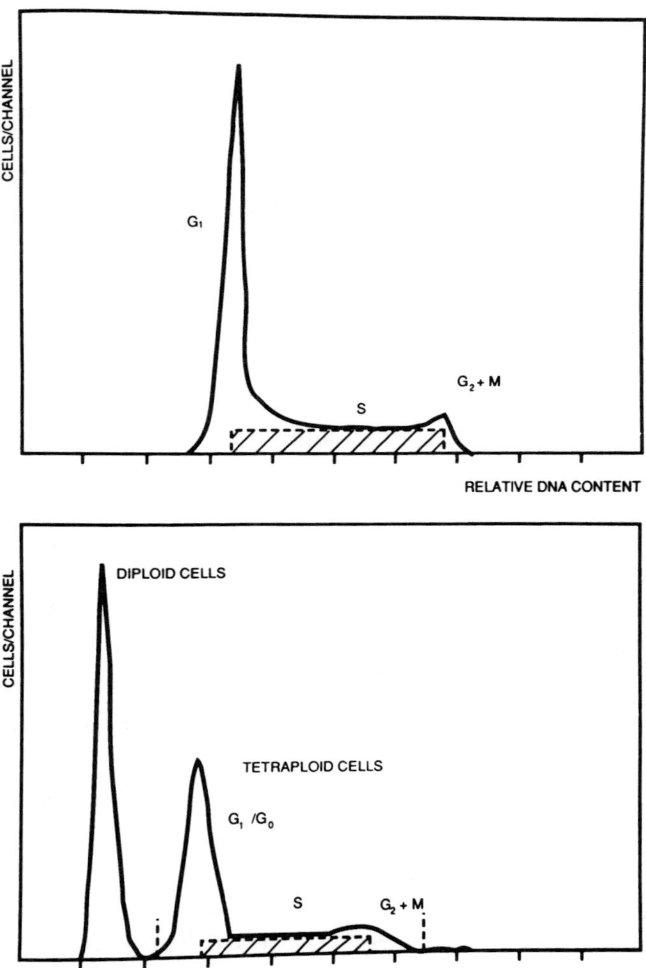


Fig. 1. The *upper panel* shows a diploid histogram. SPF is calculated by fitting a rectangle to represent the S phase. The left-hand side of the rectangle is set at the peak channel of $G_1$ and the right-hand side by the peak channel of $G_2$. The height of the rectangle is defined by determining the mean number of cells/channel in the central region of the S phase well away from either $G_1$ or $G_2$; for a 1023-channel display we would determine this average in the central 40 channels. The *lower panel* shows a DNA histogram for a tetraploid tumor. The tumor $G_1$ peak has twice the normal amount of DNA (DI = 2.0). SPF is calculated by a modified version of the rectangular method.

### 3.4.1. BrdUrd Staining Protocol

1. Centrifuge 1–2 × $10^6$ cell aliquots in alcohol, pour off supernatant, add 1 mL 0.1 *M* HCl, and incubate for 10 min at 37°C.

2. Add 2 mL of PBS, spin, remove acid, and wash in PBS.
3. Add 20 µL of anti-BrdUrd antibody in a volume of 100 µL, and incubate at room temperature for 60 min.
4. Add 2 mL of PBS, centrifuge, wash again with PBS.
5. Add 10 µL of $F(ab)_2$ FITC-labeled antibody in a volume of 100 µL, and incubate for 30 min.
6. Wash as after the first antibody, centrifuge, and resuspend in PBS containing 50 µg of PI and 200 µg of RNase/mL.
7. Measure total DNA content, BrdUrd labeling, and light scatters on ~$10^4$ cells (*see* **Note 4**).

## 4. Notes

A general problem in writing this chapter has been to describe methods that are generally applicable to all or even most circumstances. A problem with DNA and multiparametric flow cytometry methods is the large number of variations in technique described in the literature. As an example, even within one laboratory it is not always possible to have one technique for tissue disaggregation. I have tried to describe methods that we have found to have wide applicability but few of the methods will work in all circumstances without some modification.

1. Tissue disaggregation. We have found that needle aspiration when carried out by a skilled operator produces a crude cell suspension with adequate cell numbers in most cases from a wide variety of fresh tissues. However, if larger cell numbers are required and sufficient tissue is available then one of a large number of enzymatic or physical/enzymatic methods described in the literature may be more appropriate. Particularly for tough fibrous tissues the precise optimal protocol can take some time to determine.
2. The use of paraffin wax-embedded material. The description in the early 1980s of a method that allowed DNA histograms to be produced from archival paraffin-embedded tissue led to an explosion in clinical DNA flow cytometric studies. This technique allows large retrospective studies to be performed rapidly and studies on rare neoplasms that would take too long to accumulate fresh tumor samples to be performed. Another advantage is that the same protocol works well on all the tissues we have tried. However, fixation of the tissue, which may have occurred many years previously, is important. Overlong fixation, particular in unbuffered formalin or fixation in Bouin's fixative or fixatives containing heavy metals, will usually prevent the use of tissue for DNA flow cytometry. In addition, excessive heating of tissue during fixation leads to poor-quality DNA histograms. The general quality of data from paraffin wax-embedded tissue is poorer than from the equivalent fresh tissue, but given good fixation we have had success rates of between 80% and 95% of archival samples in particular series.
3. Data analysis. A number of commercial programs are routinely used to analyze DNA histograms. Within each of these programs there is generally a choice of models to fit the data. As well as the type of curve fitting routine applied there are

often choices regarding factors such as subtraction of debris, manual or automatic recognition of peaks, and so forth. It is generally best to use the simplest model that appears to give a good fit to the experimental data. In practice, the better the quality of the data, the less affected by different assumptions are the answers produced. Thus, it is worth the effort to produce the best quality raw data possible with the material available.

4. BrdUrd labeling. The foregoing protocol works in our hands on most cell types but better quality data are obtained on some cells by adding pepsin (1 mg/mL) to the HCl. This modified method yields a suspension of nuclei, and a similar method with a stronger HCl solution (typically 1–2 *M*) can be used on minced, alcohol-fixed samples of solid tissue. This obviates the need for a separate disaggregation procedure. For some cell types stronger acid or a longer exposure time may work better even if single cells are treated. The aim of the acid treatment is to denature a small amount of the DNA to allow access of the anti-BrdUrd antibody. If too much denaturation occurs, then the BrdUrd staining may be good but a poor DNA profile will be achieved because PI stains only double-stranded DNA.

## References

1. Camplejohn, R. S. (1994) The measurement of intracellular antigens and DNA by multiparametric flow cytometry. *J. Microsc.* **176,** 1–7.
2. Gong, J., Li, X., Traganos, F., and Darzynkiewicz, Z. (1994) Expression of $G_1$ and $G_2$ cyclins measured in individual cells by multiparametric flow cytometry: a new tool in the analysis of the cell cycle. *Cell Prolif.* **27,** 357–371.
3. Darzynkiewicz, Z., Juan, G., Li, X., Gorczyca, W., Murakami, T., and Traganos, F. (1997) Cytometry in cell necrobiology—analysis of apoptosis and accidental cell-death (necrosis). *Cytometry* **27,** 1–20.
4. Volm, M., Bak, M., Hahn, E. W., Mattern, J., and Weber, E. (1988) DNA and S-phase distribution and incidence of metastasis in human primary lung carcinoma. *Cytometry* **9,** 183–188.
5. Feichter, G. E., Kaufmann, M., Müller, A., Haag, D., Eckhardt, R., and Goerttler, K. (1989) DNA index and cell cycle analysis of primary breast cancer and synchronous axillary lymph node metastases. *Breast Cancer Res. Treat.* **13,** 17–22.
6. Levack, P. A., Mullen, P., Anderson, T. J., Miller, W. R., and Forrest, A. P. M. (1987) DNA analysis of breast tumor fine needle aspirates using flow cytometry. *Br. J. Cancer* **56,** 643–646.
7. Vindelov, L. L. and Christensen, I. J. (1990) A review of techniques and results obtained in one laboratory by an integrated system of methods designed for routine clinical flow cytometric DNA analysis. *Cytometry* **11,** 753–770.
8. Ormerod, M. G., ed. (2000) *Flow Cytometry: A Practical Approach*, 3rd ed., IRL Press, Oxford, England, pp. 83–97.
9. Baisch, H., Goehde, W., and Linden, W. A. (1975) Analysis of PCP-data to determine the fraction of cells in the various phases of cell cycle. *Radiat. Environ. Biophys.* **12,** 31–39.

10. Camplejohn, R. S., Ash, C. M., Gillett, C. E., Raikundalia, B., Barnes, D. M., Gregory, W. M., et al. (1995) The prognostic significance of DNA flow cytometry in breast cancer: results from 881 patients treated in a single centre. *Br. J. Cancer* **71,** 140–145.
11. Wilson, G. D., Dische, S., and Saunders, M. I. (1995) Studies with bromodeoxyuridine in head and neck cancer and accelerated radiotherapy. *Radiother. Oncol.* **36,** 189–197.

# 11

# SDS-PAGE and Western Blotting Techniques

**Christine Blancher and Adam Jones**

## 1. Introduction

The goal of Western blotting, or more correctly, immunoblotting, is to identify with a specific antibody a particular antigen within a complex mixture of proteins that has been fractionated in a polyacrylamide gel and immobilized onto a membrane. Immunoblotting can be used to determine a number of important characteristics of protein antigens—the presence and quantity of an antigen, the relative molecular weight of the polypeptide chain, and the efficiency of extraction of the antigen.

Immunoblotting occurs in six stages: (1) extraction and quantification of protein samples; (2) resolution of the protein sample in sodium dodecyl sulfate-polyacrylamide denaturing gel electrophoresis (SDS-PAGE); (3) transfer of the separated polypeptides to a membrane support; (4) blocking nonspecific binding sites on the membrane; (5) addition of antibodies; and (6) detection.

Sample preparation is important for obtaining accurate separation of the proteins on the basis of molecular weight. Depending on whether an antigen is primarily extracellular, cytoplasmic, or membrane-associated different procedures might be required to prepare the sample initially. Although there are exceptions, many soluble nuclear and cytoplasmic proteins can be solubilized by lysis buffers that contain the nonionic detergent Nonidet P-40 (NP-40) and either no salt at all or relatively high concentrations of salt (e.g., 0.5 *M* NaCl). However, the efficiency of extraction is often greatly affected by pH of the buffer and the presence or absence of chelating agents such EDTA. Extraction of membrane-bound and hydrophobic proteins is less affected by ionic strength of the lysis buffer but often requires a mixture of ionic and nonionic detergents. Many methods of solubilization, particularly those that involve mechanical disruption of cells, release intracellular proteases that can digest the target

From: *Methods in Molecular Medicine, vol. 57:*
*Metastasis Research Protocols, Vol. 1: Analysis of Cells and Tissues*
Edited by: S. A. Brooks and U. Schumacher © Humana Press Inc., Totowa, NJ

protein. The susceptibility of different proteins to attack by proteases varies widely, with cell-surface and secreted proteins generally being more resistant than intracellular proteins. It is therefore important to keep the cell extracts in ice to minimize proteolytic activity. In addition, inhibitors of proteases are commonly included in lysis buffer. The sample treatments used should destroy any secondary or tertiary protein structure so these do not alter the migration of the proteins through the acrylamide matrix. Therefore, the sample preparation should solubilize and denature proteins, dissociate polypeptides, and reduce disulfide bonds. This is achieved through a combination of SDS, a reducing agent (usually β-mercaptoethanol), and heat. When comparing samples, be certain to correct for the total amount of protein in the sample. This can be done by many of the commonly used protein concentration determination methods. The Bradford method is relatively accurate for most proteins, except for small basic polypeptides such as ribonuclease or lysozyme. It is also hampered by detergent concentrations higher than about 0.2% (e.g., Triton X-100, SDS, NP-40). Commercially available colorimetric assay systems are now available (e.g., Pierce [BCA protein assay reagent] or Bio-Rad [DC protein assay]); they are easier to use and relatively cheap.

The proteins diluted in a sample buffer are separated by gel electrophoresis. Most often, the electrophoresis system of choice will be a Tris–glycine discontinuous SDS-polyacrylamide gel *(1)*. The SDS coats the proteins, providing them with a negative charge proportional to their length. When the coated sample is run on an SDS polyacrylamide gel, the proteins separate by charge and by the sieving effect of the gel matrix. Sharp banding of the proteins is achieved by using a discontinuous gel system that has stacking and separating gel layers that differ in either salt concentration, pH, acrylamide concentration, or a combination of these.

Once separated, antigens are transferred (immobilized) onto a membrane that binds the proteins nonspecifically. Transfer usually is achieved by placing the membrane in direct contact with the gel and then placing this sandwich in an electric field to drive the proteins from the gel onto the membrane *(2)*. Electrophoretic transfer can be achieved either by complete immersion of a gel-membrane sandwich in a buffer (wet transfer) or by placing the gel-membrane sandwich between absorbent paper soaked in transfer buffer (semidry transfer). For the wet transfer, the sandwich is placed in a buffer tank with platinum wire electrodes. For the semidry transfer, the gel-membrane sandwich is placed between carbon plate electrodes. Apparatus for both types of transfers are available commercially.

Before the blot can be processed for antigen detection, it is essential to block the membrane to prevent nonspecific adsorption of the immunological reagents. The most commonly used blocking solution that is compatible with nearly all detection systems is nonfat dried milk *(3)*, which could be associated with Tween-20 to reduce eventual excessive background *(4)*.

Virtually all Western blots are probed in two stages. An unlabeled antibody specific to the target protein is first incubated with the membrane in the presence of blocking solution. The membrane is then washed and incubated with the secondary reagent: an antiimmunoglobulin that carries a reporter group, that is either radiolabeled or coupled to an enzyme such as horseradish peroxidase or alkaline phosphatase. After further washing, detection is achieved by a colourimetric reaction, enhanced chemiluminescence or autoradiography, depending on the nature of the conjugate.

## 2. Materials

### 2.1. Extraction and Quantification of Protein Samples

1. The lysis buffers that are commonly used to prepare extracts of mammalian cells for immunodetection are shown in the following list. In the absence of any information about the target antigen, we recommend trying the triple- and single-detergent lysis buffer before attempting more specialized methods of extraction. The properties of commonly used protease inhibitors are summarized in **Table 1**.
   a. Triple-detergent lysis buffer: 50 m*M* Tris-HCl, pH 8.0, 150 m*M* NaCl; 0.02% sodium azide; 0.1% SDS; 100 µg/mL of phenylmethylsulfonyl fluoride (PMSF); 1 µg/mL of aprotinin; 1 µg/mL of pepstatin A; 1 µg/mL of leupeptin; 1% NP-40; 0.5% sodium deoxycholate.
   b. Single-detergent lysis buffer: 50 m*M* Tris-HCl, pH 8.0; 150 m*M* NaCl; 0.02% sodium azide, 100 µg/mL of PMSF; 1 µg/mL of aprotinin; 1 µg/mL of pepstatin A; 1 µg/mL of leupeptin; 1% Triton X-100 or 1% NP-40.
   c. High-salt lysis buffer: 50 m*M* *N*-2-hydroxyethylpiperazine-*N*′-2-ethanesulfonicacid (HEPES), pH 7.0; 500 m*M* NaCl; 1% NP-40; 100 µg/mL of PMSF; 1 µg/mL of aprotinin; 1 µg/mL of pepstatin A; 1 µg/mL of leupeptin.
   d. No-salt lysis buffer: 50 m*M* HEPES, pH 7.0; 1% NP-40; 100 µg/mL of PMSF; 1 µg/mL of aprotinin; 1 µg/mL of pepstatin A; 1 µg/mL of leupeptin.
2. Ice-cold phosphate-buffered saline (PBS): Mix 11.5 g of 80 m*M* of disodium hydrogen orthophosphate anhydrous, 2.96 g of 20 m*M* of sodium dihydrogen orthophosphate, and 5.84 g of 100 m*M* sodium chloride in 1 L of distilled water. Adjust the pH to 7.5.
3. Rubber policemen for scraping cell monolayers.
4. Either a sonicator or a homogenizer with immersible tip, or 23-gauge hypodermic needle, to shear chromosomal DNA to a lower viscosity in the protein samples.
5. Protein quantification:
   a. Preparation of a series of protein samples for a standard curve: Use a protein as similar in its properties to your sample as possible (i.e., if doing antibody concentrations, use a purified antibody). If your sample is unknown, use antibody. Bovine serum albumin (BSA) gives a value about twofold higher than its weight for Bradford dye binding assays ***(5)***, but it is adequate for Lowry-based assays ***(6)***. The dilution of the protein standard should be made from 0.2 mg/mL to approx 1.5 mg/mL. For best results, the standards should always be prepared in the same buffer as the protein sample.

**Table 1**
**Properties of Commonly Used Protease Inhibitors**

| Protease inhibitor | Enzyme reactivity | Work solution | Stock solution |
|---|---|---|---|
| Aprotinin (Trasylol) | Kallikrein, trypsin, plasmin, chymotrypsin, plasminogen activator, elastase. | 1–2 μg/mL | 10 mg/mL in 0.01 *M* HEPES, pH 8.0 |
| Leupeptins (supplied as a 3:1 mixture of propionyl and acetyl derivatives) | Plasmin, trypsin, chymotrypsin, papain, thrombin, cathepsin B,H,L. | 1–2 μg/mL | 10 mg/mL in water |
| Pepstatin A | Pepsin, renin, cathepsin D. | 1 μg/mL | 1 mg/mL in ethanol |
| Antipain | Cathepsins A,B,D, papain, plasmin, trypsin, thrombin, calpain. | 1–2 μg/mL | 1 mg/mL in water |
| PMSF (phenylmethyl sulfonyl fluoride) | Chymotrypsin, trypsin, thrombin. | 100 μg/mL | 1.74 mg/mL (10 m*M*) in isopropanol |
| TLCK (tosyllysine chloromethyl ketone) | Trypsin, papain. | 50 μg/mL | 1 mg/mL in 0.05 *M* sodium acetate, pH 5. |
| TPCK (tosylphenylalanine chloromethyl ketone) | Chymotrypsin, papain. | 100 μg/mL | 3 mg/mL in ethanol |
| EDTA | Metalloproteases. | 1–10 m*M* | 0.5 *M* in water |

b. Bradford dye concentrate: Dissolve 100 mg of Coomassie brilliant blue R250 in 50 mL of 95% ethanol. Add 100 mL of concentrated phosphoric acid. Add distilled water to a final volume of 200 mL. The dye is stable at 4°C for at least 6 mo. This dye concentrate is also available commercially from Bio-Rad.

6. Plastic or glass cuvets with 1 cm path length matched to a laboratory spectrophotometer. If analyzing multiple samples it is easier to perform microplate assays using microtiter plates and microplate reader.

**Caution:**

Sodium azide is poisonous. It should be handled with great care, gloves should be worn while working with it, and solutions containing it should be clearly marked. PMSF is extremely destructive to the mucous membranes of the respiratory tract, the eyes, and the skin. PMSF may be fatal if inhaled, swallowed, or absorbed through the skin. In case of contact, immediately flush eyes or skin with copious amounts of water. Discard contaminated clothing.

## 2.2. Resolution of the Protein Sample in SDS-PAGE

Two formats of gels are often used, minigels (7 × 10 cm) and large gels (16 × 16 cm). For most of the analytical and some preparative fractionation techniques the minigel is the best choice.

1. Bio-Rad Mini-PROTEAN vertical gel electrophoresis apparatus and casting equipment. This apparatus is commonly used and combines fast separations with easy setup. It allows the running at the same time of two sandwich gels of variable thickness and number of wells.
2. For each sandwich gel: Inner glass plate and outer glass plate, two spacers, and a Teflon comb of the same thickness (0.5–1.5 mm).
3. Power pack.
4. 29% w/v Acrylamide/1% w/v *bis*-acrylamide stock solution, use as supplied by manufacturer.
5. SDS: Prepare a 10% w/v stock solution in deionized water and store at room temperature.
6. 0.5 *M* Tris-HCl, pH 6.8: Dissolve 6.05 g of Tris base in 40 mL of distilled water. Adjust to pH 6.8 by slow addition of 1 *M* HCl. Make volume up to 100 mL with distilled water.
7. 1.5 *M* Tris-HCl, pH 8.8: Dissolve 90.9 g of Tris base in 350 mL of water; adjust the solution to pH 8.8 by slow addition of 1 *M* HCl. Make volume up to 500 mL with distilled water.
8. 10% w/v Ammonium persulfate (APS): Use freshly made solution or prepare 10 mL and store aliquots of 1 mL at –20°C for several months (store at 4°C for 1 wk maximum).
9. *N*,*N*,*N*′,*N*′-Tetramethyl ethylene diamine (TEMED): Use as supplied by the manufacturer.
10. SDS sample buffer: 50 m*M* Tris-HCl, pH 6.8, 2% w/v SDS, 5% v/v glycerol, 1% v/v β-mercaptoethanol, 0.01% w/v bromophenol blue. Make up 10 mL of 6X sample buffer by mixing 6 mL of 0.5 *M* Tris-HCl, pH 6.8, 1.2 g of SDS, 3 mL of glycerol, 600 µL of β-mercaptoethanol; of 6 mg bromophenol.
11. Water-saturated butanol solution: Combine 100 mL of *n*-butanol and 5 mL of distilled water in a bottle. Shake well. Use top layer for overlaying gels. Store at room temperature.
12. Prestained protein molecular weight markers.
13. Tris-glycine electrophoresis buffer: 25 m*M* Tris, pH 8.3; 250 m*M* glycine (electrophoresis grade); 0.1% SDS. A 5X stock solution can be made by dissolving 15.1 g of Tris base, 94 g of glycine, 10 g of SDS; make up volume to 1 L with water. Dilute with deionized water for 1X working solution.

    **Caution:** Acrylamide and *bis*-acrylamide are potent neurotoxins and are absorbed through the skin. Their effect is cumulative. Polyacrylamide is considered to be nontoxic, but it should be treated with care because it may contain small quantities of unpolymerized material.

### 2.3. Staining SDS-Polyacrylamide Gels with Coomassie Brilliant Blue

1. Coomassie brilliant blue solution: Dissolve 0.25 g of Coomassie brilliant blue R250 in 90 mL of methanol—$H_2O$ (1:1 v/v) and 10 mL of glacial acetic acid. Filter the solution through a Whatman no. 1 filter to remove any particulate matter.
2. Destaining solution: Mix 90 mL of methanol—$H_2O$ (1:1 v/v) and 10 mL of glacial acetic acid.

### 2.4. Transfer of the Separated Polypeptides to a Membrane Support

1. Electrophoretic transfer apparatus, either a wet system (Trans-blot cell, including: buffer tank, two gel holder cassettes, four fiber pads, modular electrode assembly) or a semi-dry electrophoretic transfer cell.
2. Power pack.
3. Sheets of absorbent filter paper (Whatman 3MM or equivalent).
4. Transfer membrane (nitrocellulose filter) or preferably polyvinylidene fluoride (PVDF).
5. Transfer buffer (for semidry or wet system): 25 m*M* Tris base, 190 m*M* glycine, 15% methanol (for 5 L, dissolve 15.1 g of Tris base and 72 g of Glycine in 4 L of distilled water, add 750 mL of methanol; and make up to 5 L with distilled water). Store at 4°C.

### 2.5. Staining the Blot with Ponceau S

1. Ponceau S solution: 0.1% w/v Ponceau S, 5% v/v acetic acid, in water. Available commercially from Sigma.

### 2.6. Blocking Nonspecific Binding Sites on the Membrane

1. Blocking solution: 5% w/v nonfat dry milk (e.g., Marvel) and 0.05% Tween-20, in PBS. Store at 4°C only for 1 or 2 d.
2. Platform shaker.

### 2.7. Addition of Antibodies

1. Blocking solution: 5% w/v nonfat dry milk (e.g., Marvel) and 0.05% Tween-20, in PBS. Store at 4°C only for 1 or 2 d.
2. Monoclonal or polyclonal primary antibody (against antigen of interest). The appropriate amount of primary antibody should be determined empirically in pilot experiments. The recommended test dilutions are: polyclonal antibodies, 1:100 to 1:5000; supernatants from cultured hybridoma cells, undiluted to 1:100; ascitic fluid from mice bearing hybrid myelomas, 1:1000 to 1:10,000.
3. Washing buffer: 0.2% Tween in PBS.
4. Secondary antibody against primary antibody linked to either horseradish peroxidase for chromogenic assays or alkaline phosphatase for chemiluminescent assays. Used at the dilution recommended by the supplier (usually 1:1000).
5. Heat-sealable plastic bags (Sears Seal-A-Meal or equivalent) and heating bag sealer.

### *2.8. Detection*

Some researchers use $^{125}$I-labeled Protein A and autoradiography to detect the bound primary antibody. Although this method is sensitive, there are methods as sensitive that avoid the hazards of radioactivity. These systems rely on the use of second antibodies that recognize common features of primary antibody and also carry a reporter enzyme. The reporter enzyme, usually alkaline phosphatase (AP) or horseradish peroxidase (HRP), catalyze reactions that generate either a colored precipitate or chemiluminescence.

Enhanced chemiluminescent (ECL) detection of proteins on membrane was first introduced by Amersham a few years ago and has revolutionized Western blotting. Since then it has become the method of choice for the detection of proteins using directly or indirectly labeled HRP antibody conjugates. Now several suppliers offer very good chemiluminescent detection systems with new improved sensitivity that allow detection and quantification of very low protein signals: Amersham "ECL" and "ECL Plus" Western blotting detection reagents, Biolabs "Phototope-HRP Western blot detection kit," and Pierce "SuperSignal" and "SuperSignal Ultra" chemiluminescent substrate for Western blotting.

Equipment needed are an X-ray film cassette, a roll of SaranWrap® (or similar), a timer, and autoradiography film recommended for chemiluminescence detection (e.g., Hyperfilm ECL, Amersham).

## 3. Methods

### *3.1. Extraction and Quantification of Protein Samples*

#### *3.1.1. Lysis of Cultured Mammalian Cells*

Cells should be harvested as quickly as possible and all reagents kept in ice.

1. Wash cells twice with cold PBS.
2. Using 2 mL of PBS for a 35-mm dish and 4 mL of PBS for a 100-mm dish, harvest cells into a 5-mL round-bottom tube by scraping the plate with a rubber policeman. Cells growing in suspension should be concentrated by centrifugation (1000*g* for 5 min) and then transferred to a 5-mL round-bottom tube before washing twice with PBS.
3. Pellet the cells by centrifuging at 3000*g* for 10 min at 4°C. Tip off the PBS overlying the pellet and aspirate the last drops carefully.
4. Add 400—500 µL of lysis buffer and vortex-mix to homogenize (*see* **Notes 1–4**).
5. Shear the chromosomal DNA by sonication or homogenization (Ika, Ultra-turrax T8 disperser) for 30 s to 2 min depending on the power output of the apparatus. This should be sufficient to reduce the viscous lysate to manageable levels. The shearing can also be achieved by passing the sample three or four times through a 23-gauge hypodermic needle.
6. Samples can be stored at –20°C for a few months or –70°C for longer (*see* **Note 5**).

### 3.1.2. Protein Quantification

Several methods can be employed to determine the concentration of proteins, the method of choice depending upon the constituents of the storage buffers and the purity of the protein.

1. UV detection is the quickest of all methods for quantifying protein solutions. The absorbance at 280 nm is due primarily to the presence of tyrosine and tryptophan.

   This method is used for pure protein solutions in buffers that do not interfere with this wavelength.

   Take an aliquot of the protein sample and the standards, and read the absorbance at 280 nm against the blank (lysis buffer).

   For antibodies and BSA, use the table below to calculate the concentration. A very rough approximation for other proteins is 1 absorbance unit is equal to 1 mg/mL.

   | Protein | $A_{280}$ (for 1 mg/mL) |
   |---|---|
   | IgG | 1.35 |
   | IgM | 1.2 |
   | BSA | 0.7 |

   If the protein solution is contaminated with nucleic acids, read the absorbance vs a suitable control at 280 nm and 260 nm, and calculate the approximate concentration using the equation:

   $$\text{Protein concentration (in mg/mL)} = (1.55 \times A_{280})-(0.76 \times A_{260}).$$

2. Bradford assay: Dilute the concentrated dye binding solution 1:5 with distilled water. Filter if any precipitate develops.
   Add 5 mL of diluted dye binding solution to each protein sample, standard, and blank (lysis buffer). Allow the color to develop for at least 5 min but not longer than 30 min. The red dye will turn blue as it binds protein. Read absorbance at 595 nm. The standard curve will be linear between about 20 and 150 µg/mL.
3. Several colorimetric assays are available commercially (e.g., Pierce, bicinchoninic acid [BCA] protein reagent and Bio-Rad, DC protein assay). Both assay conditions are based on a modification of the Lowry method *(6,7)*. They offer the advantage of being faster (15–30 min) and easier to perform than the Lowry method, compatible with most ionic and nonionic detergents, linear for working range from 20 to 2000 µg/mL, and adaptable to microtiter plates.

## 3.2. Resolution of the Protein Sample in SDS-PAGE

1. Wash the glass plates and wipe with alcohol prior to use.
2. Make a sandwich of one large and one small glass plate separated by a spacer at both sides and lock into the casting stand with the smaller plate foremost. Ensure that the spacers and plates are all level at the bottom (*see* **Note 6**).
3. Either pouring directly from a universal or using a Pasteur pipet, pour some water between the plates of the sandwich. This will identify any leaks before pouring the gel. If the system is leak free pour out the water and begin preparing the gel.

4. The gel is poured in two layers. A separating gel first, which fills all but the top 1 cm of the sandwich, followed by a stacking gel (*see* **Note 7**). The stacking gel is made the same for all proteins but different final percentages of acrylamide are used to separate proteins of different molecular weights in the separating gel (*see* **Table 2**). As an approximate guide use 5% gels for 60–200 MW proteins (high-range molecular weight markers), 10% gels for 16–70 MW proteins (mid-range molecular weight markers), and 15% gels for 12–45 MW proteins (low-range molecular weight markers) (*see* **Note 8**).
5. Add the 10% APS and the TEMED immediately prior to pouring the gel, as it will polymerize rapidly (*see* **Note 9**). Leave a small amount of gel in the universal after pouring, this can be used as a guide to when the main gel has set.
6. Pour a thin (2–5 mm) layer of butanol on top of the separating gel. This ensures a clean interface between separating and stacking gels.
7. Once the separating gel has set, usually within 20 min at room temperature, pour off the butanol and now prepare and pour the stacking gel. Again pour immediately after adding 10% APS and TEMED. Pour the stacking gel to the top of the sandwich and insert the Teflon comb immediately. For both gels be careful to avoid air bubbles. If any appear they can usually be teased to the edge of the gel and burst with a hypodermic needle.

   This gel will polymerize within 40 min, again this can be checked by comparing with the small residual amount of gel left in the universal.
8. When comparing samples, aim to load a constant quantity and volume of protein in all wells. This will vary depending on total supply of protein but 50–100 µg is a reasonable quantity (*see* **Note 10**). Dilute the protein samples in lysis buffer to the required concentration, in order to load as small a volume as possible, and then mix with the 6X SDS sample buffer to obtain a 1X final solution.
9. Denature the proteins in the sample by heating to 100°C for 10 min, then leave at room temperature while loading.
10. While heating the samples, remove the gel sandwich from the casting frame and attach to the electrophoresis apparatus.
11. Fill the tank and electrophoresis apparatus with Tris–glycine electrophoresis buffer.
12. Carefully remove the Teflon comb from the sandwich, and using a 1-mL Gilson pipet wash all wells with some of the electrophoresis buffer. This removes any unpolymerized gel that could interfere with loading.
13. Load one well at either edge with the prestained molecular weight markers. Then continue with samples. Record position of samples in relation to the marker.
14. Attach the electrophoresis apparatus to an electric power supply (the positive electrode should be connected to the bottom buffer tank). Apply a voltage of 8 V/cm height of the gel. After the dye front has moved into the resolving gel, increase the voltage to 15 V/cm height and run the gel until the bromophenol blue reaches the bottom of the resolving gel (about 1 or 2 h for a small gel). Then turn off the power supply.
15. Remove the sandwich plates from the electrophoresis apparatus and place them on a paper towel. To disassemble, slide one spacer almost completely out so that only one corner remains between the glass plates. Rotating the spacer will now separate the plates.

**Table 2**
**Solutions for Preparing, Resolving, and Stacking Gels for SDS-PAGE**

| | Component volumes (mL) per gel mold volume of | | | | | | | |
|---|---|---|---|---|---|---|---|---|
| Solution components | 5 mL | 10 mL | 15 mL | 20 mL | 25 mL | 30 mL | 40 mL | 50 m |
| 6% | | | | | | | | |
| $H_2O$ | 2.6 | 5.3 | 7.9 | 10.6 | 13.2 | 15.9 | 21.2 | 26.5 |
| 30% Acrylamide mix | 1.0 | 2.0 | 3.0 | 4.0 | 5.0 | 6.0 | 8.0 | 10.0 |
| 1.5 *M* Tris, pH 8.8 | 1.3 | 2.5 | 3.8 | 5.0 | 6.3 | 7.5 | 10.0 | 12.5 |
| 10% SDS | 0.05 | 0.1 | 0.15 | 0.2 | 0.25 | 0.3 | 0.4 | 0.5 |
| 10% APS | 0.05 | 0.1 | 0.15 | 0.2 | 0.25 | 0.3 | 0.4 | 0.5 |
| TEMED | 0.004 | 0.008 | 0.012 | 0.016 | 0.02 | 0.024 | 0.032 | 0.04 |
| 8% | | | | | | | | |
| $H_2O$ | 2.3 | 4.6 | 6.9 | 9.3 | 11.5 | 13.9 | 18.5 | 23.2 |
| 30% Acrylamide mix | 1.3 | 2.7 | 4.0 | 5.3 | 6.7 | 8.0 | 10.7 | 13.3 |
| 1.5 *M* Tris, pH 8.8 | 1.3 | 2.5 | 3.8 | 5.0 | 6.3 | 7.5 | 10.0 | 12.5 |
| 10% SDS | 0.05 | 0.1 | 0.15 | 0.2 | 0.25 | 0.3 | 0.4 | 0.5 |
| 10% APS | 0.05 | 0.1 | 0.15 | 0.2 | 0.25 | 0.3 | 0.4 | 0.5 |
| TEMED | 0.003 | 0.006 | 0.009 | 0.012 | 0.015 | 0.018 | 0.024 | 0.03 |
| 10% | | | | | | | | |
| $H_2O$ | 1.9 | 4.0 | 5.9 | 7.9 | 9.9 | 11.9 | 15.9 | 19.8 |
| 30% Acrylamide mix | 1.7 | 3.3 | 5.0 | 6.7 | 8.3 | 10.0 | 13.3 | 16.7 |
| 1.5 *M* Tris, pH 8.8 | 1.3 | 2.5 | 3.8 | 5.0 | 6.3 | 7.5 | 10.0 | 12.5 |
| 10% SDS | 0.05 | 0.1 | 0.15 | 0.2 | 0.25 | 0.3 | 0.4 | 0.5 |
| 10% APS | 0.05 | 0.1 | 0.15 | 0.2 | 0.25 | 0.3 | 0.4 | 0.5 |
| TEMED | 0.002 | 0.004 | 0.006 | 0.008 | 0.01 | 0.012 | 0.016 | 0.02 |
| 12% | | | | | | | | |
| $H_2O$ | 1.6 | 3.3 | 4.9 | 6.6 | 8.2 | 9.9 | 13.2 | 16.5 |
| 30% Acrylamide mix | 2.0 | 4.0 | 6.0 | 8.0 | 10.0 | 12.0 | 16.0 | 20.0 |
| 1.5 *M* Tris, pH 8.8 | 1.3 | 2.5 | 3.8 | 5.0 | 6.3 | 7.5 | 10.0 | 12.5 |
| 10% SDS | 0.05 | 0.1 | 0.15 | 0.2 | 0.25 | 0.3 | 0.4 | 0.5 |
| 10% APS | 0.05 | 0.1 | 0.15 | 0.2 | 0.25 | 0.3 | 0.4 | 0.5 |
| TEMED | 0.002 | 0.004 | 0.006 | 0.008 | 0.01 | 0.012 | 0.016 | 0.02 |
| 15% | | | | | | | | |
| $H_2O$ | 1.1 | 2.3 | 3.4 | 4.6 | 5.7 | 6.9 | 9.2 | 11.5 |
| 30% Acrylamide mix | 2.5 | 5.0 | 7.5 | 10.0 | 12.5 | 15.0 | 20.0 | 25.0 |
| 1.5 *M* Tris, pH 8.8 | 1.3 | 2.5 | 3.8 | 5.0 | 6.3 | 7.5 | 10.0 | 12.5 |
| 10% SDS | 0.05 | 0.1 | 0.15 | 0.2 | 0.25 | 0.3 | 0.4 | 0.5 |
| 10% APS | 0.05 | 0.1 | 0.15 | 0.2 | 0.25 | 0.3 | 0.4 | 0.5 |
| TEMED | 0.002 | 0.004 | 0.006 | 0.008 | 0.01 | 0.012 | 0.016 | 0.02 |

| )lution components<br>% Stacking gels | 1 mL | 2 mL | 3 mL | 4 mL | 5 mL | 6 mL | 8 mL | 10 mL |
|---|---|---|---|---|---|---|---|---|
| $H_2O$ | 0.56 | 1.1 | 1.7 | 2.2 | 2.8 | 3.4 | 4.5 | 5.6 |
| 30% Acrylamide mix | 0.17 | 0.33 | 0.5 | 0.67 | 0.83 | 1.0 | 1.3 | 1.7 |
| ).5 *M* Tris, pH 6.8 | 0.25 | 0.5 | 0.75 | 1.0 | 1.25 | 1.5 | 2.0 | 2.5 |
| 10% SDS | 0.01 | 0.02 | 0.03 | 0.04 | 0.05 | 0.06 | 0.08 | 0.1 |
| 10% APS | 0.01 | 0.02 | 0.03 | 0.04 | 0.05 | 0.06 | 0.08 | 0.1 |
| TEMED | 0.001 | 0.002 | 0.003 | 0.004 | 0.005 | 0.006 | 0.008 | 0.01 |

Modified from Sambrook et al., 1989 *(8)*.

16. The gel can now be stained with Coomassie brilliant blue or used for Western blotting.

### 3.3. Staining SDS-Polyacrylamide Gels with Coomassie Brilliant Blue

1. Immerse the gel in at least 5 vol of staining solution and place on a slowly rotating platform for a minimum of 4 h at room temperature.
2. Remove the stain and save it for future use. Destain the gel by soaking it in destaining solution on a slowly rocking platform for 4–8 h, changing the destaining solution three or four times.
3. The more thoroughly the gel is destained, the smaller the amount of protein that can be detected by staining with Coomassie brilliant blue. Destaining for 24 h usually allows as little as 0.1 µg of protein to be detected in a single band.
   An alternative quicker method of staining or destaining can be achieved by the following methods:
   - Staining and destaining at higher temperatures (45°C) (*see* **Note 11**).
   - Including a few grams of an anion-exchange resin or a piece of sponge in the destaining buffer. This absorbs the stain as it leaches from the gel.
   - Destaining electrophoretically in apparatuses that are available commercially.
4. After destaining, gels may be stored indefinitely in water, in a sealed plastic bag without any diminution in the intensity of staining.

### 3.4. Transfer of the Separated Polypeptides to a Membrane Support

Two types of electrophoresis apparatus are available for electroblotting.

#### 3.4.1. Semidry Electrophoretic Transfer (9)

Transfer of proteins from a gel to a membrane using a semidry method gives even and rapid transfer and does not require a large power source. It can also be adapted to handle stacks of gel-membrane sandwiches (up to six). In the original description, different buffers were described for the anode and cathode, but recently it has been demonstrated that a simple buffer system using diluted Tris–glycine SDS-polyacrylamide gel running buffer is just as effective and much easier.

1. Rinse electrode plates of the semidry apparatus with distilled water and wipe off any beads of liquid that adhere to them with Kimwipes® or other absorbent tissues.
2. Wearing gloves, cut six sheets of absorbent paper (Whatman 3MM or equivalent) and one sheet of membrane (PVDF or nitrocellulose) to the size of the gel. If the paper overlaps the edge of the gel, the current will short-circuit the transfer and bypass the gel, preventing efficient transfer (*see* **Notes 12–15**).
3. Soak the PVDF membrane for 15 s in 100% methanol, then immerse it with the absorbent paper in the transfer buffer for at least 5 min or, float the nitrocellulose membrane on the surface of deionized water, and allow it to wet from beneath by capillary action. Then, submerge the membrane in the water for at least 5 min to displace trapped air bubbles. Soak the six pieces of absorbent paper in transfer buffer for at least 30 s.
4. Wearing gloves, set up the transfer apparatus as follows:
   - Lay the bottom plate of the apparatus (the anode) flat on the bench, graphite side up.
   - Place on the electrode three sheets of absorbent paper that have been soaked in transfer buffer. Stack the sheets one on top of the other so that they are exactly aligned. Using a glass pipet as a roller, squeeze out any air bubbles.
   - Place the membrane on the stack of absorbent paper. Make sure the filter is exactly aligned and that no air bubbles are trapped between it and the absorbent paper.
   - Transfer briefly the SDS-polyacrylamide gel produced in **Subheading 3.2.** to a tray of deionized water, and then place it exactly on the top of the membrane.
   - Place the final three sheets of absorbent paper on the gel, again making sure that they are exactly aligned and that no air bubbles are trapped (*see* **Notes 16–19**).
5. Carefully place the upper electrode (the cathode) on the top of the stack, graphite side down. Connect the electrical leads (positive or red leads to the bottom graphite electrode) and commence transfer. Running time is 45 min to 1.5 h with a current of 0.8 mA/cm$^2$ of gel. Avoid extending the running time, as this can cause drying out (*see* **Note 20**).
6. After transfer, disconnect the power source. Carefully disassemble the apparatus. Mark the membrane to allow orientation (usually by snipping off the lower left-hand corner, the number one lane).
7. The polyacrylamide gel can now be stained with Coomassie brilliant blue (*see* **Subheading 3.3.**) to verify transfer. The prestained markers can serve as internal markers for transfer and molecular weight.
8. Let the blot air-dry (1 h at 37°C or 2 h at room temperature) to improve the protein binding and then process the membrane for staining with Ponceau S (to assess the quality of transfer) (*see* **Subheading 3.5.**) or blocking (*see* **Subheading 3.6.**) as appropriate.

### *3.4.2. Wet Electrophoretic Transfer* (2)

For submerged or wet transfers, no one set of transfer conditions offers complete and even transfer of proteins with good retention on the membrane. However, most of the common transfer techniques are adequate.

1. Wearing gloves, cut four sheets of absorbent paper (Whatman 3MM or equivalent) and one sheet of membrane (PVDF or nitrocellulose) to the size of the gel (*see* **Notes 12–15**).
2. Soak the PVDF membrane for 15 s in 100% methanol, then immerse it with the absorbent paper in the transfer buffer for at least 5 min.
   Or, float the nitrocellulose membrane on the surface of deionized water, and allow it to wet from beneath by capillary action. Then, submerge the membrane in the water for 2 min. Move the membrane to soak in transfer buffer for 5 min. Wet the absorbent paper by soaking in transfer buffer.
3. Immerse the fiber pads in transfer buffer to ensure they are thoroughly soaked. Be careful to exclude air bubbles.
4. Assemble the transfer sandwich. One half of the gel holder cassette is usually black (this will be on the cathode side). Building the sandwich in the following order, with the black plate on the bottom, will ensure that the negatively charged proteins will migrate from the gel toward the anode and thus onto the transfer membrane. Keep all the components wet and make sure the sandwich is tightly assembled.
   - Place a fiber pad on the bottom of the cassette holder.
   - Place two sheets of absorbent paper that have been soaked in transfer buffer. Stack the sheets one on top of the other so that they are exactly aligned. Using a glass pipet as a roller, squeeze out any air bubbles.
   - Place the SDS-polyacrylamide gel produced in **Subheading 3.2.**, on top of the absorbent paper. Then put the membrane on top of the gel. Make sure the filter paper is exactly aligned and that no air bubbles are trapped between layers.
   - Place two sheets of absorbent paper on the top of the stack. Then put a fiber pad on top of the absorbent paper (*see* **Notes 16–19**).
5. Lock the sandwich in the cassette holder. Then place it in the tank blotting apparatus so that the black side of the cassette holder with the gel is facing the cathode (–). Add enough transfer buffer to the blotting apparatus to cover the cassette holder.
6. Transfer for 4–18 h. Thicker gels and higher molecular weight proteins require longer transfers. The following conditions have been optimized to permit transfer and retention of a very wide range of molecular weight proteins.

   For proteins >100,000 MW on a 15 × 15 × 0.1 cm gel, transfer at 28 V for 1 h, then at 84 V for 14–16 h.

   For proteins <100,000 MW on a 15 × 15 × 0.1 cm gel, transfer at 63 V for 4–16 h.

   In all transfers, the temperature will rise substantially during the run, and it is essential to use a cooling coil or to run the transfer in a cold room to avoid the generation of gas bubbles in the sandwich (*see* **Note 20**).
7. After transfer, disconnect the power source. Remove the cassette holder from the blotting apparatus. Open the cassette holder and remove the fiber pad and filter paper with forceps. Mark the membrane to allow orientation (usually by snipping off the lower left-hand corner, the number one lane).
8. The polyacrylamide gel can now be stained with Coomassie brilliant blue (*see* **Subheading 3.3.**) to verify transfer. The prestained markers can serve as internal markers for transfer and molecular weight.

9. Let the blot air-dry (1 h at 37°C or 2 h at room temperature) to improve the protein binding and then process the membrane for staining with Ponceau S (to assess the quality of transfer) (*see* **Subheading 3.5.**) or blocking (*see* **Subheading 3.6.**) as appropriate.

### 3.5. Staining the Blot with Ponceau S

Ponceau S is applied in an acidic aqueous solution. Staining is rapid but not permanent; the red stain will wash away in subsequent processing. Because the binding is reversible, the stain is compatible with most antigen visualization techniques. Therefore, Ponceau S can be used routinely to verify gel loading and transfer, and to locate molecular weight markers (if not using prestained marker).

1. If the membrane has been dried, float it on the surface of a tray of deionized water and allow it to wet from beneath by capillary action. Then, submerge the filter in the water for at least 5 min to displace trapped air bubbles.
2. Transfer the filter to a tray containing a solution of Ponceau S stain. Incubate the filter for 1–5 min with gentle agitation.
3. When the bands are visible, wash the membrane in several changes of deionized water at room temperature. Process until the color of the membrane background becomes almost white.
4. Mark the positions of proteins used as molecular weight standards with waterproof black ink if not using a prestained marker.
5. Then process the membrane for blocking (*see* **Notes 21** and **22**).

### 3.6. Blocking Nonspecific Binding Sites on the Membrane

Place membrane in a shallow tray that is slightly larger than the membrane or in a sealed plastic bag with 50–100 mL of blocking buffer. Gently agitate on a platform shaker for 2–3 h at room temperature (*see* **Notes 23–28**).

### 3.7. Addition of the Antibodies (see Notes 29–35)

1. Dilute the primary antibody in blocking buffer according to manufacturer's guidelines or as suggested in **Subheading 2.7., item 2.** Incubate the membrane with this primary antibody solution for 1–2 h at room temperature or overnight in a cold room, with continuous agitation on a platform shaker (*see* **Note 36**). This stage is best performed in a heat-sealed plastic bag or even, if antibody is scarce by putting 1 mL of diluted antibody in the center of a small glass plate and laying the membrane onto this. In the last method, ensure that the side of the membrane that was in contact with the gel is now facing the antibody. Cover the glass plate with a sheet of parafilm slightly bigger than the membrane to prevent drying.
2. Cut open the plastic bag or remove the membrane from the glass plate, and discard the blocking solution and antibody. Place the membrane in a large tray and wash it 4× (10 min each time) with 200 mL of washing buffer on a platform shaker.
3. Immediately incubate the membrane with the secondary immunological reagent. Dilute the secondary antibody according to manufacturer's guidelines in blocking buffer and repeat the process described in **step 1** for the primary antibody.

4. Cut open the plastic bag or remove the membrane from the glass plate, and discard the blocking solution and antibody. Place the membrane in a large tray and wash it once for 15 min and then 4× (5 min each time) with 200 mL of washing buffer, on a platform shaker.
5. A final wash should be performed in PBS without Tween, as this may interfere with certain chromogenic assays.

### 3.8. Detection

1. Remove membrane from the final wash and perform the enhanced chemiluminescent detection according to the manufacturer's intructions. This protocol will generally involve:
   - Mixing of at least two different substrates and short incubation with the membrane under gentle incubation.
   - Draining the membrane of excess detection reagent, wrapping it in SaranWrap, and gently smoothing out air pockets.
   - Placing the membrane, protein face up, in the film cassette.
   - Switching off the lights of a dark room and carefully placing a sheet of autoradiography film on top of the membrane, closing the cassette and exposing for 15 s.
   - Removing the film and developing it. On the basis of its appearance, estimate how long to continue the exposure of a second piece of film. Second exposure can vary from 1 min to 1 h (*see* **Notes 37–41**).
2. The membrane can be stripped and reprobed several times to either clarify or confirm results or when small or valuable samples are being analyzed ***(10)***. Sequential reprobing of membranes with a variety of antibodies is possible following the steps in the following subheading. To strip the membrane wash initially in distilled water for 5–10 min. Wash in 0.2 *M* sodium hydroxide for 5 min, then wash again in distilled water. Process to the blocking of the membrane as previously.

## 4. Notes

1. The cells are suspended in lysis buffer to prevent the formation of an insoluble mass when the SDS sample buffer is added. This step should be carried out as quickly as possible, using ice-cold lysis buffer, to minimize proteolytic degradation. Most types of mammalian tissue can be rapidly teased apart with forceps or cut into small pieces with scissors or scalpel beneath the surface of the lysis buffer.
2. Mammalian tissues can be equally well dispersed mechanically and then dissolved directly in SDS-gel-loading buffer. Mammalian cells in tissue culture may be lysed gently with detergents as described in **Subheading 3.1.1.** or alternatively lysed directly in SDS sample buffer, if the target antigen is resistant to this type of extraction.
3. Aprotinin is a basic polypeptide of 58 amino acids that aggregates if repeatedly frozen and thawed. The stock solution should be stored in small aliquots at –20°C. Each aliquot should be discarded after use.
4. PMSF is inactivated in aqueous solutions. The rate of inactivation increases with pH and is faster at 25°C than at 4°C. The half-life of a 20 μ*M* aqueous solution of

PMSF is about 35 min at pH 8.0. This means that solutions of PMSF can be safely discarded after they have been rendered alkaline (pH >8.6) and stored for several hours at room temperature.

5. After thawing, samples that have been stored at –20°C should be centrifuged at 12,000*g* for 5 min at 0°C in a microfuge. This removes aggregates of cytoskeletal elements.
6. Prepare numerous gels. Wrap unused gels in a moistened paper towel and plastic wrap. Store at 4°C for up to 1 wk.
7. Stacking gel length should be 1 cm from well bottom to top of separating gel.
8. Band resolution can be improved by doubling the salt concentration in stacking and separating gels, but the gel must be run at lower voltages.
9. Use fresh ammonium persulfate and high-quality acrylamide. Prepare clean plates prior to mixing acrylamide and work fast.
10. The lowest amount of an average-sized protein that can be detected by Western blotting is approx 1–5 ng. Approximately 100 μg of total cellular protein can be applied to a lane in a 0.75-mm-thick SDS-polyacrylamide gel without overloading.
11. Microwave methods of staining and destaining gels may shorten time considerably, but vapors are harmful. Hot solvents may harm the oven, and the gel can fracture. Avoid excessive heating of gels.
12. It is important to wear gloves when handling the gel, 3MM papers, and membrane. Oils and secretions from the skin will prevent the transfer of proteins from the gel to the filter.
13. The PVDF membrane is hydrophobic and offers a uniformly controlled pore structure with high binding capacity for proteins. When compared to a nitrocellulose membrane, it has improved handling characteristics and staining capabilities, increased solvent resistance, and a higher signal-to-noise ratio for enhanced sensitivities.
14. Antigens may bind differently to different types of membranes. To maximize detection sensitivity, test different types of membranes with the particular antigen–antibody combination to be used.
15. Occasionally lot-to-lot variability may be seen with some membranes.
16. After the transfer step, membranes can be wrapped in plastic and stored at 4°C for up to 3 mo. But once subsequent detection procedures are started, do not let the membrane dry.
17. 0.1% w/v SDS in the transfer buffer may improve transfer efficiency of large proteins (>100 kDa MW).
18. Methanol improves binding of smaller proteins to membrane.
19. Low transfer pH (<8) or high amounts (>0.1% w/v) SDS in the gel or buffer can lead to inefficient binding of the antigen.
20. If the Western blot transfer efficiency is low, try longer transfer times or higher transfer voltages.
21. Ponceau S staining could be used as well on a blocked and immunologically probed membrane, allowing a useful rescue or verification of old blots.
22. Because the pink-purple color of Ponceau S is difficult to capture photographically, the stain does not provide a permanent record of the experiment.

23. Membranes may be left in the blocking solution overnight in a cold room if more convenient.
24. Elimination of Tween-20 from buffers can lead to increased background.
25. Greater than 0.05% Tween-20 in buffers may elute proteins from PVDF membranes.
26. Some blocking proteins may reduce antigen recognition by certain antibodies. If this is suspected, try a different blocking protein (BSA, gelatin or casein, use at 1% w/v in PBS–Tween).
27. Check blocking agent for residual alkaline phosphatase activity because it could lead to high background when using an alkaline phosphatase-based detection system.
28. Drying the membrane is claimed to improve the retention of proteins on the filter and to reduce the nonspecific binding of antibodies on the membrane during subsequent processing. However, it may also result in further denaturation and consequent alteration in immunoreactivity. Drying may therefore be advantageous for some protein–antibody combinations and disadvantageous for others. This can be established empirically for the target proteins of interest.
29. Some antibodies (particularly monoclonal) recognize epitopes that may be buried or denatured when the antigen is bound to surfaces such as nitrocellulose. This effect may be enhanced when blotting protein antigens from SDS-containing gels, possibly eliminating antibody recognition.
30. When using a low-titer antibody, increase signal by eliminating Tween-20 in buffers, increasing the incubation time, or increasing the concentration of the primary antibody solution.
31. Backgrounds will usually increase when using conjugated second antibodies at dilutions lower than 1:2500.
32. A few IgGs and other immunoglobulins (e.g., IgM) will tend to stick nonspecifically to membranes, resulting in high background.
33. Localized background can result from primary antibody recognition of epitopes shared by other protein species in the sample or a primary antiserum containing a mixture of antibodies with multiple specificity. The latter problem is sometimes overcome by preadsorbing the antisera to remove the crossreacting antibodies.
34. Improperly stored antibodies will lose activity over time, making detection results inconsistent. Primary antisera should be stored in aliquots at –20°C to avoid freeze–thaw cycles and prevent contamination. Secondary antibodies should be stored at 4°C or in aliquots at –20°C. 0.02% w/v sodium azide or 0.05% w/v thimerosal can be added as a preservative. Azide will inhibit HRP activity.
35. As a general rule, as large a volume as possible of washing buffer should be used each time.
36. Incubation times and temperatures will vary and should be optimized for each antibody. The conditions indicated are recommended starting points.
37. The volume of the detection solution should be sufficient to cover the membranes. The final volume required is about 0.125 mL/cm$^2$ membrane.
38. Drain off excess detection reagent by holding the membrane vertically and touching the edge of the membrane against a tissue paper. Gently place the membrane, protein side down, on the SaranWrap. Close the SaranWrap to form an envelope, avoiding pressure on the membrane.

39. Ensure that there is no free detection reagent in the film cassette; the film must not get wet.
40. Do not move the film while it is being exposed.
41. If background is high the membrane may be washed twice for 10 min with wash buffer and redetected following the same method with slight loss of sensitivity. If overexposure occurs because of high light emission resulting from high target antigen concentration, leave blots in the cassette for 5–10 min before reexposing to film.

## References

1. Laemmli, E. K. (1970) Cleavage of structural proteins during the assembly of the head of bacteriophage T4. *Nature* **227,** 680–685.
2. Towbin, H., Staehelin, T., and Gordon, J. (1979) Electrophoretic transfer of proteins from polyacrylamide gels to nitrocellulose sheets: Procedure and some applications. *Proc. Natl. Acad. Sci. USA* **76,** 4350–4354.
3. Johnson, D. A., Gautsch, J. W., Sportsman, J. R., and Elder, J. H. (1984) Improved technique utilizing non-fat dry milk for analysis of proteins and nucleic acids transferred to nitrocellulose. *Gene Analyt. Techn.* **1,** 3–8.
4. Batteiger, B., Newhall, W. J., and Jones, R. B. (1982) The use of Tween 20 as a blocking agent in the immunological detection of proteins transferred to nitrocellulose membranes. *J. Immunol. Methods* **55,** 297–307.
5. Bradford, M. M. (1976) A rapid and sensitive method for the quantitation of microgram quantities of protein utilizing the principle of protein-dye binding. *Analyt. Biochem.* **72,** 248–254.
6. Lowry, O. H., Rosebrough, N. J., Farr, A. L., and Randall, R. J. (1951) Protein measurement with Folin phenol reagent. *J. Biol. Chem.* **193,** 265–275.
7. Peterson, G. L. (1983) Determination of total protein. *Methods Enzymol.* **91,** 95–119.
8. Sambrook, J., Fritsch, E. F., and Maniatis, T. (1989) *Molecular Cloning: A Laboratory Manual*, 2nd ed. Cold Spring Harbor Laboratory Press, Cold Spring Harbor, NY.
9. Kyhse-Andersen, J. (1984) Electroblotting of multiple gels: A simple apparatus without buffer tank for rapid transfer of proteins from polyacrylamide to nitrocellulose. *J. Biochem. Biophys. Methods* **10,** 203–209.
10. Kaufmann, S. H., Ewing, C. M., and Shaper, J. H. (1987) The erasable Western blot. *Analyt. Biochem.* **161,** 89–95.

# 12

# Assessment of Gelatinases (MMP-2 and MMP-9) by Gelatin Zymography

**Marta Toth and Rafael Fridman**

## 1. Introduction

The invasion and metastasis of tumor cells has been shown to require proteolytic activity in order to degrade components of the extracellular matrix (ECM). The hydrolysis of the ECM appears to facilitate tumor cell migration contributing to the metastatic dissemination of malignant cells *(1)*. A major group of proteases that has been directly associated with tumor metastasis is the matrix metalloproteinases (MMPs), a family of endopeptidases known to cleave many ECM proteins *(1)*. The MMPs are multidomain proteases that contain a zinc atom in the active site and are produced in a latent inactive form (zymogen) *(2)*. Acquisition of enzymatic activity requires cleavage of the inhibitory N-terminal domain *(3)*. Thus, generation of the active form usually occurs concomitantly with a decrease in molecular mass and exposure of the active site. Once activated, all the MMPs are specifically inhibited by a group of endogenous protease inhibitors known as the tissue inhibitors of metalloproteinases (TIMPs), which bind to the active site inhibiting catalytic activity *(4)*.

The gelatinases MMP-2 (gelatinase A, EC 3.4.24.24) *(5)* and MMP-9 (gelatinase B, EC 3.4.24.35) *(6)* are two members of the MMP family that have been extensively studied owing to their consistent association with tumor invasion and metastasis. It has been reported that the expression and activity of gelatinases are elevated in many malignant human tumors and correlate with tumor progression *(7,8)*. MMP-2 and MMP-9 were originally described as type IV collagenases because of their ability to promote the hydrolysis of collagen IV *(5,6)*, a major component of basement membranes and a major structural barrier for tumor cell invasion. Both enzymes can also cleave a variety of ECM

From: *Methods in Molecular Medicine, vol. 57:*
*Metastasis Research Protocols, Vol. 1: Analysis of Cells and Tissues*
Edited by: S. A. Brooks and U. Schumacher © Humana Press Inc., Totowa, NJ

proteins but they are extremely efficient in hydrolyzing denatured collagen I (gelatin) and consequently they are referred to as gelatinases *(9)*. The ability of MMP-2 and MMP-9 to degrade denatured collagen I was developed into a relatively easy yet powerful technique to detect their presence in biological samples. This technique, known as gelatin zymography, identifies gelatinolytic activity in biological samples using sodium dodecyl sulfate (SDS)-polyacrylamide gels impregnated (copolymerized) with gelatin ***(10,11)***. This is basically an adaptation of the methodology for SDS-polyacrylamide gel electrophoresis (PAGE) described in Chapter 11 by Blancher and Jones. To maintain enzymatic activity, the samples are electrophoresed under nonreducing conditions. After removal of the SDS from the gel by Triton X-100 and incubation in a calcium-containing buffer, the partially renatured enzymes can degrade the gelatin leaving a cleared zone that can be detected after staining of the gel. In the presence of SDS the enzymes are denatured exposing their active site, which permits both the latent and active forms of the gelatinases to exhibit gelatinolytic activity after the partial renaturation. Furthermore, SDS disrupts the noncovalent interactions between gelatinases and TIMPs, allowing for the detection of gelatinase activity independently of the presence of TIMPs. Detection of gelatinase activity in latent enzymes and/or in enzymes derived from samples rich in TIMPs is a major paradox of gelatin zymography. Under physiological conditions, latent gelatinases do not possess enzymatic activity *(3,9)*. Likewise, in the presence of TIMPs, gelatinase activity is usually inhibited *(4)*. Thus, gelatin zymography cannot provide definite information on the net MMP-2- and MMP-9-dependent proteolytic activity present in a given sample because it does not take into account endogenous inhibitors. Gelatin zymography, however, is a useful qualitative tool for the detection and analysis of the level and type of the gelatinases expressed in different cell types/tissues at any given time and/or after different treatments. For example, it is possible to determine which gelatinases are expressed in tumor cells with various degrees of invasive potential whether they are derived from established cancer cell lines or from tumor biopsies. Furthermore, the regulation of gelatinase expression in vitro in response to a variety of factors can be studied. It should be pointed out, however, that owing to the enzymatic nature of the method and the many variables involved, zymography is too crude to be used as a quantitative technique (discussed in **Note 1**).

Another important feature of the gelatin zymography technique is the ability to assess the activation status of the enzymes. Because the active forms of the gelatinases do not normally possess the inhibitory N-terminal domain (~10 kDa), the molecular mass of the species detected can be used, in some cases, as an indicator of the degree of zymogen activation in the sample. Therefore, it is possible to correlate the presence of active forms with the invasive characteristics of a given cell/tumor type. It is also possible to study the factors that play a role in gelatinase activation. However, as will be discussed in **Note 2**, discrimination between active and latent gelatinase species using gelatin zymography is prone to serious pitfalls. Nevertheless, owing to the simplicity and sensitivity of the gelatin zymography it is the method of choice to assess expression of gelatinases in tumor cells. Here we provide a detailed protocol on how to set up the gelatin zymography technique and how to use it in samples relevant for the study of tumor cell invasion and metastasis. We have also provided hints for the interpretation of results.

## 2. Materials

1. Disposable plastic gel casting cassettes, 1.0 mm thick, 10 × 10 cm.
2. Acrylamide–*bis*-acrylamide stock: Prepare 200 mL of 30% w/v acrylamide, 0.8% w/v *bis*-acrylamide solution in distilled water ($dH_2O$) (*see* **Note 3**). Use a surgical mask when weighing acrylamide powder. Store the acrylamide solution in a dark bottle at 4°C where it is stable for at least 6 mo. Unpolymerized acrylamide is a neurotoxin; **always handle with gloves**!
3. Separating gel buffer stock: Prepare 200 mL of 1.88 *M* Tris-HCl buffer, pH 8.8. Autoclave and store it at room temperature. It is stable for at least 6 mo.
4. Stacking gel buffer stock: Prepare 200 mL of 1.25 *M* Tris-HCl buffer, pH 6.8. Autoclave and store it at room temperature. It is stable for at least 6 mo.
5. 1% w/v Gelatin (Sigma, St. Louis, MO, cat. no. G-8150): Dissolve the gelatin in 5 mL of $dH_2O$ and heat the solution at 60°C in a water bath for at least 20 min; mix well. Make sure that the gelatin is completely dissolved. Cool down the gelatin solution to room temperature before use. Prepare it fresh.
6. 20% w/v SDS: Prepare 200 mL of 20% SDS in $dH_2O$. Use a surgical mask when weighing SDS. Store it at room temperature. It is stable for 1 yr.
7. 10% w/v Ammonium persulfate (APS): Prepare 1–5 mL of 10% APS in $dH_2O$, depending on the number of gels to be prepared. Store it at 4°C for no longer than 2 wk.
8. *N*, *N*, *N*', *N*'-tetramethylethylenediamine (TEMED): Store it in a dark bottle at 4°C.
9. Running buffer stock (10X): Prepare 1 L of 0.25 *M* Tris base and 1.92 *M* glycine, pH 8.3. The pH should be correct without adjusting. Store it at room temperature. It is stable for months.
10. Running buffer: Dilute the running buffer stock 10× with $dH_2O$ to make 1 L and supplement with 5 mL of 20% w/v SDS to a final concentration of 0.1% w/v. Store it at room temperature. It is stable for months.
11. Sample buffer (4X): Prepare 10 mL of 250 m*M* Tris-HCl, pH 6.8, 40% v/v glycerol, 8% w/v SDS; and 0.01% w/v bromophenol blue. Store it at –20°C in 0.5-mL aliquots. Before use, warm it up to dissolve the SDS.
12. Renaturing solution stock (10X): Prepare 200 mL of 25% v/v Triton X-100 in $dH_2O$. Store it at room temperature. It is stable for months.
13. Developing buffer stock (10X): Prepare 1 L of 500 m*M* Tris-HCl, pH 7.8, 2 *M* NaCl, 50 m*M* $CaCl_2$, and 0.2% v/v Brij 35. Store it at 4°C for 6 mo.

14. Staining solution: Prepare 1 L of 0.5% w/v Coomassie blue R-250, 5% v/v methanol, and 10% v/v acetic acid in $dH_2O$. Filter. Store it at room temperature. This solution is reusable.
15. Destaining solution: Prepare 1 L of 10% v/v methanol, 5% v/v acetic acid in $dH_2O$. Store it at room temperature for months.
16. Phosphate-buffered saline (PBS): Prepare 1 L of 10 m*M* phosphate buffer, pH 7.1; 137 m*M* NaCl; and 2.7 m*M* KCl. Autoclave and store it at room temperature. It is stable for months.
17. Lysis buffer: Prepare 100 mL of 25 m*M* Tris-HCl, pH 7.5; 100 m*M* NaCl; and 1% v/v Nonidet P-40 (NP-40). Store it at 4°C for 6 mo. Right before use add protease inhibitors: 10 μg/mL aprotinin, 2 μg/mL leupeptin, and 4 m*M* benzamidine.
18. Tris-buffered saline (TBS): Prepare 1 L of 50 m*M* Tris-HCl, pH 7.5; 150 m*M* NaCl. Store it at 4°C for months.
19. TBS-CM: Prepare 200 mL of TBS supplemented with 1 m*M* $CaCl_2$ and 1 m*M* $MgCl_2$. Store it at 4°C and use within a week.
20. TBS-B: Prepare 50 mL of TBS containing 5 m*M* $CaCl_2$ and 0.02% v/v Brij-35. Store it at 4°C for 1 mo or until a visible precipitate appears.
21. TBS-CM-Triton X-114: Prepare 50 mL of 1.5% v/v Triton X-114 in TBS-CM. Store it at 4°C. It is stable for months. Right before use, add protease inhibitors: 10 μg/mL aprotinin, 2 μg/mL leupeptin, and 4 m*M* benzamidine.
22. Gelatin-agarose beads (Sigma cat. no. G-5384): Wash the beads (the amount needed for the experiment) twice with TBS-B before use to remove preservative solution. Make a 50% suspension of gelatin–agarose beads in TBS-B.

## 3. Methods

### *3.1. Gel Preparation*

The following protocol is for the preparation of eight gels (10% polyacrylamide–0.1% w/v gelatin) (*see* **Notes 4** and **5**).

1. Separating gel: Mix 17.8 mL of $dH_2O$, 5 mL of 1% w/v gelatin, 16.6 mL of acrylamide–*bis*-acrylamide stock, 10 mL of separating gel buffer stock, 0.25 mL of 20% w/v SDS; and 150 μL of 10% w/v APS in a 125-mL filtration flask at room temperature. Degas the solution for approx 5 min.
2. Stacking gel: Mix 11 mL of $dH_2O$, 2 mL of acrylamide–*bis*-acrylamide stock, 2 mL of stacking gel buffer stock, 0.1 mL of 20% w/v SDS, and 75 μL of 10% w/v APS in a 125-mL filtration flask at room temperature. Degas the solution for approx 5 min.
3. Add 30 μL of TEMED to the separating gel solution to initiate polymerization. Swirl the solution rapidly without causing bubble formation or aeration.
4. Immediately, pipet 6.2 mL of separating gel solution into each cassette avoiding the formation of bubbles.
5. Carefully overlay the separating gel solution with $dH_2O$ up to the top of the cassette using a syringe. Do not disturb the surface of the separating gel solution.
6. Let the gel polymerize for at least 1 h at room temperature. Polymerization is complete when a discrete line of separation can be noted between the gel and the water overlay.

7. Decant the overlay water from the separating gel.
8. Immediately, add 10 µL of TEMED to the stacking gel solution, swirl rapidly, and pipet the solution on top of the polymerized separating gels until it reaches the top of the front plate.
9. Rapidly, insert the appropriate combs (usually 10 wells) into the liquid stacking gel, making sure that no bubbles remain trapped under the comb. Let the stacking gel polymerize at room temperature (about 30–60 min).
10. Store the gels (comb on top) in a sealed plastic bag or container containing 1X running buffer to keep them moist. Store them at 4°C for up to 2–3 wk (*see* **Note 5**).

### 3.2. Sample Preparation

#### 3.2.1. Serum-Free Conditioned Media

The gelatinases are secreted enzymes *(9)*. Therefore, in cultured cells, a significant part of the gelatinase pool is found in the media. Because serum contains gelatinases, it is necessary to prepare serum-free conditioned media for gelatin zymography (*see* **Fig. 1**).

1. Grow the cells to be tested to approx 80% confluence in complete growth media.
2. Wash the cell monolayer with sterile PBS or serum-free media to remove the serum completely.
3. Incubate the cells with serum-free media at 37°C for at least 12–16 h (*see* **Note 6**).
4. Collect the media and centrifuge (400*g*, 5 min at 4°C) to remove cells and debris. Keep the supernatant.
5. Mix 75 µL of the clarified supernatant with 25 µL of 4X sample buffer; vortex-mix. Let the sample stand at room temperature for 10–15 min. Do not heat the sample! Load 30 µL per lane. If the intensity of the gelatinolytic bands is very low, the conditioned media can be concentrated using commercially available concentrators (10 kDa cutoff) or subjected to gelatin-agarose purification as follows:
6. In a microcentrifuge tube, mix 1 mL of clarified conditioned media with 30 µL of the gelatin–agarose bead suspension and rotate at 4°C for at least 1 h.
7. Centrifuge (16,000*g*) the tubes in a microcentrifuge (approx 1–2 min) and carefully aspirate the supernatant.
8. Wash the beads (containing the bound gelatinases), at least twice, with 1 mL of cold TBS-B.
9. After the last wash, carefully aspirate the supernatant. Add 20 µL of 1X sample buffer to the beads to elute the bound enzymes. Do not heat the samples! Spin the tubes and load the supernatants on the gel.

#### 3.2.2. Cell Lysates

In addition to being secreted into the media, the gelatinases are also cell associated, intracellularly and extracellularly. Intracellular enzymes represent gelatinases undergoing different stages of posttranslational processing whereas the extracellular gelatinases represent enzymes that are associated with the cell surface/matrix. Analysis of cell lysates therefore can provide valuable infor-

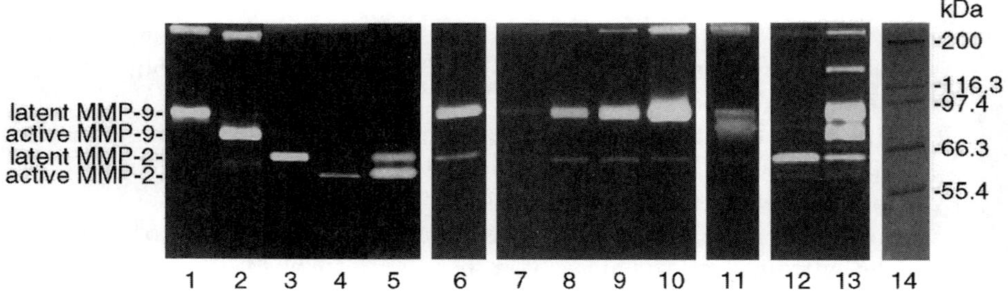


Fig. 1. Gelatin zymogram of biological samples from different sources. *Lanes 1–5*, purified recombinant MMP-9 (*lane 1*, latent form; *lane 2*, activated form) and MMP-2 (*lane 3*, latent form; *lane 4*, active form; *lane 5*, mixture of active and latent forms). Note the presence of the MMP-9 dimer form (~200 kDa). *Lane 6*, serum-free conditioned media of HT1080 fibrosarcoma cells. *Lanes 7–10*, serum-free conditioned media of nonmalignant breast epithelial MCF10A cells untreated (*lane 7*) or treated with increasing concentrations (10, 25, and 50 ng/mL) of tumor necrosis factor-α (TNF-α) (*lanes 8–10*). Note the obvious induction of MMP-9 in response to TNF-α compared to untreated cells. *Lane 11*, cell lysate of TNF-α treated MCF10A cells. Note the presence of the intracellular, lower molecular mass (~83–85 kDa), precursor (partially glycosylated) form of latent MMP-9. *Lanes 12* and *13*, tissue extracts (30 μg of protein) from the benign (*lane 12*) and carcinoma (*lane 13*) section of a breast biopsy. Note the induction of MMP-9 and a 130–140-kDa band of unknown origin in the carcinoma. Also note the appearance of a approx 80–85 kDa form in the carcinoma. Note that it is difficult to determine whether this is active MMP-9 or the intracellular precursor form. *Lane 14*, unreduced molecular weight standard (Novex).

mation on the rate of synthesis, cell surface–matrix association and activation status of surface-associated gelatinases. Cell lysates are prepared as follows:

1. Wash the cells twice with cold PBS.
2. Add cold lysis buffer. Use 2 mL of lysis buffer per 150-mm dish to obtain a final total protein concentration of approx 2–3 mg/mL.
3. Scrape the cells into the lysis buffer with a rubber policeman, collect the lysate, and incubate it on ice for at least 15 min.
4. Vortex-mix and centrifuge (16,000*g*) for 20 min at 4°C in a microcentrifuge. Collect the supernatant and measure protein concentration.
5. Mix 75 μL of the supernatant with 25 μL of 4X sample buffer. Do not heat the sample. Load up to 30 μL per lane. In the minigel system described here, do not load more than 40 μg of total protein per lane, as excessive amount of protein will disrupt band separation and resolution (*see* **Note 7**). If the level of gelatinases is low, phase partition with Triton X-114 can be used to concentrate the cell-associated gelatinases.

### 3.2.3. Phase Extraction with Triton X-114

Triton X-114 is a biphasic detergent that is soluble in aqueous buffers at 4°C but above 20°C samples containing Triton X-114 separate into two distinct phases (aqueous and detergent). In general, hydrophilic proteins allocate to the aqueous phase and hydrophobic proteins to the detergent phase. However, anomalous distributions of proteins have been reported. Being hydrophilic in nature, gelatinases are found mostly in the aqueous phase. However, gelatinases (most evident with MMP-2) have also been found in the detergent phase ***(12,13)***, suggesting a strong association with the plasma membranes. Phase partition is carried out as follows:

1. Wash the cells twice with cold PBS. During all the following steps, keep samples on ice unless otherwise stated.
2. Add 2 mL of cold TBS-CM-Triton X-114 solution containing protease inhibitors per 150-mm dish. This should yield approx 2–3 mg of total protein per milliliter.
3. Scrape the cells into the solution and transfer to a tube.
4. Incubate the extract on ice for at least 15 min and then centrifuge (16,000*g*, 20 min) at 4°C.
5. Collect the supernatant, transfer to a new tube and incubate it for 2 min in a 37°C water bath.
6. Centrifuge (16,000*g*) the sample at room temperature for approx 5 min to obtain the lower detergent and upper aqueous phases.
7. Carefully collect and transfer the aqueous phase into a new tube without disturbing the detergent phase. Cool down the tubes briefly. Keep the tube with the detergent phase on ice.
8. Add 30 μL of the gelatin-agarose beads to 500 μL of the aqueous phase and proceed as described in **Subheading 3.2.1., steps 6–9**. Load the samples onto the gel.
9. Add 2 vol of TBS-B and 1 vol of 4X sample buffer to the detergent phase (4X dilution) (*see* **Note 8**). Do not heat the sample. Load 30 μL per lane.

### 3.2.4. Preparation of Tissue Extracts

Fresh tissue biopsies derived from tumor samples are an important source for examining gelatinase expression during tumor progression **(Fig. 1)**.

1. Cut the tissue of interest (~50 mg) into small pieces. Remove any visible fat.
2. Add approx 500 μL of cold lysis buffer with protease inhibitors (*see* **Subheading 2., item 17**).
3. Homogenize the tissue on ice with a pestle (Kontes, Vineland, NJ, cat. no. 749520-0000) in a microcentrifuge tube (*see* **Note 9**).
4. Centrifuge (16,000*g*) the homogenate for 10 min at 4°C. Collect the supernatant and measure the protein concentration.
5. Adjust the protein concentration to 1 μg per μL of 1X sample buffer. Load onto the gel equal amounts of protein per lane (*see* **Note 7**).

### 3.2.5. Gelatinase Standards

It is important to include gelatinase standards in each zymogram to accurately determine the type and activation status of the enzyme(s) expressed in a given sample. Conditioned medium from HT1080 human fibrosarcoma cells (American Type Culture Collection, CCL-121) is optimal because it contains both MMP-2 (72 kDa) and MMP-9 (92 kDa) **(Fig. 1)**. Conditioned media obtained from HT1080 cells treated for 16 h with phorbol ester (100 n*M*) or concanavalin A (10 µg/mL) contain active MMP-2 (62 kDa) and can be used as reference for active MMP-2 *(7)*. If available, purified natural or recombinant enzymes can be used **(Fig. 1)** (*see* **Note 10**).

1. Dilute the purified gelatinases to a final concentration of 1 ng/µL of 1X sample buffer. Do not heat the sample.
2. Load 1–5 ng of the enzymes per lane.

## 3.3. Running and Developing of Gel

1. Gently pull the comb out from the stacking gel and peel off the tape from the bottom of the cassette. Place the cassette into the gel apparatus and fill the buffer chambers with 1X running buffer.
2. Load the samples and run the gel at constant voltage (125 V, starting current should be approx 30–40 mA/gel) until the bromophenol blue tracking dye reaches the bottom of the gel (approx 90 min). These running conditions will prevent overheating of the gel (*see* **Note 4**).
3. Carefully remove the gel from the cassette and place it in plastic tray containing 100 mL of renaturing solution. Incubate the gel for 30 min at room temperature with gentle agitation.
4. Decant the solution and rinse the gel at least once with 300 mL of $dH_2O$.
5. Incubate the gel at room temperature for an additional 30 min in 100 mL of developing buffer with gentle agitation.
6. Decant the developing buffer and replace it with 100 mL of fresh developing buffer. Incubate the gel at 37°C for approx 16 h in a closed tray (*see* **Note 11**).
7. Decant the developing buffer and stain the gel in staining solution for at least 1 h or until the gel is uniformly dark blue. The staining solution can be collected and used again. However, it will require a longer staining time.
8. Destain the gel with destaining solution until areas of gelatinolytic activity appear as clear sharp bands over the blue background (*see* **Note 7**).

## 4. Notes

1. Regarding the use of gelatin zymography for quantitative assessments: Although gelatin zymography has been claimed to be a quantitative technique, major problems arise during the process of method standardization. As with any quantitative assay, a reproducible and accurate standard is essential. However, owing to the many variables involved in the assay including (a) variations in the amount

of copolymerized gelatin; (b) nature, source, and loading of the sample; (c) incubation time and temperature; (d) washing conditions; (e) staining and destaining conditions; and (f) source and condition (latent, active) of the standards, it is unreliable to establish reproducible assay conditions when analyzing multiple samples.

2. Latent vs active species: Because gelatin zymography is widely used to study gelatinase activation, it is important to be aware of the pitfalls. Indeed, the complexity of the activation process and the inherent limitations of SDS-PAGE make identification of active species by gelatin zymography not straightforward. Gelatinases are usually activated in a sequential process involving generation of inactive intermediate species. Therefore, if good band separation and mass determination are not achieved, the intermediate species can be easily confused with the active forms. In addition, owing to differences in glycosylation between the precursor (intracellular) and the mature (secreted) form of latent MMP-9 *(16)*, the intracellular precursor form of MMP-9 (~85 kDa) can be easily mistaken for the active species. This is prone to occur when the samples examined are derived from cell lysates or tissue extracts, all of which may contain the intracellular precursor enzyme (**Fig. 1**). Inclusion, in the same gel, of active enzymes (*see* **Subheading 3.2.5.**) as standards and/or reliable unreduced molecular weight markers (Mark 12 from Novex, cat. no. LC5677) will aid in discriminating between inactive (intermediate, precursor) and active species. When in doubt, additional methods including immunoblot analysis with antibodies recognizing latent or active species, inhibitor trapping ($\alpha$2-macroglobulin) *(3)* and/or enzymatic assays are required to determined the activation status of the gelatinases.
3. For reliable electrophoresis results, it is crucial to prepare a good acrylamide–*bis*-acrylamide stock solution from quality reagents. Ready-to-use 30% acrylamide–*bis*-acrylamide 37.5:1 (2.6% C) solution can be purchased from Bio-Rad Laboratories (Hercules, CA; cat. no. 161-0158), allowing reproducible results and removing the need for the handling of acrylamide powder.
4. We commonly use 10% polyacrylamide gels for separating gelatinases. However, the percentage and the thickness of the separating gel can be varied depending on the aim of the separation. For instance, to better visualize the dimeric form of MMP-9 (~200 kDa) (**Fig. 1**) and/or to obtain a better resolution of closely opposed bands (latent and active forms), a lower percentage (7–8%) of polyacrylamide solution can be used. However, the gelatinolytic bands will be less sharp. Alternatively, the gel can be run for an additional 30 min after the tracking dye have reached the bottom of the gel.
5. It is very convenient to prepare the gels in advance since the polymerized gels can be stored at 4°C for 2–3 wk without any effects on resolution. Avoid bacterial contamination of buffers and solutions as bacterial proteases may result in the appearance of nonspecific gelatinolytic bands. This can be minimized by sterile filtration of buffers and stock solutions and storage at 4°C, as indicated.
6. Preparation of conditioned media: Be aware that long incubation times (48–72 h) of cells in serum-free media can affect viability. Incubate the cells in a minimum amount of serum-free media (~50–60% of the amount of growth media) to obtain more concentrated samples.

7. Gelatinolytic bands: Under optimal conditions, gelatinolytic bands should be sharp and well defined (**Fig. 1**). This is generally true with samples derived from conditioned media or with purified enzymes. Cell lysates and tissue extracts, in contrast, may produce distorted bands owing to the high protein content. This is resolved by minimizing the amount of protein loaded per lane or concentrating the sample using phase partition and agarose-beads purification as described in **Subheading 3.2.3.** The sharpness and resolution of the bands also depend on the time of staining and destaining of the gels. Because other proteases besides MMPs can cleave gelatin, it is important to ascertain the nature of the enzymes detected. Incubating the gel in developing buffer containing 20 m*M* EDTA will cause the disappearance of bands produced by metalloproteases. By comparing the molecular weight of the bands in the samples with known gelatinase standards and the use of EDTA, a reliable assessment of the nature and molecular mass of the enzymes can be made. Other major MMPs showing gelatinolytic activity are stromelysin 1 (MMP-3) and interstitial collagenase (MMP-1) ***(3)***. However, the intensity of the bands produced by these enzymes is significantly lower than that elicited by gelatinases.
8. Triton X-114 extraction and detergent phase: Owing to the high sensitivity of gelatin zymography, presence of gelatinases in the detergent phase may be the result of contamination with enzymes coming from the aqueous phase. To minimize contamination, repeat again the extraction by adding ice-cold TBS-CM to the detergent phase. Vortex-mix and repeat phase partition as described in **Subheading 3.2.3., steps 5–7**. Discard the upper phase and continue as described in **Subheading 3.2.3., step 9**.
9. Preparation of tissue extracts: The amount of lysis buffer can be varied depending on the source of the tissues. Be aware that owing to inherent differences in tissue structure, between and within specimens, protein extractability may vary ***(11)***. Therefore, the results should be carefully interpreted! After centrifugation, the tissue homogenates may contain floating lipids. Repeat the centrifugation to obtain a clear homogenate.
10. Gelatinase standards: Purified gelatinases can now be obtained from commercial sources. To activate purified gelatinases, incubate the enzymes in 1 m*M* *p*-aminophenylmercuric acetate (APMA) (**Toxic!**). This should be prepared from a fresh stock of 10 m*M* APMA in 50 m*M* NaOH and then diluted 10-fold in TBS-B. Incubate the purified latent enzymes (2 ng/µL) with APMA in a 37°C water bath. MMP-2 is readily activated to the 62-kDa form after approx 30 min while MMP-9 is partially activated to the 82-kDa form after 2 h. Be aware that presence of TIMPs in the enzyme preparation will inhibit or slow the activation process causing generation of intermediate inactive forms (64 kDa for MMP-2 and 85 kDa for MMP-9) ***(4)***. On the other hand, long incubation times with APMA or TIMP-free gelatinases will cause the appearance of low molecular mass active forms—a 45-kDa form for MMP-2 ***(14)*** and a 67-kDa form for MMP-9 ***(15)***.
11. The time of incubation of the gel in the developing buffer is critical. Because the presence of the gelatinolytic bands is the result of enzymatic activity,

varying the incubation time will affect the size of the bands. For most conditions, overnight incubation will provide optimal resolution and reproducible results. Therefore, for better resolution, it is preferable to accordingly increase or decrease the amount of sample loaded into the gel rather than changing the incubation time.

## Acknowledgments

The authors would like to thank the members of our laboratory: Margarida Bernardo, David Gervasi, Martin Pietila, and Zhong Dong for suggestions and tips. This work has been supported by a grant from the National Institutes of Health (CA-61986) to R. F.

## References

1. Liotta, L. A., Steeg, P. S., and Stetler-Stevenson, W. G. (1991) Cancer metastasis and angiogenesis: an imbalance of positive and negative regulation. *Cell* **64,** 327–336.
2. Birkedal-Hansen, H., Moore, W. G., Bodden, M. K., Windsor, L. J., Birkedal-Hansen, B., DeCarlo, A., and Engler, J. A. (1993) Matrix metalloproteinases: a review. *Crit. Rev. Oral Biol. Med.* **4,** 197–250.
3. Nagase, H. (1997) Activation mechanisms of matrix metalloproteinases. *Biol. Chem.* **378,** 151–160.
4. Murphy, G. and Willenbrock, F. (1995) Tissue inhibitors of matrix metalloendopeptidases. *Methods Enzymol.* **248,** 496–510.
5. Collier, I. E., Wilhelm, S. M., Eisen, A. Z., Marmer, B. L., Grant, G. A., Seltzer, J. L., et al. (1988) H-ras oncogene-transformed human bronchial epithelial cells (TBE-1) secrete a single metalloprotease capable of degrading basement membrane collagen. *J. Biol. Chem.* **263,** 6579–6587.
6. Wilhelm, S. M., Collier, I. E., Marmer, B. L., Eisen, A. Z., Grant, G. A., and Goldberg, G. I. (1989) SV40-transformed human lung fibroblasts secrete a 92-kDa type IV collagenase which is identical to that secreted by normal human macrophages. *J. Biol. Chem.* **264,** 17,213–17,221.
7. Stetler-Stevenson, W. G. (1994) Progelatinase A activation during tumor cell invasion, *Invas. Metastas.* **14,** 259–268.
8. Himelstein, B. P., Canete-Soler, R., Bernhard, E. J., Dilks, D. W., and Muschel, R. J. (1994) Metalloproteinases in tumor progression: the contribution of MMP-9. *Invas. Metastas.* **14,** 246–258.
9. Murphy, G. and Crabbe, T. (1995) Gelatinases A and B. *Methods Enzymol.* **248,** 470–484.
10. Heussen, C. and Dowdle, E. B. (1980) Electrophoretic analysis of plasminogen activators in polyacrylamide gels containing sodium dodecyl sulfate and copolymerized substrates. *Anal. Biochem.* **102,** 196–202.
11. Woessner, J. F., Jr. (1995) Quantification of matrix metalloproteinases in tissue samples. *Methods Enzymol.* **248,** 510–528.
12. Azzam, H. S., Arand, G., Lippman, M. E., and Thompson, E. W. (1993) Association of MMP-2 activation potential with metastatic progression in human breast cancer cell lines independent of MMP-2 production. *J. Natl. Cancer Inst.* **85,** 1758–1764.

13. Lewalle, J. M., Munaut, C., Pichot, B., Cataldo, D., Baramova, E., and Foidart, J. M. (1995) Plasma membrane-dependent activation of gelatinase A in human vascular endothelial cells. *J. Cell. Physiol.* **165,** 475–483.
14. Fridman, R., Fuerst, T. R., Bird, R. E., Hoyhtya, M., Oelkuct, M., Kraus, S., et al. (1992) Domain structure of human 72-kDa gelatinase/type IV collagenase. Characterization of proteolytic activity and identification of the tissue inhibitor of metalloproteinase-2 (TIMP-2) binding regions. *J. Biol. Chem.* **267,** 15,398–15,405.
15. O'Connell, J. P., Willenbrock, F., Docherty, A. J., Eaton, D., and Murphy, G. (1994) Analysis of the role of the COOH-terminal domain in the activation, proteolytic activity, and tissue inhibitor of metalloproteinase interactions of gelatinase B. *J. Biol. Chem.* **269,** 14,967–14,973.
16. Toth, M., Gervasi, D. C., and Fridman, R. (1997) Phorbol ester-induced cell surface association of matrix metalloproteinase-9 in human MCF10A breast epithelial cells. *Cancer Res.* **57,** 3159–3167.

# II

# Genetic Aspects of Metastasis

# 13

# *In Situ* Hybridization to Localize mRNAs

**Richard Poulsom**

## 1. Introduction

*In situ* hybridization (ISH) to localize sites of expression of mRNA is widely applicable to studies of invasion and metastasis in human pathology specimens and tissues from experimental animals or cell cultures. ISH can provide crucial information about where a specific gene is expressed, before suitable antisera can be made. The ICRF *In Situ* Hybridization Service has hybridized more than 44,000 sections of principally formalin-fixed, paraffin wax-embedded tissues, with a high success rate. Our preferred method for screening for mRNAs in routinely fixed paraffin wax-embedded materials uses $^{35}$S-labeled riboprobes, and has been developed over several years from the method described by Senior and colleagues *(1)*. It may seem unfashionable to use an isotopic method, but $^{35}$S provides the best balance of specificity, sensitivity, and cost; most importantly, unlike nonisotopic ISH, it gives reproducible and easily interpretable results without the need to adjust the method to individual cell types within sections. For alternative methodology for fluorescent *in situ* hybridization (FISH) techniques, please refer to Chapter 14 by Goker and Shipley.

The protocol described here gives good results for the vast majority of routine surgical specimens, and useful data from about half of the blocks sampled to date from the discarded archive of a General Hospital Pathology Department (avoiding postmortem specimens). With the economies of scale, reagent costs are approximately US $11 (UK £6.50) per slide.

Comparison of different published protocols for ISH shows that steps regarded as essential by one group may not be used by another group. Nevertheless, there are some common principles and checkpoints, and only one needs to be wrong for the entire experiment to fail.

From: *Methods in Molecular Medicine, vol. 57:*
*Metastasis Research Protocols, Vol. 1: Analysis of Cells and Tissues*
Edited by: S. A. Brooks and U. Schumacher © Humana Press Inc., Totowa, NJ

The key stages are:

### 1.1. Probe Design and Production

Riboprobes are single-stranded RNA molecules made by in vitro transcription using one of the DNA-directed RNA polymerases (SP6, T3, or T7), nucleoside triphosphates (including [$^{35}$S]UTP) and a suitable DNA template. Templates are lengths of double-stranded DNA that contain sequences for RNA polymerase binding and initiation, followed by a sequence that is specific to the target RNA. To make a riboprobe capable of binding to an mRNA the strand complementary to the target mRNA must be made, that is, an RNA version of the noncoding strand. Specific cDNA fragments can be amplified by reverse-transcriptase polymerase chain reaction (RT-PCR) using primers containing RNA polymerase sequences, for example *(2)*. More commonly, templates are made from plasmid DNA that has been linearized with a restriction endonuclease. Cloning cDNA fragments into plasmids is a convenient way to prepare large amounts of good quality template. Key features of an ideal plasmid for making an antisense riboprobe are shown in **Fig. 1** and explained in **Note 1**. Template DNA must be of good quality, as explained in **Note 2**.

### 1.2. Preparation of Tissues

The results in **Fig. 2** were all obtained using paraffin wax-embedded blocks of tissues fixed with a formalin fixative and processed routinely as if for pathological evaluation. Even 30-yr-old blocks may work well. Postmortem specimens should be avoided as they may not have intact mRNA (check with a β-actin probe; *see* **Note 3**). Frozen tissues may give good mRNA labeling, but are awkward to store and handle, and give poor morphology. Cell cultures can be studied as cytospins or directly if grown on glass slides.

### 1.3. Permeabilization of Tissues

This step is to enable the probe to gain access to the target and is standard for paraffin wax-embedded blocks. This step is usually not required for frozen sections and cell cultures.

### 1.4. Hybridization

The riboprobe is encouraged to base pair with homologous complementary sequences in conditions designed to speed up the rate of association, yet minimize nonspecific binding to other sequences and proteins.

### 1.5. Washing

Following hybridization, RNase A is used to reduce background by degrading unhybridized riboprobe and cleaving imperfect hybrids. The resultant fragments are then washed out at higher stringency.

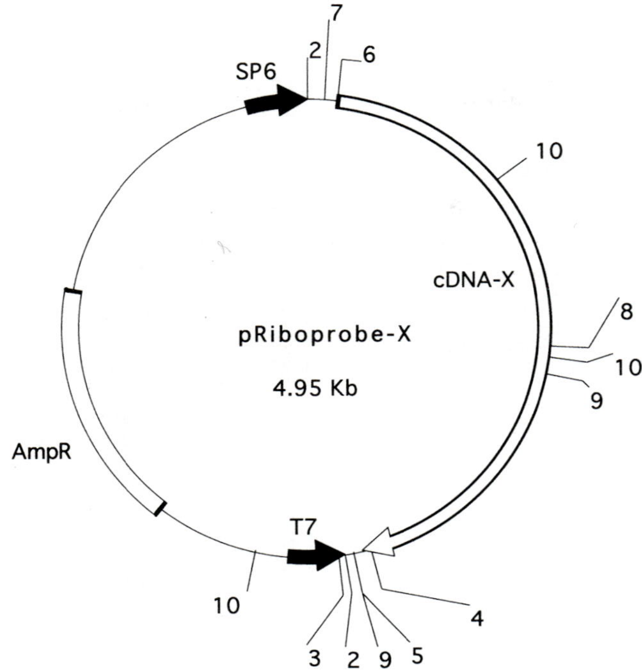


Fig. 1. Features of an ideal plasmid for making a riboprobe. *1*, Multicopy plasmid containing antibiotic resistance gene used during propagation in bacteria. The less complex members of the pGEM series (Promega) are preferred to pBluescript (Stratagene), or pCR series vectors (Invitrogen). *2–2*, Extent of the multiple cloning site. *3*, T7 RNA polymerase will give an antisense probe in this particular construct. *4*, 3' End of the partial cDNA insert to be used as a riboprobe template; note poly $(A)^+$ tract has been removed beforehand. *5*, Minimal vector sequence between 3 and 4 and avoiding GC-rich content (*see* **Note 1**). *6*, 5' End of the partial cDNA insert. *7*, Potential restriction endonuclease site for forming template, but probe would be too long to use without partial hydrolysis, and would still contain domain 8. *8*, Sequence conserved between gene family members or an Alu repeat. *9, 9*, Potential restriction endonuclease sites for forming template, except that it cuts between the T7 RNA polymerase site and the partial cDNA insert. *10, 10, 10,* Restriction endonuclease site chosen to linearize plasmid into template. It creates 5' overhangs and one of the three fragments bears the T7 RNA polymerase site and approx 1 kb of cDNA.

### *1.6. Autoradiography*

Slides are coated with a thin layer of sensitive photographic emulsion, dried, then kept for 2–70 d to allow a latent image to form. A permanent image made up of silver grains is then developed, a counterstain used to reveal the tissue architecture, and a glass coverslip applied to protect the completed slide.

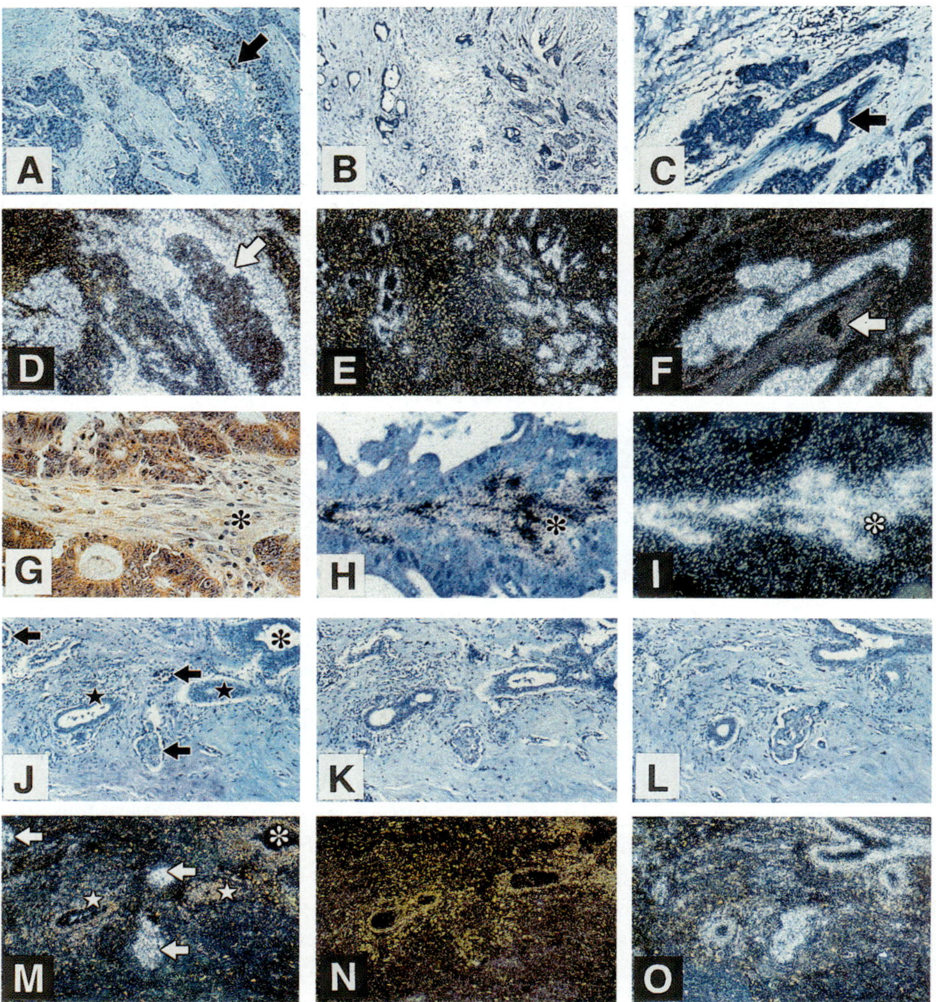


Fig. 2. Micrographs showing localization of mRNAs in routine histology blocks. Excepting **G**, all brightfield images are accompanied by their reflected-light darkfield counterpart, obtained simply by switching from conventional illumination to epi-illumination (*see* **Note 7**). **(A,D)** Localization of VEGF mRNA in breast cancer using a 204-base riboprobe, exposing for 11 d *(31)*. Only the most intense clusters of silver grains bordering the necrotic region (arrowed) are visible in brightfield; using reflected light, expression is seen to be widespread in tumor epithelium. No expression is detected in the stromal cells (golden nuclei). **(B,E)** The mRNA encoding the membrane receptor HER4 *(32)* is easily detected in invasive breast cancer (and not in stromal tissues) using a 1-kb riboprobe and exposing for 10 d. **(C,F)** HER4 mRNA *(32)* is present throughout the carcinoma cells and not the epithelium of the duct (arrowed) or inflammatory cells (*top center* and *bottom left*). **(G,H,I)** Gelatinase A expression in

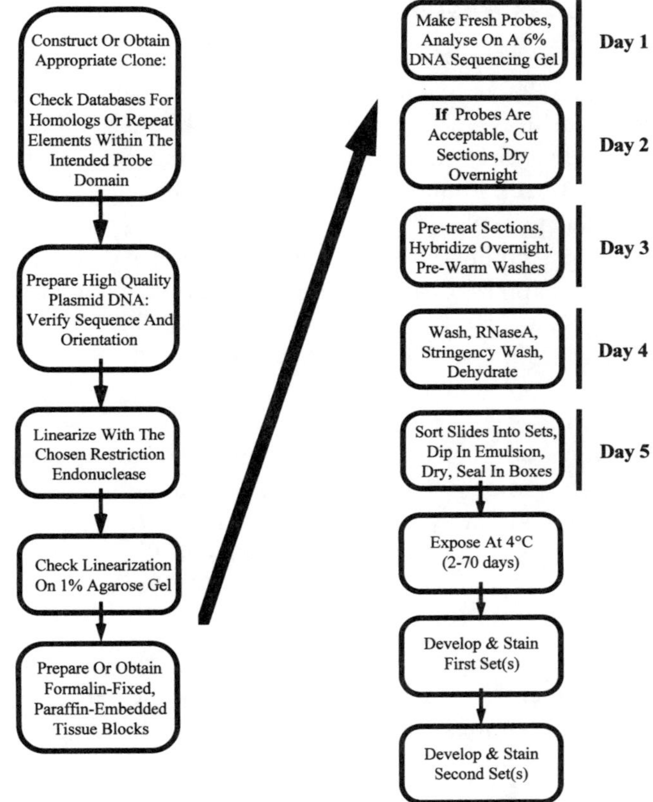


Fig. 3. Flowchart showing the principal steps involved in *in situ* hybridization.

---

colorectal cancer *(10)* appears predominantly within the tumor epithelium by immunohistochemistry (brown reaction product in G), whereas the mRNA appears confined to cells within the stromal cores (asterisks, pink desmoplastic stroma in **H**). In **I**, autoradiographic silver grains are white in reflected light; other reflecting structures are colored. **(J–O)** Three pairs of micrographs showing localization of mRNAs in near serial sections of a breast cancer specimen as part of a screening programme to identify differentially expressed genes (*see* Chapter 20 by Kao and Francia). All slides were processed together and developed after 11 d exposure. **(J,M)** Candidate gene *11AN1* is seen to be overexpressed in invasive (arrowed) vs benign (star) breast epithelium. The epithelium lining the duct (asterisk) expresses the mRNA at a low level. **(K,N)**. In contrast, expression of candidate *12AT2* cannot be detected in any cell type at this exposure time. **(L,O)**. The positive control probe shows that β-actin mRNA was detectable at various signal levels in all regions of this tissue.

### 1.7. Interpretation of Results

Long exposures produce dense clusters of black silver grains visible with any microscope at moderate magnification (100–200× overall). Exposure times can be reduced considerably if reflected-light darkfield conditions are used; clusters of even a few grains can be appreciated at moderate magnification and a true assessment of background graininess can be made.

ISH is an excellent technique for revealing the cell types that express the highest levels of the target, compared to surrounding tissue, and can readily identify just a few cells that express a target at moderate levels, even when a Northern blot is negative *(3)*. Conversely, if a target is expressed widely at moderate levels (e.g., glyceraldehyde-3-phosphate dehydrogenase [GAPDH] mRNA), ISH may give the impression of an "uninterpretable background," even though a Northern blot of homogenized tissue would be positive.

The entire ISH procedure, from labeling probes and cutting sections through to developing, takes about 1 wk (**Fig. 3**), plus autoradiographic exposure time of about 10 d for a moderately abundant mRNA target. For convenience and economy slides are processed in batches of 100 or 200.

## 2. Materials

Ensure that you are aware of the health and safety requirements involved in working with all reagents, as many require considerable care. We use Milli-Q water for all solutions unless specified, although other high-quality water should be satisfactory.

### 2.1. Template Preparation

1. Nuclease-free water; water shaken with 0.1% vol diethyl pyrocarbonate, then autoclaved.
2. TE buffer: 10 m*M* Tris-HCl, pH 8.0, and 1 m*M* EDTA. Store at room temperature after autoclaving.
3. Phenol–Chloroform–Isoamyl alcohol: (Life Technologies cat. no. 15593-031), stored as directed.
4. Chloroform–Isoamyl alcohol: (Sigma, cat. no. C-0549) stored at 4°C.
5. 7.5 *M* Ammonium acetate: (Sigma, cat. no. A 1542) 14.44 g made up to 25 mL with water. Store at room temperature after autoclaving.

### 2.2. Probe Labeling (In Vitro Transcription)

1. 5X Transcription buffer: Supplied with RNA polymerases purchased from Promega. Store at –20°C.
2. RNase inhibitor: 20U/µL (Roche), store at –20°C.
3. Dithiothreitol (DTT, Sigma, cat. no. D 9779): For 1 *M* dissolve 19.25 g and make up to 125 mL with water. Prepare 100 m*M* and 10 m*M* stocks and autoclave. Store single-use aliquots in microfuge tubes at –20°C for up to 1 yr.

4. AGC mix: ATP, GTP, CTP each at 6.25 m*M* in water. Prepare from Roche stocks and store 25-µL reusable aliquots in microfuge tubes at –20°C for up to 6 mo.
5. [$^{35}$S]UTP at 40 mCi/mL, 800 Ci/mmol (Amersham, cat. no. SJ 40383 or DuPont, cat. no. NEG 039H).
6. DNase I: RNase-free grade from Roche or Promega. Store at –20°C.
7. Chromaspin-30 columns: Diethylpyrocarbonate (DEPC)-equilibrated type from Clontech.
8. Ribosomal RNA: Dispense DEPC-treated water directly into the vial (Sigma, cat. no. R 5502; 25 U, 17.7 U/mg) to achieve a final concentration of 10 mg/mL. Usually ~283 µL per vial. Store 25-µL aliquots in microfuge tubes at –20°C for up to 1 yr.

### 2.3. Cutting and Preparing Paraffin Sections

1. APES-treated microscope slides: Place slides (purchased precleaned with frosted ends that can be marked with a pencil) in metal slide racks and soak overnight in 10% Decon-90. Wash in hot running tap water for 60–90 min, then twice in water. Bake slides at 180°C (approx 4 h) in the slide rack covered with aluminum foil. Baked slides can be stored for months before treating by immersing the rack for about 10 s in *freshly* prepared 3-aminopropyl-ethoxysilane (APES: Sigma, cat. no. A3648, 2% v/v in acetone AR grade) in a glass dish. Excess APES is removed by dipping twice in acetone, then twice in DEPC-treated water. Racks of slides wrapped in foil and stored at room temperature retain their adhesive ability for several weeks.
2. Rehydrating alcohols + DEPC: Used before permeabilization. Prepare 500 mL each of a series of ethanol solutions using high-quality ethanol and DEPC-treated water: 100%, 80%, 60%, and 30% ethanol. Add 0.1% vol of DEPC but *do not* autoclave. Use for only approx 200 slides.
3. Phosphate-buffered saline (PBS): For example, dissolve one PBS tablet (Sigma, cat. no. P4417) in 400 mL of water. For PBS+ treat with 0.1% vol of DEPC, then autoclave.
4. Proteinase K: Dissolve 200 mg of Proteinase K (use Sigma, cat. no. P 4914) in 20 mL of TE buffer to make a 10 mg/mL stock. Store as 1-mL aliquots in microfuge tubes at –20°C. Just before use add 1 ml to 500 mL of 37°C TE buffer to make the working solution of 20 µg/mL.
5. Glycine/2X PBS: 0.2% Glycine (e.g., Sigma, cat. no. G4392) dissolved in 2X PBS, then autoclaved.
6. 4% PFA in PBS: Add 20 g of paraformaldehyde (Sigma, cat. no. P 6148) to 500 mL of PBS, and warm carefully to approx 70°C in a fume hood. If necessary use a few drops of 1 *M* NaOH to dissolve resistant solids. Cool on ice to approx 20°C. Make fresh each time.
7. Triethanolamine buffer: 100 m*M*; dissolve 37.5 g of triethanolamine (Sigma, cat. no. T1502) in water and make up to 2 L with DEPC-treated water. Autoclave and store for up to 3 mo.
8. Acetylation buffer: Within seconds before use, add 1.25 mL of acetic anhydride (Sigma, cat. no. A 6404) to 500 mL of 0.1 *M* triethanolamine buffer.

9. Dehydrating alcohols + DEPC: Used before hybridization. Prepare 500 mL each of a series of ethanol solutions using high-quality ethanol and DEPC-treated water: 30%, 50%, 70%, 90%, and 100% ethanol. Add 0.1% vol of DEPC but *do not* autoclave. Use for only approx 200 slides.

### 2.4. Hybridization

1. RNase-free coverslips: Rinse glass coverslips in 70% alcohol, dry, then bake at 180°C in 50-mL beakers covered in aluminum foil.
2. Ribosomal RNA: As described in **Subheading 2.2., item 8**.
3. Formamide (deionized): For hybridization buffer only. Take 50 mL of formamide and 2.5 g of ion-exchange resin (Bio-Rad AG 501-×8, 20–50-mesh fully regenerated). Stir in fume hood cupboard overnight. Filter using a no. 1 filter paper to remove resin. Store at –20°C in small aliquots (~1 mL).
4. 10X Salts solution: Dissolve 14.2 g of $Na_2HPO_4$ in stages in 100 mL of water and adjust to pH 6.8. Add 176.2 g of NaCl, then 100 mL of 1 *M* Tris-HCl and 250 mL of 0.2 *M* EDTA and mix. Make up to final volume of 1 L with water. Do not autoclave or filter. Store at room temperature.
5. Dextran sulfate solution: Dissolve 50 g of dextran sulfate (Sigma, cat. no. D8906) into 100 mL of autoclaved water in a sterile bottle at 80°C (water bath). Store 1-mL aliquots in microfuge tubes at –20°C.
6. Denhardt's and salts solution: Add 5 mL of 10X salts solution directly to the vial (Sigma, cat. no. D9905; store at 0–5°C), as it contains a powder that is very hard to dissolve. Transfer to a 50-mL plastic tube and dissolve in a total of 25 mL of 10X salts solution. Store 300-μL aliquots in microfuge tubes at –20°C.
7. Hybridization buffer: For 1 mL combine the following in a microfuge tube in order: 100 μL of Denhardt's and salts solution, 500 μL of formamide (deionized), 30 μL of rRNA, 200 μL of dextran sulfate solution (prewarmed to reduce viscosity), 10 μL of 1 *M* DTT. Mix well. May be stored at –20°C for a few months.

### 2.5. Washing

1. Formamide wash buffer: Add 250 mL 10X salts solution to 1250 mL of formamide (nondeionized; BDH, Poole, Dorset). Make up to 2.5 L with DEPC-treated water.
2. TNE Buffer: For RNase A digestion. Combine 146 g of NaCl; 50 mL of 1 *M* Tris-HCl, pH 7.6; 25 mL of 0.2 *M* EDTA, pH 7.5; and make up to 5 L with water. Adjust to pH 7.2–7.6.
3. Stock RNase A: Dissolve 500 mg of ribonuclease A (Sigma, cat. no. R5503) in water (*not* DEPC treated); store 1-mL aliquots in microfuge tubes at –20°C.
4. 20X saline sodium citrate (SSC): Dissolve 175.3 g of NaCl and 88.23 g of sodium citrate in 750 mL of water. Adjust to pH 7.0 if necessary then make up to 1 L. Autoclave, then store at room temperature.
5. Ammonium acetate, 5 *M* solution: 96.25 g made up to 250 mL with water. Autoclave, then store at room temperature.
6. Dehydrating alcohols + acetate: Used before autoradiography. Prepare 500 mL each of a series of ethanol solutions using high-quality ethanol and DEPC-treated water: 30%, 50%, 70%, 90%, and 100% ethanol containing 30 mL of 5 *M* ammonium acetate in 500 mL each solution. Use for only approx 200 slides.

### 2.6. Autoradiography

1. Emulsion: (Ilford K5 in gel form, size A). The preferred pack size is 50 mL. Keep at 4°C in original lightproof packing. Once opened use within 3–4 wk. Do not reuse diluted emulsion.
2. D-19 developer: Purchased from Kodak, diluted and stored as described by manufacturer.
3. Stop solution: 1% v/v glacial acetic acid in water. Use once only. Store up to 6 mo.
4. Fixer: 30% w/v Sodium thiosulfate in water. Use once only. Store up to 6 mo.
5. Dilute Giemsa solution: For example, BDH cat. no. 350864X; dilute 1:5 with water, then filter. May be reused. Store up to 3 mo or until staining becomes weak.
6. Synthetic mountant, for example, DPX, BDH cat. no. 36125-2B.

## 3. Methods

### 3.1. Template Preparation

1. To a microfuge tube, add 50 µg of plasmid DNA (approx 1 mg/mL of solution in TE or water), 35 µL of 10X restriction endonuclease buffer appropriate to the chosen restriction enzyme and nuclease-free water to 340 µL. Mix, then remove 3.5 µL as a "predigest" sample to be kept at –20°C for later analysis.
2. Add 10 µL of the chosen restriction enzyme (10 U/µL), mix, and incubate at the temperature specified by the enzyme supplier for 2 h. Mix, then remove 3.5 µL as a "post-digest" sample to be kept at –20°C for later analysis.
3. Add 350 µL of phenol–chloroform, shake gently to emulsify for 2 min, then centrifuge at 13,000*g* for 5 min in a benchtop microcentrifuge.
4. Transfer the upper (aqueous) phase to a fresh microfuge tube, taking care to avoid the interphase.
5. Repeat **steps 3** and **4**.
6. Add 300 µL of chloroform–isoamyl alcohol, shake gently to emulsify for 2 min, then centrifuge at 13,000*g* for 5 min.
7. Transfer the upper (aqueous) phase to a fresh microfuge tube, taking care to avoid the interphase.
8. Add 100 µL of 7.5 *M* ammonium acetate, mix, then add 750 µL of ethanol and mix again.
9. Pellet the DNA at 13,000*g* for 15 min.
10. Discard the supernatant, then add 1 mL of 70% ethanol.
11. Pellet the DNA at 13,000*g* for 15 min.
12. Discard all of the supernatant by careful pipeting, and allow the pellet to air-dry (about 15 min).
13. Add 25 µL of 10 m*M* Tris; 0.1 m*M* EDTA, pH 7.6; and dissolve the template DNA (37°C, 10 min).
14. Remove 1 µL as a "final template sample." The stock template solution may be stored at –20°C for several years.
15. Assess the efficiency of plasmid cleavage production by conventional 1% agarose ethidium bromide gel electrophoresis of the three samples taken.

16. Measure the concentration of the final template solution or estimate this by comparing the fluorescence of its band with those in the pre- and post-digestion samples. If necessary adjust the concentration of the final template solution to approx 1 mg/mL.

### 3.2. Probe Labeling (In Vitro Transcription)

1. To a microfuge tube at room temperature, add in order:

   2.5 µL of 5X transcription buffer
   1.0 µL of RNase inhibitor
   0.7 µL of DTT (100 m*M*)
   2.0 µL of AGC mix
   2.4 µL of stock template solution/nuclease-free water (to give 1 µg of template DNA)
   3.5 µL of [$^{35}$S]UTP
   0.4 µL (6–8 U) of appropriate RNA polymerase from Promega

2. Incubate at 37–40°C, 60 min.
3. Destroy template by adding DNase I (1 U, 1 µL) and incubating at 37°C, 15 min. During this time prepare a Chromaspin-30 column by centrifugation as described by the manufacturer.
4. Add DTT (10 m*M*, 25 µL) and carrier ribosomal RNA (10 mg/mL, 1.5 µL) to the reaction mixture tube; mix well.
5. Remove a 1-µL sample, dilute with 50 µL water and add 3 mL scintillant to estimate initial $^{35}$S present.
6. Add bulk of reaction mixture to top of gel in a Chromaspin-30 column.
7. Centrifuge at 700*g* for 3 min at 15°C, collecting eluate into a new tube containing DTT (100 m*M*, 4 µL) and RNase inhibitor (1 µL).
8. Mix well and remove 1 µL for dilution with 50 µL of water and 3 mL of scintillant to estimate total $^{35}$S incorporated.
9. Assess riboprobe quality by standard 6% polyacrylamide sequencing gel electrophoresis of duplicate aliquots of 400,000 cpm riboprobes for 40 min, dry gel, and expose to X-ray film for 1–2 h. Meanwhile, store riboprobes at –20°C until use within a few days.
10. Discard columns and vials as appropriate for $^{35}$S waste.

### 3.3. Cutting and Preparing Paraffin Wax-Embedded Sections

1. Wearing gloves, cut sections at 4 µm using a microtome with disposable blades.
2. Float sections using a clean brush or forceps onto DEPC-treated water, fresh daily, degassed before use if necessary.
3. Collect sections onto APES-treated microscope slides.
4. Dry sections overnight in a 40°C oven in card slide trays protected from dust.
5. Dewax in fresh xylene + 0.1% DEPC for 8 min, with two changes. *(***Alcohols and xylene are not autoclaved!***)*

6. Rehydrate by sequential immersion in the rehydrating alcohols + DEPC, 5 min each.
7. Rinse in PBS+.
8. Permeabilize tissue with Proteinase K at 37°C for 10 min.
9. Rinse in glycine–2X PBS for 5 min to block the proteinase.
10. Rinse in PBS+ for 5 min.
11. Post-fix in 4% PFA in PBS for 20 min.
12. Rinse in PBS+, 3× for 5 min each.
13. In a fume hood, immerse slides in 500 mL of acetylation buffer. Mix well for 10 min.
14. Wash in PBS+ for 5 min, 3×.
15. Dehydrate by sequential immersion in the dehydrating alcohols + DEPC, 5 min each.
16. Air-dry. The specimens are now ready for hybridization.

### 3.4. Hybridization

1. Add riboprobe/water (*see* **Note 4**) to hybridization buffer in a microfuge tube, heat at 80°C for 1 min, mix well, then centrifuge briefly to reduce aerosols, then chill the mix on ice.
2. Pipet 20 µL onto each section (the mix is viscous so volumes are approximate).
3. Gently lower an RNase-free coverslip onto each slide until it makes contact with the hybridization buffer, then allow the buffer to spread under the weight of the coverslip.
4. Place the slides in, for example, plastic slide-mailing boxes each humidified with blotting paper saturated with 1X salts and 50% formamide. Place slide-mailing boxes into a lunch box humidified in the same way.
5. Seal the lunch boxes with polyvinyl chloride (PVC) tape and incubate overnight at 55°C.

### 3.5. Washing

Prewarm all solutions before use.

1. Place the first wash buffers in water baths ready for use the following day: Per 25–50 slides, 5 L of TNE buffer at 37°C, 2 L of formamide wash buffer at 55°C.
2. In a fume hood, open the lunch box containing the slides.
3. Remove the slides and place in a slide rack immersed in 500 mL of formamide wash buffer at 55°C on a rocking table. Use a filter paper at the bottom of the box to catch the coverslips.
4. Transfer the slides to a new rack, making sure each coverslip has come off, and wash again in 500 mL of formamide wash buffer at 55°C, three changes over a total of 3–4 h, using a shaking water bath.
5. Remove all traces of the formamide wash buffer using nine changes of 500 mL of TNE buffer at 37°C, shaking over approx 45 min.
6. During the TNE washes thaw 1 mL of stock RNase A. Boil for 2 min, then add to 500 mL of TNE.
7. Incubate the slides with the RNase A solution at 37°C for 1 h in a plastic lunch box. Keep specific containers for this step. Dispose of RNase-contaminated gloves.
8. Wash slides in 2X SSC for 30 min at 65°C with agitation, 2×.

9. Wash slides in 0.5X SSC for 30 min at 65°C with agitation (*see* **Note 5**).
10. Pass slides through graded ethanols (30%, 50%, 70%, 90%, and 100% ethanol) all containing 0.3 *M* ammonium acetate.
11. Air-dry, cover to avoid dust. The specimens are now ready for autoradiography.

### 3.6. Autoradiography (see Note 6)

1. In a darkroom using a 902 filter and a 15 W bulb, heat a water bath to 42°C.
2. Cool a metal plate on ice.
3. Add 2.4 mL of 5 *M* ammonium acetate (*see* **Subheading 2.5., item 5**) and 22.6 mL of water to a cut-down 100 mL measuring cylinder or beaker in the 45°C bath. Add Ilford K5 emulsion until volume is 40 mL and leave at least 10 min to allow emulsion to melt, then use a glass rod to stir the solution thoroughly. Leave for 20 min.
4. Stir slowly again. Dip a test slide, wipe the back on a paper towel, then place on the cooled plate to chill the emulsion layer. Hold the slide up to the safelight; there should be no bubbles (or allow emulsion to rest for another 5 min) and the layer of emulsion should be smooth and even with no streakiness. If not, mix again and retest.
5. Dip slides one at a time by dipping vertically into the warm emulsion until the tissue section is immersed. In one smooth movement lift the slide clear of the emulsion and rest the end on the dipping vessel for approx 10 s to recover the excess. Wipe the back of the slide with a paper towel before placing on the cooled plate.
6. Allow slides to dry for about 2.5 h, in total darkness if possible, until the surface of the emulsion is hard when scratched with a finger nail.
7. Place dry slides in a wooden slide rack/holder/box (or plastic but not metal) and add silica gel in a packet. Seal into lightproof black plastic bag.
8. Store in a 4°C refrigerator to expose emulsion (2–70 d as appropriate).
9. Develop sets of exposed slides in safelight conditions by immersion in preprepared Kodak D-19 developer at ~18°C for 4 min, agitate each minute (use glass or plastic slide carriers).
10. Immerse in 400 mL of stop solution for 30 s.
11. Immerse in tap water for 30 s.
12. Immerse in 400 mL of fixer, two changes of 4 min each.
13. Wash extensively in many changes (preferably running) cold tap water over 1 h. Slides can now be exposed to light.
14. Counterstain tissues by immersing racked slides in dilute stock Giemsa solution for 3–4 min.
15. Wash out excess stain with tap water for 30 s.
16. Air-dry, then mount in DPX under glass coverslips.
17. When the DPX is set, clean the back of the slides with a hard backed blade, then front and back with 70% ethanol and paper tissue to remove all traces of emulsion and grease.

### 3.7. Interpretation

If possible, arrange to view the sections with a microscope that allows rapid switching between conventional illumination and reflected light (*see* **Note 7**).

Under conventional illumination with ×100 overall magnification and moderate exposure times only the most intense clusters of black silver grains can be seen (**Fig. 2A,** arrow), unless much higher magnification were used and the field of view was limited to perhaps 100 or 200 cells. In contrast, looking at the same slide in reflected light (**Fig. 2D**), much lower densities of silver grains are seen easily. More importantly, the quality of background obtained can be assessed in moments. Sections that in conventional illumination appear to have no labeling (**Fig. 2B**) may in fact reveal significant patterns of mRNA expression in reflected light (**Fig. 2E**). With Giemsa counterstain, the nuclei of cells without nearby grains appear greenish-yellow or golden in reflected light, and the threshold for deciding whether a cell or cell type is labeled can be chosen while looking at thousands of cells and associated extracellular matrices (**Fig. 2B,C,E,F**). Extracellular matrices are colored blue or even pink (**Fig. 2C,H,J,K,L**) if containing elastin in conventional illumination, and in reflected light (**Fig. 2F,I,M,N,O**) take on a variety of colors easily discriminated from white reflecting silver grains.

Switching between conventional illumination and reflected-light dark field conditions is an extremely useful way to assess whether there are any significant patterns of silver grains on the section or if significant "expression" is occurring in specific regions.

Certain cell types have reputations for nonspecific binding. Paneth cells can be misleading *(4,5)*. Eosinophils frequently show nonspecific binding despite the fact that their granules are rich in RNases *(6)*; treatment with chromotrope 2R may help *(7)*. Stratified squamous epithelium in the skin can give high background, and melanin can confuse, although in reflected light melanin is brownish, unlike the bright silver grains.

Routine use of sense strand controls is of dubious value. Sense strand probes made from the same plasmid as the antisense probe contain different domains of vector sequence. These will have a different propensity to bind nonspecifically *(8,9)*. Careful attention to probe design (*see* **Note 1** and **Fig. 1**) should reduce the likelihood that your probes contain promiscuous regions.

Consider whether there may be related mRNA targets that the antisense probe should also recognize (*see* **Note 1**).

The principal concern should be with the specificity of signals obtained with the antisense probe. Was the stringency high enough? When riboprobes are used cross-species, with low-stringency washes, it is possible that some signals derive from homologous targets, for example, other members of a multigene family.

Patterns of expression derived from using several riboprobes on near serial sections can be compared and used to establish that specific, differential patterns of labeling occurred. To establish that there was hybridizable mRNA within all cell types within a section, use a probe to β-actin (**Fig. 2L,O**). Signals from β-actin mRNA are usually strongest over vascular smooth muscle cells and the centers of inflammatory cell aggregates, and are variable between other cell types. In contrast, the levels of signal produced by a probe to the "housekeeping gene" GAPDH mRNA are less variable and thus difficult to distinguish from a high background.

Sites of expression of an mRNA can conflict with expectations. For example, immunohistochemistry for gelatinase A (matrix metalloproteinase-2 [MMP-2]) in blocks of colorectal **(Fig. 2G)** or breast carcinoma indicates that this metalloproteinase is expressed principally in the tumor epithelium, yet the mRNA is localized principally to stromal cells **(Fig. 2G,H,I,** asterisks) with low or undetectable levels of gelatinase A mRNA in the tumor epithelium ***(10,11)***. This apparent conflict may be due to rapid secretion of MMP-2 protein from the actively synthesizing stromal cells and uptake by the epithelium ***(12)***.

The density of silver grains over cells can be scored semiquantitatively, as is often done for immunohistochemical staining intensity (0, 1+, 2+, 3+) ***(13)***. Automated grain-counting equipment can be used to determine the extent of labeling of cells or other areas of interest, and may give similar results as manual counting techniques ***(14)***.

Riboprobe ISH offers greater specificity and sensitivity than other ISH methods because more labels can be hybridized specifically to each target. Nonisotopic riboprobe methods may be more sensitive but their value in experiments intended to screen for sites of expression is severely limited by the fact that peak sensitivity is dependent upon the cell type and fixation conditions. With such protocols, permeabilization conditions usually need to be "titrated" for individual blocks and even cell types within a block ***(15)***. This problem is particularly evident where PCR amplification is used to increase the sensitivity ***(16,17)***. Alternative signal amplification methods, using biotin tyramide deposition to increase the number of signal-forming sites derived from each probe, have been incorporated into protocols for localizing two mRNAs simultaneously ***(18)***.

## 4. Notes

1. The sequence used to make a probe should not contain intron sequence (which will bind to all nuclei and increase background labeling), common repetitive elements such as Alu repeats or poly(A)$^+$ tracts (which could increase background by labeling all cells weakly), or domains that are highly conserved between members of a family (which will reduce the specificity of your probe).

   Do not try to use riboprobes "across" distant species; if you lower the stringency enough to get signals, some may be from undesired homologous targets. Domains from the 3' untranslated region are useful for making probes ***(10,11,19)***; they may be less conserved and also label more intensely and give less background because they are generally GC-poor, so fragments remaining after RNase digestion will wash off at lower stringency.

   Carry out a database search to see what the intended probe is homologous to and if necessary linearize the template closer to the RNA polymerase or choose a different

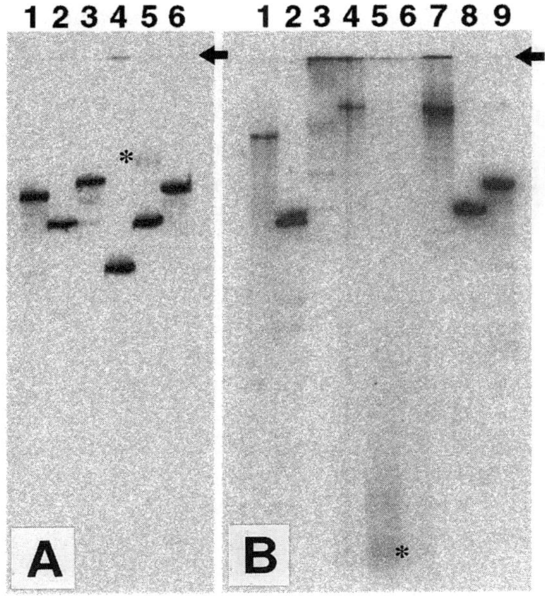


Fig. 4. Good probes and bad probes. Autoradiographs to assess riboprobe quality. Approximately 400,000 cpm of each probe were separated on 6% polyacrylamide DNA sequencing gels (Life Technologies S2 apparatus; 80 W, 40 min), dried, and exposed to Kodak X-AR film at room temperature for approx 40 min. **(A)** Some good probes as used in an experiment (Chapter 20 by Kao and Francia), and **Fig 2**. *Lane 1*, 11AN1; *lane 2*, 12AN1; *lane 3*, 12AT2; *lane 4*, 38GT2 shows some undesired banding at the top of the gel from incomplete linearisation of plasmid; *lane 5*, 40GT shows some undesired banding at the top of the gel and an additional band (asterisk) larger than the predicted size making this probe unsuitable; *lane 6*, human β-actin control riboprobe ***(19)*** is excellent. **(B)** *Lane 1*, mouse FGF8 probe is acceptable although it shows some random smearing perhaps due to nicked plasmid DNA; *lane 2*, rat β-actin probe is good with virtually all of the radioactivity concentrated in one place; *lane 3*, bovine collagen X probe was expected to be longer than 2 kb and might not enter this gel; *lanes 4–6*, unsatisfactory bovine riboprobes that show evidence for the plasmid linearising incompletely, weak or absent specific bands, and RNase degradation (asterisk); *lane 7*, human Defensin 5 riboprobe with evidence for incomplete linearization of plasmid and some nicking (smear above principal band); *lane 8*, human Defensin 6 riboprobe appears excellent; *lane 9*, human β-actin control riboprobe ***(19)*** is good (some banding evident).

region of cDNA. As a rough guide, avoid stretches of perfect match longer than 15 bases. With careful selection of probe sequence, ISH can allow detailed studies of even closely related mRNAs ***(19–21)***. Minimize the amount of GC-rich sequence derived from the multiple cloning site of plasmids, as this can increase background ***(8,9)***, or use a control vector-only riboprobe to assess this irrelevant binding ***(22)***.

There is increasing awareness of regulatory transcripts that should be detected by a sense probe ***(23–25)***.

2. Traditional alkaline lysis methods give good quality plasmid DNA. Qiagen and Wizard preparations are acceptable if the DNA is protein and RNase free. CsCl gradient DNA is often UV-nicked, giving short transcripts. Starting with small amounts of plasmid (<20 μg) often results in poor quality templates. Linearize plasmid DNA with up to fivefold excess of restriction enzyme that produces 5'-protruding or blunt ends to make an island of DNA bearing the appropriate RNA polymerase site with part or all of the cDNA insert. It does not matter if the enzyme cuts several times, as long as the desired island is created. Do not proceed to make riboprobes if the gel (**Subheading 3.1., step 15** and **Fig. 4**) shows incomplete linearization of plasmid DNA. Intact plasmids yields "wraparound" transcripts, owing to the polymerase repeatedly circling the plasmid, that are made very efficiently and will reduce the final signal-to-noise ratio. Nonspecific initiation from 3'-protruding ends (e.g., after use of *Kpn*I, *Pst*I, *Pvu*I, *Sac*I, *Sph*I) has been reported; if there are no alternative restriction sites use Klenow to blunt such templates by adding 5 U of Klenow DNA polymerase to the transcription reaction *before* adding any nucleotides or RNA polymerases; incubate at 22°C for 15 min, then add the other components.

   Use [$^{35}$S]UTP, rather than ATP, GTP, or CTP, which need higher molarities to label efficiently. Synthesis with double labels ([$^{35}$S]UTP and [$^{35}$S]CTP together for high specific activity) is inefficient. Moderate specific activity [$^{35}$S]UTP (~800 Ci/mmol at 40 mCi/mL) makes probes with approx 3 out of 4 U residues labeled, and allows efficient synthesis of long probes at ~$1.3–1.7 \times 10^9$ cpm/μg of RNA. Never add unlabeled UTP to the transcription reaction.

   $^{33}$P is a more expensive alternative giving results with similar spatial resolution ***(2)***, but no advantages in background graininess when half-life is taken into account, or improvement in sensitivity when emulsion is kept thin to maintain resolution ***(26)***. $^{3}$H offers excellent spatial resolution, but requires lengthy exposure times.

   Exceeding 10 U of RNA polymerase per pmol of template (1 μg of a 3-kb vector contains approx 0.5 pmol of template, thus needs just 5 U of enzyme) is thought to allow transcription also from other RNA polymerase sites in the template preparation.

   Do not cut sections for an ISH run until confident that the probes are acceptable. Incorporation of $^{35}$S should be >40%, but is no guide to probe quality, which should be assessed on a 6% polyacrylamide DNA sequencing gel. If transcripts are not of the expected length, consider the possibility of incorrect orientation of the insert, relative to the RNA polymerase site used. Good probes (**Fig. 4A**; lanes 1,2,3,6) are characterized by a single band of the anticipated size (size standards can be made using Promega's control templates). Secondary structures in individual templates give characteristic fingerprints of less abundant shorter fragments. Nicked DNA templates from poor plasmid preparations give smears of highly variable sizes, for example, **Fig 4B**, lane 1. RNase contamination often gives a smear of fragments low down the gel (**Fig. 4B**; lanes 4–6), but less severe contamination gives a complex pattern of fragments and such probes should be discarded. Further examples are shown in **ref.** ***27***. When RNase contamination is detected, it is usually best to discard all current labeling reagents and start with new stocks.

Hydrolysis of probes is often recommended, but this is wasteful of probe and does not improve the final signal-to-noise ratio. Short riboprobes work, but the best signal-to-noise ratios are obtained with longer probes, 0.3–1.0 kb (even 1.6 kb can work well), nonhydrolyzed and with permeabilization of sections carried out as described.

Do not use riboprobes that have been made more than a few days earlier (reexamination on a 6% polyacrylamide DNA sequencing gel usually reveals some degradation).

3. Many formalin-based fixatives work well, for example, neutral buffered formalin (NBF), formal saline, phenol formalin, formal calcium. Limit fixation to not more than one overnight. For small pieces of gut at least, perfusion fixation does not seem to give better results than immersion. Glutaraldehyde fixed tissues are too resistant to protease digestion. Avoid the coagulating fixatives Carnoy's *(28)* and methacarn, and citrus oil substitutes for xylene during processing. Tissues fixed in the presence of mercury produce autoradiographic artefacts.

   Chill paraffin blocks face up on ice as needed, but do not use any block face treatments to assist cutting. Discard the first five sections from the block, even if you have used it recently. Lift sections onto the floating-out bath with a clean brush. Sections of pancreas seem able to leak RNases into the bath but using DEPC-treated water in the bath appears to help.

   Sections dried onto APES-treated slides at 37°C overnight are best used for ISH within a week, otherwise the signal strength is reduced. Label slides with pencil. For cell cultures, consider Nunc multiwell slides, as the effects of different culture conditions can be compared on a single slide *(29)*. Cells on slides should be fixed with 4% PFA in PBS, or in NBF for a few hours, then rinsed in PBS, dehydrated with ascending alcohols, and air-dried. Use slides fresh if possible, or store slides dry at –70°C for several weeks.

   Protease treatment seems unnecessary with frozen sections or cultured cells. Postfixation in 4% PFA in PBS ensures inactivation of the proteinase (and of any RNases). Acetylation is thought to reduce background *(30)*. Sections should be permeabilized, dehydrated, and hybridized on the same day.

   For each block of tissue use a section hybridized to a positive control probe, usually β-actin *(19)*, and two sections of each "query" probe, to allow two exposure times.

4. The volume of hybridization buffer needed is approx 20 µL per section that can be covered completely by a 22 × 42 mm coverslip. For 100 µL of probe mix use 84 µL of hybridization buffer with 16 µL of probe–water. Use approx 1 × $10^6$ cpm probe per 20 µL of final probe mix. Arrange slides in a fume hood in groups for specific probes on a sheet of paper or purpose-made plastic boxes (Slide Show, BioGenex). Coverslips marked with a permanent marker are easier to see at **Subheading 3.5., step 4**.

   Hybridization is reasonably stringent at 55°C. The combination of a relatively high temperature and a low ion concentration is aimed at reducing the strength of binding between probe and tissue RNA. Bovine ribosomal RNA is used during hybridization to reduce nonspecific binding. Humidification with formamide and salts as used in the hybridization buffer minimizes evaporation and edge effects, but chambers should be filled and emptied in a fume hood.

5. Posthybridization RNase treatment and stringency washings are key steps affecting the signal-to-noise ratio, and the certainty with which the signal can be considered specific.

   The repeated post-RNase washes are important. Remember that gloves spread RNases by contact. There is no advantage in including mercaptoethanol in the stringency washes that remove remaining short probe segments.

   The use of higher stringency washes (0.1 × SSC at 65°C after **Subheading 3.5., step 7**) may be more reassuring if discrimination between highly homologous targets is needed, but specific signals will be weaker.
6. The darkroom must be dark; even indicator lamps on equipment or switched power points may expose emulsion. Test the room by leaving out dipped slides (or X-ray film) part shielded with aluminum foil for 4 h, develop, then compare the background graininess. Expose slides away from high-energy radioisotopes.

   Avoid problems by using only high-quality solvents and changing xylene and alcohols frequently.

   Forty milliliters of prepared emulsion is enough for 150 slides if a dipping chamber is used (e.g., Amersham, cat. no. RPN 39). Sections that were mounted away from the frosted end need less emulsion. Lead "doughnuts" (Sigma, cat. no. L2775) help keep the emulsion chambers steady in the water bath. Do not leave the metal plate on ice while dipped slides are drying or condensation will ruin the emulsion.

   Anticipate exposure times of 2–5 d for extremely abundant mRNAs, 7–10 d for β-actin and other abundant mRNAs, 21–35 d for moderately abundant mRNAs. For a new target with new tissue blocks develop the positive control set of slides hybridized for β-actin mRNA and a set of the test riboprobe after 10 d; if the β-actin set have worked well but there are no patterns or background graininess apparent for the test set, try developing the second set after approx 30 d.

   Giemsa's stain is good for discriminating many tissue structures and cell types. *In situ* slides stained with hematoxylin and eosin look "abnormal" because proteinase digestion increases the eosinophilia.
7. Reflected-light darkfield illumination is popular for metallurgy and can be produced using an epi-illumination lamp and objectives with an outer prism sheath that directs the light down on to the specimen. The micrographs in **Fig. 2** were taken on a relatively inexpensive Olympus BH-2 with 100 W tungsten epi-illumination lamp housing (BH2-UMA and BH2-HLSH) and ×10 and ×20 Neo S Plan IC objective lenses.

   Several alternative systems are available using epifluorescence microscopes (with or without cross-polarizing filters to obscure most elements of the tissue section). Condenser block darkfield systems cannot achieve good results if there are dense clusters of silver grains because scattered light illuminates the object as well.

   Plastic film coverslips (used by some automated coverslipping machines) produce poorer darkfield images.

   RNase pretreatment of sections as negative controls can be demanded by editors, even though it is likely that residual RNase could degrade the riboprobe rather than eliminate the RNA target. It should not be difficult to find a block of tissue that has no hybridizable mRNA.

## Acknowledgments

I am grateful to those who developed the basis of the method *(1)*; to Rosemary Jeffery, Jan Longcroft, and Len Rogers who carried out the techniques with such care; and the pathologists Gordon Stamp, Andrew Hanby, and Nick Wright who have helped me learn much using this technique. This work was supported by the Imperial Cancer Research Fund.

## References

1. Senior, P. V., Critchley, D. R., Beck, F., Walker, R. A., and Varley, J. M. (1988) The localisation of laminin mRNA and protein in the postimplantation embryo and placenta of the mouse: an in situ and immunohistochemical study. *Development* **104,** 431–446.
2. Lu, L. H. and Gillett, N. A. (1994) An optimised protocol for in situ hybridization using PCR-generated $^{33}$P-labeled riboprobes. *Cell Vision* **1,** 169–176.
3. Chinery, R., Poulsom, R., Rogers, L. A., Jeffery, R. E., Longcroft, J. M., Hanby, A. M., and Wright, N. A. (1992) Localisation of intestinal trefoil-factor mRNA in rat stomach and intestine by hybridization in situ. *Biochem. J.* **285,** 5–8.
4. Garrett, K. L., Grounds, M. D., and Beilharz, M. W. (1992) Nonspecific binding of nucleic acid probes to Paneth cells in the gastrointestinal tract with in situ hybridization. *J. Histochem. Cytochem.* **40,** 1613–1618.
5. Lacasse, J. and Martin, L. H. (1992) Detection of CD1 mRNA in Paneth cells of the mouse intestine by in situ hybridization. *J. Histochem. Cytochem.* **40,** 1527–1534.
6. Domachowske, J. B., Dyer, K. D., Adams, A. G., Leto, T. L., and Rosenberg, H. F. (1998) Eosinophil cationic protein/RNase 3 is another RNase A-family ribonuclease with direct antiviral activity. *Nucleic Acids Res.* **26,** 3358–3363.
7. Patterson, S., Gross, J., and Webster, A. D. (1989) DNA probes bind non-specifically to eosinophils during in situ hybridization: carbol chromotrope blocks binding to eosinophils but does not inhibit hybridization to specific nucleotide sequences. *J. Virol. Methods* **23,** 105–109.
8. Witkiewicz, H., Bolander, M. E., and Edwards, D. R. (1993) Improved design of riboprobes from pBluescript and related vectors for in situ hybridization. *BioTechniques* **14,** 458–463.
9. Blödorn, B., Brück, W., Rieckmann, P., Felgenhauer, K., and Mäder, M. (1998) A method for preventing artefactual binding of cRNA probes to neurons caused by in situ hybridization. *Analyt. Biochem.* **255,** 95–100.
10. Poulsom, R., Pignatelli, M., Stetler-Stevenson, W. G., Liotta, L. A., Wright, P. A., Jeffery, R. E., et al. (1992) Stromal expression of 72kDa type IV collagenase (MMP-2) and TIMP-2 mRNAs in colorectal neoplasia. *Am. J. Pathol.* **141,** 389–396.
11. Poulsom, R., Hanby, A. M., Pignatelli, M., Jeffery, R. E., Longcroft, J. M., Rogers, L., and Stamp, G. W. H. (1993) Expression of gelatinase A and TIMP-2 mRNAs in desmoplastic fibroblasts in both mammary carcinomas and basal cell carcinomas of the skin. *J .Clin. Pathol.* **46,** 429–436.

12. Ohtani, H., Tabata, N., and Nagura, H. (1995) Immuno-electron microscopic localization of gelatinase A in human gastrointestinal and skin carcinomas: difference between cancer cells and fibroblasts. *Jpn. J. Cancer Res.* **86,** 304–309.
13. Dorudi, S., Hanby, A.M., Poulsom, R., Northover, J., and Hart, I. R. (1995) Level of expression of E-cadherin mRNA in colorectal cancer correlates with clinical outcome. *Br. J. Cancer* **71,** 614–616.
14. Alison, M., R., Poulsom, R., Jeffery, R., Anilkumar, T. V., Jagoe, R., and Sarraf, C. E. (1993) Expression of hepatocyte growth factor mRNA during oval cell activation in the rat liver. *J. Pathol.* **171,** 291–299.
15. Yao, R., Sugino, I. K., Greulich, K. M., Ishida, M., Weier, H.-U., and Zarbin, M. A. (1998) Optimization of non-isotopic in situ hybridization: detection of the Y-chromosome in paraformaldehyde-fixed, wax-embedded cat retina. *Exp. Eye Res.* **66,** 223–230.
16. Boshoff, C., Schulz, T. F., Kennedy, M. M., Graham, A. K., Fisher, C., Thomas, A., et al. (1995) Kaposi's sarcoma-associated herpes virus infects endothelial and spindle cells. *Nature Med.* **1,** 1274–1278.
17. Mee, A. P., Dixon, J. A., Hoyland, J. A., Davies, M., Selby, P. L., and Mawer, E. B. (1998) Detection of canine distemper virus in 100% of Paget's disease samples by in situ-reverse transcriptase-polymerase chain reaction. *Bone* **23,** 171–175.
18. Berger, U. V. and Hediger, M. A. (1998) Comparative analysis of glutamate transporter expression in rat brain using differential double in situ hybridization. *Anat. Embryol.* **198,** 13–30.
19. Zandvliet, D. W. J., Hanby, A. M., Austin, C. A., Marsh, K. L., Clark, I. B. N., Wright, N. A., and Poulsom R. (1996) Analysis of foetal expression sites of human type II DNA topoisomerase $\alpha$ and $\beta$ mRNAs by in situ hybridization. *Biochim. Biophys. Acta* **1307,** 239–247.
20. Gilham, D. E., Cairns, W., Paine, M. J. I., Modi, S., Poulsom, R., Roberts, G. C. K., and Wolf, C. R. (1997) Metabolism of MPTP by cytochrome P4502D6 and the demonstration of 2D6 mRNA in human foetal and adult brain by in situ hybridization. *Xenobiotica* **27,** 111–125.
21. Poulsom, R., Hanby, A. M., Lalani, E.-N., Hauser, F., Hoffman, W., and Stamp, G. W. H. (1997) Intestinal trefoil factor (TFF3) and pS2 (TFF1) but not spasmolytic polypeptide (TFF2) mRNAs are co-expressed in normal, hyperplastic, and neoplastic human breast epithelium. *J. Pathol.* **183,** 30–38.
22. Dunne, J., Hanby, A. M., Poulsom, R., Jones, T., Sheer, D., Chin, W. G., et al. (1995) Molecular cloning and tissue expression of FAT, the human homologue of the *Drosophila* fat gene that is located on chromosome 4q34-q35 and encodes a putative adhesion molecule. *Genomics* **30,** 207–223.
23. Lipman, D. J. (1997) Making (anti)sense of non-coding sequence conservation. *Nucleic Acids Res.* **25,** 3580–3583.
24. Moorwood, K., Charles, A. K., Salpekar, A., Wallace, J. I., Brown, K. W., and Malik, K. (1998) Antisense WT1 transcription parallels sense mRNA and protein expression in fetal kidney and can elevate protein levels in vitro. *J. Pathol.* **185,** 352–359.

25. Takeda, K., Ichijo, H., Fujii, M., Mochida, Y., Saitoh, M., Nishitoh, H., et al. (1998) Identification of a novel bone morphogenetic protein-responsive gene that may function as a non-coding RNA. *J. Biol. Chem.* **273,** 17079–17085.
26. Longcroft, J. M., Jeffery, R. E., Rogers, L. A., Hanby, A. M., and Poulsom, R. (1994) $^{33}$P is a poor alternative to $^{35}$S for routine isotopic in situ hybridization. *Cell Vision* **1,** 90–91.
27. Poulsom, R., Longcroft, J. M., Jeffery, R. E., Rogers, L. A., and Steel, J. H. (1998) A robust method for isotopic riboprobe in situ hybridization to localise mRNAs in routine pathology specimens. *Eur. J. Histochem.* **42,** 121–132.
28. Urieli-Shoval, S., Meek, R. L., Hanson, R. H., Ferguson, M., Gordon, D., and Benditt, E. P. (1992) Preservation of RNA for in situ hybridization: Carnoy's versus formaldehyde fixation. *J. Histochem. Cytochem.* **40,** 1879–1885.
29. Hodivala, K. J. and Watt, F. M. (1994) Evidence that cadherins play a role in the downregulation of integrin expression that occurs during keratinocyte terminal differentiation. *J. Cell Biol.* **124,** 589–600.
30. Hayashi, S., Gillam, I. C., Delaney, A. D., and Tener, G. M. (1978) Acetylation of chromosome squashes of drosophila melanogaster decrease the background in autoradiographs from hybridization with [$^{125}$I]-labeled RNA. *J. Histochem. Cytochem.* **26,** 677–679.
31. Lee, A. H. S., Dublin, E. A., Bobrow, L. G., and Poulsom, R. (1998) Invasive lobular and invasive ductal carcinoma of the breast show distinct patterns of vascular endothelial growth factor expression and angiogenesis. *J. Pathol.* **185,** 394–401.
32. Srinivasan, R., Poulsom, R., Hurst, H. C., and Gullick, W. J. (1998) Expression of the c-erbB-4/HER4 protein and mRNA in normal human foetal and adult tissues and in a survey of nine solid tumor types. *J. Pathol.* **185,** 236–245.

# 14

# Fluorescence *In Situ* Hybridization

**Hakan Goker and Janet Shipley**

## 1. Introduction

*In situ* hybridization describes the annealing of a labeled nucleic acid to complementary nucleic acid sequences in a fixed target (e.g., chromosomes, free nuclei, nuclei in tissue sections, and DNA) followed by visualisation of the location of the probe. Since its development about 30 yr ago *(1,2)*, it has transformed into a highly effective and rapid technique for uses such as characterizing chromosome aberrations, gene mapping, and marker ordering as well as expression studies.

All *in situ* hybridization originally used radioactively labeled probes, and methodology for *in situ* hybridization using radioactive probes is covered in Chapter 13 by Poulsom. The strict regulations on radioactivity, long exposure times, and some practical difficulties with the use of radioactive labels limited the wide application of the technique. In the 1980s, several methods using non-radioactive labeling were developed *(3–7)*. The ease and effectiveness of fluorescence methods (fluorescence *in situ* hybridization [FISH]) in particular have now almost rendered the radioisotopic techniques obsolete. FISH has been developed to incorporate chromosome painting and the analysis of the whole genome for aberrations using the approaches of comparative genomic hybridization (CGH; described in detail in Chapter 15 by Roylance), multifluor FISH (M-FISH), and spectral karyotyping (SKY).

In FISH, essentially the probes are labeled either directly or indirectly with various fluorochrome dyes such as fluorescein isothiocyanate (FITC) and tetramethyl rhodamine that fluoresce at different wavelengths when excited by ultraviolet (UV) light. Most commonly deoxyuridine 5'-triphosaphate (dUTP) is conjugated directly to the fluorochrome or to biotin or digoxygenin and used in the place of dTTP in a labeling reaction. Following hybridization, washing off excess unbound probe at specific stringencies, indirectly labeled probes are detected; the location of probes

From: *Methods in Molecular Medicine, vol. 57:*
*Metastasis Research Protocols, Vol. 1: Analysis of Cells and Tissues*
Edited by: S. A. Brooks and U. Schumacher © Humana Press Inc., Totowa, NJ

can be seen under an epifluorescence microscope. The advantage of fluorescence detection for *in situ* hybridization includes the ability to use multiple fluorochromes to allow simultaneous specific detection of multiple probes. In this way the linear order of probes can be determined on metaphase chromosomes, interphase nuclei, and released chromatin. FISH also enables numerical (deletion/gain) ***(8)*** and structural (translocation/inversion) aberrations to be detected.

This chapter deals with DNA probes hybridized to DNA targets primarily for localization of genes and markers and refining cytogenetic analysis. We describe the methods of one or more color FISH on metaphase chromosomes, interphase nuclei, nuclei in paraffin wax-embedded sections, and released chromatin. M-FISH ***(9)*** and SKY ***(10)*** probes and analysis systems are currently commercially available from Vysis and Applied Spectral Imaging Ltd. As these companies provide detailed protocols, the methodologies are not described here. For CGH, please refer to Chapter 15 by Roylance.

### 1.1. Probes

A wide variety of probe types are used to hybridize to target molecules in different preparations for a variety of FISH applications. Probes contain sequences homologous to either specific repetitive or unique regions of the genome.

Probes that detect tandemly repeated sequences for centromere and telomere visualization (α-satellites, β-satellites, telomere probes) are useful to deduce aneuploidy and aberrations in tumor material either in metaphase chromosomes or in some cases interphase nuclei. Chromosome specific paints are probes that hybridize to the whole or part of the chromosome and are generally derived from chromosomes sorted on a fluorescence-activated cell sorter or chromosomes microdissected from fixed preparations. These are commercially available and the many unique regions in the probe bind specifically to targets along the length of the chromosome while the nonspecific repetitive regions are suppressed through the addition of unlabeled nonspecific repetitive DNA such as Cot-1 DNA. This approach is known as chromosomal *in situ* suppression (CISS) or competitive *in situ* suppression ***(11,12)***. Suppression of nonspecific repeats is also required for unique sequence probes that contain such elements, which is likely for any genomic probe greater than a few kilobases (kb) in size. Unique sequence probes include cloned cDNA or more usually genomic DNA which vary according to the cloning vector's capacity to accept different lengths of sequence. The following are the approximate insert sizes of the most commonly used probes:

Plasmids, 500 bp–2 kbp
Phages, 5–10 kbp
Cosmids, 30–40 kbp
P1-artificial chromosomes (PACs), 130–150 kbp ***(13)***
Bacterial artificial chromosomes (BACs) 100 kbp–1 Mbp ***(14)***
Yeast artificial chromosomes (YACs) 100 kbp–2 Mbp ***(15)***

The degree of homology between the probe and target, and potentially less homologous targets, is a consideration in determining the stringency of hybridization and washing. The standard conditions are fairly stringent but, for example, for cross species hybridizations the stringency may be lowered. Another consideration for unique sequence probes is the size of the insert as this impacts on the amount of probe required. As the insert size gets smaller, the detection of signals gets more difficult. Mapping of probes smaller than 2–3 kbp is difficult and needs statistical analysis. Stronger signals may be obtained through indirect detection and extra rounds of signal amplification following labeling with molecules such as biotin.

As the amount of probe is important, it is essential for accurate measurement of the DNA concentration. In our laboratory, DNA concentration is measured using a fixed wavelength fluorometer, which is designed specifically for the accurate quantitation of low concentrations of DNA. DNA quantitation is based on the binding of *bis*-benzimidazole, commonly known as Hoechst 33258, to DNA. The emission and detection peaks of the fluorometer are optimized for detecting the bound form of the dye. The binding of Hoechst 33258 to DNA is unaffected by buffers commonly used to extract DNA from whole cells, low levels of detergents, high salt concentrations, and most importantly by RNA contamination, as RNA does not compete with DNA for binding to Hoechst 33258.

### 1.2. One or more Color FISH on Metaphase Chromosomes

FISH on metaphase chromosomes can be used to localize a new probe to a specific chromosome band, to reveal characteristic chromosomal abnormalities such as translocations, to check the localization of a potentially chimeric probe (e.g., yeast artificial chromosomes), or to determine the linear order of probes on a chromosome. In relation to cancer and metastasis, metaphase FISH can be used to characterize specific chromosomal aberrations in tumor cell cultures and cell lines. To help a karyotype-untrained eye, the probe can be cohybridized to a chromosome with a differentially labeled marker or centromere specific probe. For linear ordering there should be at least 1–3 Mb distance between the two probes ***(16,17)*** for clear visualization, and as chromatin folding may cause problems with relatively close probes several metaphases should be analyzed or a higher resolution approach taken.

### 1.3. One or more Color FISH on Interphase Nuclei

Interphase FISH can be applied to cases where metaphase material is not available, for example, fresh or frozen solid tissue tumors and archival material embedded in paraffin wax. Because chromatin in interphase nuclei is less condensed than in metaphase chromosomes, mapping resolution is higher. Interphase FISH analysis of many nuclei can be used to determine probe order in the 100–1000 kb range. The signal patterns change according to chromatin folding, the degree of condensation, and whether the cells have gone through the DNA synthesis stage of the cell cycle when they were fixed in methanol–acetic acid (*see* **Subheading 3.1.**). If the cells are captured before cell division, each chromosome will have only one chromatid; thus one would observe hybridization signals as single dots. After replication of DNA during the S-phase, the probe will hybridize to two sister chromatids per chromosome, giving signal doublets.

### 1.4. FISH on Chromatin Fibers

The resolution of FISH can be increased even higher by using free chromatin fibers that are fixed on a glass slide. This enables ordering of probes that are overlapping or only a few kilobases apart. The distances in released chromatin can be estimated by using the ratio between the known length of a probe in kilobases and the length of the probe signal in micrometers as an internal ruler ***(18)***. Statistical analysis is required to eliminate errors due to inconsistent chromatin decondensation and DNA breaks. Rearrangements can also be analyzed using specific probes (e.g., if a gene sequence is inserted into the exons or introns of another gene) ***(19)***.

Chromatin can be released both from fixed cells ***(18)*** and fresh/frozen cells ***(20)***. Both methods are described in this chapter and variations of the releasing agents will be discussed in the Notes.

### 1.5. M-FISH and SKY

Multifluor FISH (multiplex FISH) and SKY are technological developments that have only recently become available, allowing independent recognition of multiple chromosomes simultaneously. Previously, only a few fluorochromes were available for FISH, thus the number of probes that could be tested per experiment by chromosome painting was limited. In 1990, Nederlof et al. ***(21)*** described a method for detecting more than three target DNA sequences using only three fluorochromes. This technique was named "combinatorial labeling" ***(21–23)*** or "ratio labeling" ***(24)***. A probe can be labeled with one or more fluorochrome-conjugated nucleotides and detected with one or more optical filters. The number of useful Boolean combinations of N fluorochromes is $2^N$–1. Following this idea, Speicher et al. developed epifluorescence filter sets and computer software to detect and discriminate 27 different DNA probes hybridized simultaneously, a method known as M-FISH ***(9)***. Twenty-two autosomes and two sex chromosomes are labeled combinatorially and hybridized to metaphase spreads. In this way, simple and complex translocations, interstitial deletions and insertions, chromosomal aneusomies, and double minute chromosomes can be identified rapidly.

SKY is a similar approach, developed around the same time, in which each human chromosome is observed in a different color ***(10)***. SKY uses a spectral imaging approach that combines Fourier spectroscopy, charge-coupled device (CCD) imaging, and optical microscopy to measure simultaneously at all points in the sample emission spectra in the visible and near-infrared spectral range. The main difference between SKY and M-FISH is therefore that in SKY the image is captured with a triple bandpass filter set that allows all dyes to be excited and measured together without image shift.

## 2. Materials

### 2.1. Preparation and Processing of Slides

#### 2.1.1. Lymphocyte Culture and Preparation of Metaphase Slides

1. Sodium heparin: Stock concentration of 1000 U/mL, stored at 4°C.
2. Phytohemagglutinin (PHA M form): Available dissolved in 5 mL of distilled sterile water, stored in 1-mL aliquots at –20°C or at 4°C for 30 d (GibcoBRL-Life Technologies cat. no. 10576-023).
3. Culture medium: RPMI 1640, L-Glutamine (200 m*M*).
4. 25-$cm^2$ Cell culture flasks.
5. Fetal bovine serum.
6. Penicillin and streptomycin solution: Stock concentration of 10.000 IU/mL.
7. Colcemid solution: 10 µg/mL, store at 4°C.
8. Hypotonic solution: 0.075 *M* potassium chloride, warm to 37°C before use.
9. Fixative solution: 1 vol glacial acetic acid to 3 vol methanol; prepare fresh and cool to 4°C before use.
10. Clean 76 mm × 26 mm glass microscope slides. Store in 500 mL of 100% ethanol and dry immediately with lint-free tissues before use.
11. Water bath at 80°C.
12. Fine-tipped plastic pipets.
13. Silica gel as desiccant.
14. Versene (cell line culture harvest): 1:5000 (1X), store at 4°C (GibcoBRL).
15. Trypsin–EDTA (cell line culture harvest): 0.5 g of trypsin and 0.2 g of EDTA per liter of Modified Puck's Saline A (GibcoBRL).

#### 2.1.2. Cell Line Culture and Preparations of Metaphase Slides

Same as **Subheading 2.1.1.** (for growing the cells, use the appropriate cell culture media).

#### 2.1.3. Interphase Nuclei

##### 2.1.3.1. Cells Harvested in Suspension

Same as **Subheading 2.1.1.**

##### 2.1.3.2. Frozen Tissue (Touch Preps)

1. Clean 76 mm × 26 mm glass microscope slides. Store in 500 mL of 100% ethanol, and dry immediately with lint-free tissues before use.

2. Fixative solution: 1 vol glacial acetic acid to 3 vol methanol; prepare fresh and cool to 4°C before use.
3. Silica gel as desiccant.
4. L-15 (Leibovitz) with Glutamax-l (Sigma cat. no. 31415-011), store at 4°C.
5. Dulbecco's PBS: Phosphate-buffered saline (500 mL for 10 slides); store at 4°C.
6. Type H collagenase: Dissolve 1 g powder in 10 mL of L-15 and store at –20°C as 50 μL aliquots (cat. no. C-8051).
7. 1 *M* $MgCl_2$, pH 7.3: Autoclaved.
8. Formaldehyde solution: Store at 15–30°C (contains 10–14% methanol added as a stabilizer, Fisher Scientific, cat. no. F/1500/PB08). Toxic; use in fume cupboard.
9. 70%, 90%, and 100% ethanol solution series.

#### 2.1.3.3. Paraffin Wax-Embedded Sections

1. Clean 76 mm × 26 mm silane-treated glass microscope slides.
2. Histoclear.
3. 1 *M* Sodium thiocyanate: Prewarmed to 45°C prior to use.
4. Pepsin: Powdered, pepsin A 1:10,000.
5. 0.9% NaCl; adjust pH to 1.5 with 10 *M* HCl.
6. Hank's balanced salt solution.
7. 70%, 90%, and 100% ethanol solution series.

### *2.1.4. Fiber Chromatin*

#### 2.1.4.1. Chromatin Release from Fixed Cells

1. Dulbecco's PBS: Store at 4°C.
2. Chromatin releasing solution A: 0.05 M NaOH in 30% ethanol (*see* **Note 1**).
3. 24 mm × 60 mm long coverslips.
4. Methanol.
5. 70%, 90%, and 100% ethanol solution series.

#### 2.1.4.2. Chromatin Release from Fresh and Frozen Cells

1. Dulbecco's PBS: Store at 4°C.
2. Chromatin releasing solution B: 0.5% sodium dodecyl sulfate (SDS), 50 m*M* EDTA, 200 m*M* Tris-HCl, pH 7.4.
3. Fixative solution: 1 vol glacial acetic acid to 3 vol methanol; prepare fresh.

## *2.2. Denaturation of Slides*

### *2.2.1. Nonparaffin Wax-Embedded Slides*

1. Denaturation solution: 70% Formamide, 2X saline sodium citrate (SSC), make up in sterile water and adjust pH to 7.0.
2. 70%, 90%, and 100% ethanol solution series.

### *2.2.2. Paraffin Wax-Embedded Slides*

Codenaturation method is used for paraffin wax-embedded sections *(25)*. *See* **Subheading 3.2.2.**

## 2.3. Preparation of Probes

### 2.3.1. Growing Clones and Extracting DNA

#### 2.3.1.1. Yeast Artificial Chromosomes

1. YPD agar plates: YPD-agar is stored as solid. Melt agar and cool to 55°C and pour into Petri dishes; dry in tissue culture hood for 15 min before use.
2. SD broth 5X: 35 g of yeast nitrogen base without amino acids, 70 g of casamino acids, 100 g of dextrose (D-glucose). Dissolve in 1 L of distilled sterile water, filter sterilize with a 0.2 µm filter, and keep at 4°C. Dilute to 1X SD with sterile water before use. (Adenine sulfate and tyrosine [*see* **steps 3** and **4**] need to be added to 1X SD, as they would come out of solution in 5X SD.)
3. Adenine sulfate: Dissolve 55 mg of adenine sulfate in a few drops of 10 *M* NaOH and make up to 1 mL with sterile water. Filter sterilize as described above. Add 100 µL into each 100 mL of 1X SD culture.
4. Tyrosine: Dissolve 55 mg of tyrosine in a few drops of 10 *M* NaOH and make up to 1 mL with sterile water. Filter sterilize as described above. Add 100 µL into each 100 mL of 1X SD culture.
5. Ampicillin: Stock concentration of 50 mg/mL; add 100 µL into each 100 mL of 1X SD culture.
6. 250-mL Corning conical centrifuge tubes.
7. Resuspension solution: 0.9 *M* Sorbitol; 20 m*M* EDTA; 10 m*M* Tris-HCl, pH 7.5.
8. 14 *M* β-mercaptoethanol (BME).
9. Lyticase: 20 U/µL of stock concentration.
10. 76 mm × 26 mm glass microscope slides and 22 mm × 22 mm coverslips.
11. TEN–SDS solution: 9 mL of TEN (10 m*M* Tris-HCl, pH 8.0, 40 m*M* EDTA, pH 8.0, 10 m*M* NaCl), 1 mL of 10% w/v SDS, make fresh, mix well, and use immediately.
12. Proteinase K: 50 mg/mL stock concentration.
13. RNase A: 10 mg/mL stock concentration.
14. Phenol: Tris-buffered.
15. Chloroform–isoamylalcohol mix: 24 vol of chloroform to 1 vol of isoamylalcohol.
16. 3 *M* Sodium acetate: Autoclaved.
17. 100% ethanol cooled to 4°C.
18. 70% ethanol.
19. 30% Glycerol in 1X SD for making glycerol stocks.

#### 2.3.1.2. Plasmid, BAC, Cosmid, and PAC

1. LB agar plates: LB agar is stored as solid. Melt agar and cool to 55°C, add appropriate antibiotic, and pour into Petri dishes. Dry in tissue culture hood for 15 min before use.
2. LB broth: Store at 4°C until use.
3. Appropriate antibiotic.
4. Solution 1: 50 m*M* glucose, 10 m*M* EDTA, 25 m*M* Tris-HCl, pH 8.0, 100 µg/mL of RNase A, filter sterilized and cooled to 4°C.
5. Solution 2: 0.2 *N* NaOH, 1% w/v SDS, freshly prepared.

6. Solution 3: 60 mL of 5 *M* potassium acetate, 11.5 mL of glacial acetic acid, 28.5 mL of water, filter sterilized and kept at room temperature (RT). The resulting solution is 3 *M* with respect to potassium and 5 *M* with respect to acetate.
7. Phenol: Tris buffered.
8. Chloroform–isoamylalcohol mix: 24 vol of chloroform to 1 vol of isoamylalcohol.
9. 100% Isopropanol.
10. 70% Ethanol.
11. RNase A: 10 mg/mL stock concentration.

#### 2.3.2. Measuring DNA Concentration

1. TKO 100 minifluorometer: Hoefer Scientific Instruments (San Francisco, CA).
2. TKO 105 fluorometry cuvet.
3. 1 mg/mL Hoechst 33258 concentrated dye stock in water; store away from light at 4°C for up to 6 mo.
4. Standard DNA solution in water: Calf thymus DNA standard, 100 µg/mL for assay solution A and 1 mg/mL for assay solution B.
5. 10X TNE concentrated buffer: 100 m*M* Tris-HCl, 2 *M* NaCl, 10 m*M* EDTA, pH 7.4 in water. Can be stored at 4°C for up to 12 mo.
6. 1X TNE working buffer: Dilute concentrated buffer 1:10.
7. (a) Assay solution A: 10 µL of concentrated dye stock, 100 mL of working buffer (1X TNE), standard assay for measuring DNA samples between 10 ng/mL and 500 ng/mL final concentration, prepare fresh and mix well. Or, (b) assay solution B: 100 µL of concentrated dye stock, 100 mL of working buffer (1X TNE), extended range assay, for measuring DNA samples between 100 ng/mL and 2000 ng/mL final concentration; prepare fresh and mix well.

### 2.4. Labeling and Processing of Probes

#### 2.4.1. Probe Labeling

1. 1 µg of DNA solution to be labeled.
2. 10X dNTP mix including fluro-dNTP: Add and mix well the following components, store away from light at –20°C as 60 µL aliquots.

| Contents as final mix | Stock concentration | 300 µL total volume |
|---|---|---|
| 0.2 m*M* dCTP[a] | 10 m*M* | 6 µL |
| 0.2 m*M* dATP[a] | 10 m*M* | 6 µL |
| 0.2 m*M* dGTP[a] | 10 m*M* | 6 µL |
| 0.1 m*M* dTTP[a] | 10 m*M* | 3 µL |
| 0.1 m*M* fl-dUTP[b] | 1 m*M* | 30 µL |
| 50 m*M* $MgCl_2$ | 1 *M* | 15 µL |
| 100 m*M* BME | 1 *M* | 30 µL |
| 100 µg/mL BSA[c] | 1 mg/mL | 30 µL |
| 500 m*M* Tris-HCl, pH 7.8 | 1 *M* | 150 µL |
| Sterile water | | 24 µL |

[a]Ultrapure dNTP set 100 m*M* solutions, Pharmacia Biotech, cat. no. 27-2035-01, store at –20°C.

[b]Fluorescein or rhodamine conjugated dUTP (FluoroGreen, RPN2121, and FluoroRed, RPN2122, respectively, Amersham Life Science); store away from light at –20°C.

[c]Bovine serum albumin (BSA) (DNAse free) Pharmacia Biotech (cat. no. 27-8915-01); store at 4°C.

3. DNase 1 enzyme mix: 15 U/mL DNase (Promega), 50 m*M* Tris-HCl, 5 m*M* magnesium acetate, 1 m*M* BME, 50% v/v glycerol, 100 µg/mL of BSA; store at –20°C.
4. DNA polymerase 1:10 U/µL, Promega cat. no. M205A; store at –20°C.
5. 0 5 *M* EDTA, pH 7.5.
6. Sterile distilled water.

#### *2.4.2. Chromosome Paints*

Chromosome-specific paints can be obtained commercially. We use chromosome-specific paints from Cambio Laboratories, Cambridge, UK. These are prepared from degenerated oligonucleotide primed-polymerase chain reaction (DOP-PCR) amplified DNA obtained from fluorescence-activated cell sorter (FACS)-sorted chromosomes. Each manufacturer has different conditions for denaturation and application of their probes that should be followed and carried out carefully.

#### *2.4.3. Processing of Noncommercial Probes*

1. Labeled probe.
2. Sephadex G-50 soaked in TE buffer: Slowly add 30 g of Sephadex G-50 beads into 500 mL of sterile TE, pH 8.0. Make sure the powder is dispersed and TE covers all the powder in excess. Add more TE if required. Let stand overnight at room temperature. Decant the supernatant and replace with an equal amount of TE. Store at 4°C in a screw-capped bottle.
3. 1-mL Syringes.
4. Glass wool.
5. Amber (dark) colored microcentrifuge tubes.
6. TE buffer, pH 7.5: Autoclaved.
7. Human Cot-1 DNA (1 mg/mL) (Gibco BRL, cat. no. 15279-011), store at 4°C.
8. 100% Ethanol.
9. Hybridization buffer: 10% w/v Dextran sulfate, 2X SSC, 50% v/v formamide, 1% v/v Tween-20; adjust pH to 7.0, store at –20°C as ready-to-use aliquots.

### *2.5. Hybridization*

1. 22 mm × 22 mm no. 1.5 coverslips.
2. Rubber cement.
3. Humid chamber: A plastic box with a sheet of paper towel moistened with water. For paraffin FISH codenaturation method: Fill an empty Gilson pipet tip box with 250 mL of water at 4°C. Cover the tip holder with 3M filter paper.

### *2.6. Posthybridization Washes*

1. 50% v/v Formamide, 2X SSC, pH 7.0: Warm the solution in the Coplin jar to 42°C in a water bath prior to use.
2. 2X SSC pH 7.0: Autoclaved. Warm in a Coplin jar to 42°C in a water bath prior to use.
3. PBS solution with 0.001% v/v of Igepal: 500 µL of Igepal detergent in 500 mL of PBS solution. Warm to dissolve Igepal, mix well, store at 4°C.

4. 70%, 90%, and 100% ethanol series.
5. DAPI solution: Vectashield mounting medium with 4',6-diamidino-2-phenylindole (DAPI, 1.5 µg/mL, Vector Laboratories Inc., Burlingame, CA 94010).
6. 22 mm × 50 mm no. 0 coverslips.

### 2.7. Image Capture and Analysis

Epifluorescence microscope equipped with appropriate filters in combination with a cooled CCD camera and suitable computer hardware and software.

## 3. Methods

### 3.1. Preparation and Processing of Slides

#### 3.1.1. Metaphase Chromosomes from Lymphocytes

Necessary precautions should be taken before starting work with biological material. All the steps prior to fixing the cells should be carried out in a tissue culture hood.

1. Add 10 mL of fetal bovine serum, 1 mL of penicillin and streptomycin solution, and 1 mL of phytohemagglutinin (PHA) into each 100 mL of RPMI medium and mix.
2. Take 5 mL of normal or test blood from a volunteer into a tube containing 50 µL of sodium heparin, roll tube gently to mix. Five milliliters of blood will give 10 flasks of culture from which approx 200 slides can be made.
3. Aliquot 10 mL of media into each cell culture flask and add 0.5 mL of blood. Mix.
4. Gas all flasks briefly (10–20 s) with 5% $CO_2$.
5. Incubate cultures at 37°C for 72 h.
6. Between 30 and 90 min prior to harvesting (*see* **Note 2**), add 100 µL of colcemid solution and return to 37°C (colcemid solution arrests cell division).
7. Warm 0.075 *M* KCl to 37°C. Make fresh fixative and cool to 4°C.
8. After incubation with colcemid, transfer cultures to capped 10-mL centrifuge tubes and centrifuge at 500*g* for 5 min.
9. Discard supernatant using a plastic pipet and resuspend the pellet in the remaining supernatant by flicking gently.
10. Add 10 mL of prewarmed hypotonic 0.075 *M* KCl (hypotonic solution causes the cells to swell by osmosis making them burst easily when dropped onto slides), mix by inverting and incubate at 37°C for 8 min.
11. Centrifuge at 500*g* for 5 min and discard supernatant, leaving the interphase layer which contains some of the white cells.
12. Flick tube gently to resuspend the pellet. Continue flicking the tube while adding cold fixative dropwise to pellet. After carefully adding the first 5–10 drops, fill the tube up to 10 mL with fixative. Mix by inverting the tube gently. Make sure the pellet is resuspended and does not have any cell clumps, as after fixation single cells cannot be retrieved from the clump.
13. Centrifuge at 500*g* for 10 min.
14. Discard supernatant with pipet (into appropriate fixative waste, *not* into chloros).

15. Repeat **steps 12–15** up to 6 more times.
16. Resuspend the pellet in a few drops of fixative (0.5–1 mL). (After this, the fixed cells can be kept at 4°C for maximum a week or at –20°C for a month, before making slides after storage, the cells should be washed in fresh fixative at least one more time.)
17. Take a slide out of ethanol and wipe dry with lint-free tissue (e.g., Kimwipes®).
18. Hold slide over 80°C water bath for 2 s.
19. Take away from water bath and hold slide at an approx 40° angle.
20. Drop fixed cell suspension on to both halves of the moistened slide surface from a distance about 30 cm (i.e. two separate drops onto each slide). (*See* **Note 3** for alternative conditions for slide making.) Usually the quality of metaphases change from one flask to next, thus it is advised to make separate batches of slides from each flask.
21. Dry slides at room temperature.
22. Observe one slide made from each tube under a phase-contrast microscope (×20 objective lens) to assess the approximate number of metaphases and amount of cytoplasm on each half of the slide. An ideal slide is 5–10 metaphases per field with no cytoplasm around them. If there are too many metaphases on the slide, more fix can be added to the tube until the required density is achieved (*see* **Note 4** for reducing cytoplasm).
23. Repeat **steps 17–21** until required number of slides are made.
24. Slides should be stored at –20°C in a box containing desiccant (e.g., silica gel).

### *3.1.2. Metaphase Chromosomes from Cell Line Cultures*

1. Add 25 µL colcemid solution into a cell culture flask (25-cm$^2$ flasks with 5 mL of media) and incubate at 37°C for up to 2 h.
2. Warm 0.075 *M* KCl to 37°C. Make fresh fixative and cool to 4°C. **Steps 3–6** are for adherent cell cultures. For supension cell cultures go directly to **step 8**.
3. Transfer medium from flask into a 10 mL centrifuge tube, avoiding the cell layer.
4. Put 2–3 mL of versene solution into the flask, replace the cap, and swirl to wash off the serum from all the surfaces of the flask (serum stops the action of trypsin–EDTA). Discard versene.
5. Put 2–3 mL trypsin–EDTA solution into flask, swirl and incubate at 37°C for 3 min.
6. Tap flask 4–5× (observe under phase-contrast microscope if all cells are detached from the flask) and pour the contents of the centrifuge tube back into the flask (to stop the action of trypsin).
7. Transfer the suspension back into the centrifuge tube.
8. Centrifuge at 500*g* for 5 min.
   From here onward the fixation and slide making is the same as above (**Subheading 3.1.1., steps 9–24** inclusive).

### *3.1.3. Processing of Cells in Interphase for FISH*

#### 3.1.3.1. Cells Harvested in Suspension

This method is usually used in cases where tumor material cannot be cultured to obtain metaphase chromosomes, for example, pleural effusion. The same procedure is followed as in the method for fixing metaphase chromosomes.

1. Centrifuge sample at 500*g* for 5 min.
2. Follow **Subheading 3.1.1., steps 9–24**.

#### 3.1.3.2. Touch Preps from Frozen and Fresh Tissue

1. Take slides out of 100% ethanol and wipe dry with lint-free tissue.
2. (a) For frozen tissue:
   Keep frozen tissue on dry ice until use. With alcohol-cleaned forceps remove tissue from the vial and cut off a small piece with a sterile scalpel. With forceps take a tissue piece and apply gently to areas of hybridization on clean slides. Try to rotate tissue so that large clumps of thawed tissue are not deposited onto the slide. (If tissue sample is stuck in the vial, immerse vial in liquid nitrogen and break vial after freezing with pestle and mortar.)
   Go to **step 4**.
   (b) For fresh tissue:
   Touch tumor sample lightly onto slides at RT.
3. Immediately place slides in cold methanol for 20 min. Be careful not to let the slide dry out.
   Go to **step 5**.
4. Let slides air-dry for 1 min. Fix tissue for 20 min in a Coplin jar filled with fixative cooled to –20°C.
5. Transfer slides into fresh fixative at room temperature for 20 min.
6. Allow to air-dry. (When observed under a phase-contrast microscope, try to find dull gray cells which are good for hybridization.)
7. Store slides at –20°C with silica gel until processing.
   Processing:
8. Mix 50 µL of collagenase solution in 50 mL of L-15 medium and warm to 37°C in a Coplin jar.
9. Treat slides in collagenase for 5 min at 37°C.
10. Rinse in Dulbecco's PBS with 50 m*M* $MgCl_2$, pH 7.3, for 10 min at RT, twice.
11. Fix slides in 1% v/v formaldehyde in PBS with 50 m*M* $MgCl_2$, pH 7.3, for 10 min at RT. Formaldehyde is toxic and thus should be used in a fume cupboard.
12. Rinse slides in PBS with 50 m*M* $MgCl_2$ for 2 min at RT, 3×.
13. Dehydrate by processing through 70% ethanol, 90% ethanol, and 100% ethanol for 2 min each. Air-dry.
14. Slides are ready for denaturation. They can be stored at –20°C with desiccant until use.

#### 3.1.3.3. Paraffin Wax-Embedded Sections

1. 4 µm sections are cut from formalin-fixed, paraffin wax-embedded material and placed on clean 76 mm × 26 mm silane-treated glass microscope slides and left overnight at 65°C.
2. Heat slides to 100°C on a hotplate for 2 h.
3. Wash immediately in Histoclear for 10 min at RT, twice, to dewax paraffin.

4. Rinse slides in 100% ethanol for 5 min, twice, and air-dry.
5. Soak slides into prewarmed 1 *M* sodium thiocyanate for 20–30 min at 45°C to make proteins more accessible to pepsin digestion.
6. Rinse well with distilled water.
7. Incubate slides in 0.5% w/v pepsin (in 0.9% w/v NaCl, pH 1.5) for 15–20 min at 37°C with occasional shaking.
8. Wash slides in Hank's balanced salt solution for 5 min at RT.
9. Rinse with distilled water and dehydrate in 70, 90, and 100% ethanol solution series, 2 min each.
10. Slides are ready for hybridization to the probes. Please note that incubation time in **steps 5** and **7** change according to tissue type, and the thickness of the section may vary according to the tissue type.

### *3.1.4. Fiber Chromatin*

#### 3.1.4.1. Chromatin Release from Fixed Cells

1. Transfer slides of fixed cells into PBS for 1 min to rehydrate the cells. Alternatively fixed cells are dropped onto clean slides and placed in PBS before evaporation of the fixative. As the recently dropped cells are only loosely attached to the slide, avoid any agitation of the slide during this incubation step.
2. Take slides out of PBS and drain on a paper towel, without allowing the slide to dry out.
3. Immediately add 100 µL of releasing solution A on one end of slide and drag the solution to the other end using the short edge of a long coverslip (*see* **Note 1**).
4. Fix the released chromatin fibers by carefully rinsing the slide with methanol. To avoid chromatin loss, start with a minimal amount of methanol added on one end of the slide, which is held horizontally. Try to keep the methanol on the slide for as long as possible until the gelatinous bulk, chromatin, is attached to the slide.
5. Air-dry slides.
6. Dehydrate the slides by leaving them in 70%, 90%, and 100% ethanol for 2 min each.
7. Store slides at –20°C, if not used immediately.

#### 3.1.4.2. Chromatin Release from Fresh and Frozen Cells

1. Cells are touch imprinted onto one end of the slide (as described in **Subheading 3.1.3.2.**; do not immerse slides in fixative after making the touch preps) from frozen and fresh tumor samples. For cell lines, the cells are trypsinized from the flask surface (*see* **Subheading 3.1.2., steps 3–8**), washed in PBS (at –20°C), resuspended in PBS to $10^6$ cells/mL and 2 µL of cell suspension is placed on one end of the slide.
2. Air-dry the slide.
3. Immediately add 10 µL of chromatin release solution B onto touch imprinted cells or 5 µL onto dried cell solution (from cell culture) and leave for 5 min to lyse the cells.
4. Tilt the slide to allow the drop of DNA to run down the slide.
5. Air-dry the stream of DNA .
6. Fix the DNA in methanol–acetic acid fixative for 2–5 min and air-dry.
7. Store slides at –20°C with silica gel, if not used immediately.

## 3.2. Denaturation of Slides

Some of the slides need to be baked at 65–68°C for 30 min to 2 h before denaturation. They will be mentioned individually in the following methods.

### 3.2.1. Nonparaffin Wax-Embedded Slides

All slides made from fresh material, except touch imprints, are baked at 65–68°C for 2 h before denaturation. Carry out baking before proceeding with denaturation.

1. Preheat denaturation solution to 73 ± 1°C in a Coplin jar in a water bath. Make sure the temperature of the denaturation solution is above 72°C before slides are immersed.
2. Immerse the slides into denaturation solution for 2–3 min (*see* **Note 5**).
3. Transfer slides into 70% ethanol for 3 min, then into 90% ethanol for 3 min and 100% ethanol for 3 min. Change the 100% ethanol and leave slides in for an extra 3 min.
4. Put slides on to a hotplate at 40°C until probe is ready.

### 3.2.2. Paraffin Wax-Embedded Slides

*See* **Subheading 3.5.3.** for codenaturation.

## 3.3. Preparation of Probe

### 3.3.1. Growing Clones and Extracting DNA

#### 3.3.1.1. Yeast Artificial Chromosomes

1. Streak YACs from glycerol stock onto YPD agar plates and incubate inverted at 30°C for 48 h.
2. Prepare 100 mL of 1X SD with adenine sulfate, tyrosine, and ampicillin for each YAC clone.
3. Inoculate 100 μL of 1X SD mixture with a single YAC colony and incubate shaking (250 rpm) at 30°C for 72 h.
4. Take out 0.5 mL of culture to make glycerol stocks. Mix 0.5 mL of YAC culture with 0.5 mL of 30% v/v glycerol in 1X SD, and store at –80°C.
5. Centrifuge the rest of the culture at 1500*g* for 10 min at 20°C to pellet the cells.
6. Discard the supernatant and resuspend cells in 5 mL of sorbitol–EDTA–Tris-HCl mixture and 5 μL of BME. Transfer into 50-mL Falcon tubes.
7. Remove 20 μL of cells, put on a microscope slide, and place a coverslip on top of the cells to observe as negative control of cell lysis.
8. To the rest of the cell suspension add 15 μL of lyticase (300 U) and incubate with shaking at 37°C for 1–2 h until the cells are lysed.
9. Take 20 μL of cells, put on a microscope slide, and place a coverslip on top.
10. Observe the slides from **steps 7** and **9** under a phase-contrast microscope to check if the cells have lysed. Lysed cells appear dull and brown whereas intact cells appear as bright circles. If necessary continue lyticase incubation until nearly all cells are lysed.
11. Centrifuge lysed cell culture at 1500*g* for 10 min.

12. Discard supernatant and resuspend pellet in 1 mL of TEN–SDS solution using either a 5-mL or 10-mL pipet tip.
13. Add 10 μL of Proteinase K (50 mg/mL stock concentration) and incubate shaking at 55°C overnight (or at 37°C for 2 h).
14. Add 20 μL of RNase (10 mg/mL stock concentration) at 37°C for 30 min. Optional: To make DNA extraction easier, add 3 mL of water into cultures after RNase incubation.
15. Add equal volume of phenol, mix well, and centrifuge for 10 min at 1500*g*. If there is too much cell debris left, repeat this step once more.
16. Transfer top layer into a fresh Falcon tube and add equal volume of phenol–chloroform/isoamylalcohol (i.e., 1 vol culture, 0.5 vol phenol, 0.5 vol chloroform–isoamylalcohol). Mix well and centrifuge at 1500*g* for 10 min.
17. Transfer top layer into a fresh Falcon tube and add equal volume of chloroform–isoamylalcohol. Centrifuge well and spin at 1500*g* for 10 min.
18. Transfer top layer into a fresh Falcon tube and add 1/10th vol of 3 *M* sodium acetate. Mix well.
19. Immediately add 2 vol of ice-cold 100% ethanol. Mix well and leave at –20°C for 1 h or overnight.
20. Centrifuge at 1500*g* for 10 min.
21. Discard the supernant and wash with 70% ethanol and centrifuge at 1500*g* for 10 min.
22. Discard the supernant and dry the DNA pellet in a vacuum dryer.
23. Add 250 μL of sterile distilled water and dissolve the pellet using a Gilson pipet.
24. Transfer into microcentrifuge tubes, and incubate at 65°C for 15 min to make sure DNA is dissolved. After it cools down, measure DNA concentration using TKO 100 minifluorometer (*see* **Subheading 3.3.2.**).

#### 3.3.1.2. Plasmid, BAC, Cosmid, PAC DNA Extraction

1. Streak bacteria on LB-agar plates supplemented with appropriate antibiotic and incubate inverted at 37°C for overnight.
2. Inoculate a single isolated bacterial colony into 10 mL of LB-broth in a 50-mL Falcon tube, supplemented with appropriate antibiotic, and incubate shaking at 225–300 rpm at 37°C for 16 h.
3. Centrifuge the culture at 1500*g* for 10 min.
4. Discard supernatant and drain pellet by inverting tube onto a paper towel (take care that the pellet does not slide down the side of the tube).
5. Resuspend pellet well by vortex-mixing in 400 μL of ice-cold solution 1. At this stage, the cell suspension can be transferred to 1.9-mL microcentrifuge tubes.
6. Add 800 μL of solution 2 and invert tube very gently to mix the contents. Leave at room temperature for 5 min. The appearance of the suspension should change from very turbid to almost translucent.
7. Add 600 μL of solution 3 slowly while gently shaking the tube. A thick white precipitate of protein and *E. coli* DNA will form. Place tube on ice for at least 5 min.

8. Centrifuge at 1500*g* for 10 min at 4°C.
9. Remove tubes from the centrifuge and put on ice. Transfer supernatant into a fresh microcentrifuge tube. Try to avoid any white precipitate material.
10. Add an equal volume of phenol–chloroform–isoamylalcohol. Mix and centrifuge at maximum speed in a microcentrifuge for 5 min.
11. Transfer the top layer into a fresh microcentrifuge tube and add an equal volume of chloroform–isoamylalcohol, mix well and centrifuge at maximum speed in a microcentrifuge for 5 min.
12. Transfer the top layer into a fresh microcentrifuge tube.
13. Add 0.7 vol of isopropanol, mix well, and place tube on ice for at least five min or leave at –20°C overnight.
14. Centrifuge at maximum speed for 10 min at 4°C.
15. Discard supernatant and add 0.5 mL of 70% ethanol, and invert tube several times to wash the DNA pellet. Centrifuge at maximum speed for 5 min at 4°C with the DNA pellet facing toward the outside of the centrifuge.
16. Discard as much of the supernatant as possible. Occasionally, pellets will become dislodged from tube so it is better to aspirate off the supernatant carefully rather than pour it off.
17. Air-dry DNA pellet at room temperature.
18. Resuspend DNA pellet in 100 µL of water with occasional tapping. DNA can be stored at –20°C if not used immediately.
19. Incubate tube at 65°C for 15 min. After it cools down, measure DNA concentration using TKO 100 minifluorometer (*see* **Subheading 3.3.2.**).

### 3.3.2. Measuring DNA Concentration

Prepare assay solution appropriate for the expected DNA content of the extraction process.

1. Turn on TKO 100 at least 15 min before use.
2. Adjust the SCALE control knob on the TKO 100 to approx 50% sensitivity—five clockwise turns from the fully counterclockwise position.
3. Pipet 2 mL of appropriate assay solution at RT into the cuvet.
4. Zero the instrument by turning the ZERO control knob until the display reads "000." The readout will normally flicker ±3 units around the zero when set for 50% sensitivity.
5. Pipet 2 µL of the appropriate DNA reference standard into the assay solution. Mix by pipeting in and out of the cuvet with a disposable pipet. Close the lid over the cuvet well.
6. Set the scale on the reference standard. For the standard assay, turn the SCALE knob until the display reads "100," indicating 100 ng/ml final concentration, or 100 µg/mL undiluted. For the extended asay, turn the SCALE knob until the display reads "1000," indicating 1000 ng/mL final concentration, or 1000 µg/mL undiluted.

7. Repeat **steps 3–6** once or twice until the readout on the display reproducibly reads number set in **step 6**, matching the concentration of your reference standard. Between samples, empty the cuvet and drain it thoroughly by blotting upside down on a paper towel.
8. Measure your samples in the same manner as described in **steps 3–6**. Be sure to zero the instrument each time you load 2 mL of assay solution. Do not adjust the scale knob.

   Since the labeling reaction is done in 50 μL of total volume, make sure the undiluted DNA concentration is not lower than 30 ng/μL. It is advisable to check the DNA concentration and quality on a 1% agarose gel alongside DNA molecular weight markers, before proceeding to labeling reaction.

### 3.4. Labeling and Processing of Probes

All steps involving fluorescent material should be carried out in subdued light to avoid bleaching of the fluorochromes.

#### 3.4.1 FISH Probes

1. Mix in a 1.5-mL microcentrifuge tube placed on ice, 1 μg of DNA, 5 μL of fluoro-dNTP mix, 0.5 μL of DNA polymerase 1, 5 μL of DNase enzyme mix, and sterile, distilled water to 50 μL total volume. Mix by pipeting.
2. Incubate in a water bath set at 15°C in subdued light for 3 h.
3. At 40 min prior to the completion of 3 h prepare Sephadex G-50 columns for purifying the labeled probe from unincorporated nucleotides. Remove plungers from 1-mL syringes (one for each labeling reaction) and place them in lidless, conical, 10-mL centrifuge tubes. Plug syringes with glass wool with the help of the plunger up to the 0.1-mL mark. Fill syringes with TE soaked Sephadex G-50. Centrifuge at 700*g* for 4 min. Fill the syringes up with TE and centrifuge again. Repeat once more. Transfer syringes into fresh centrifuge tubes. Columns are ready to use. Do not let them dry out.
4. Add 3 μL of 0.5 *M* EDTA to stop the labeling reaction.
5. Add 50 μL of water and mix by pipeting.
6. Load labeled probe onto Sephadex G-50 columns and centrifuge at 700*g* for 4 min.
7. Transfer purified labeled probe from the centrifuge tube into amber microcentrifuge tubes.
8. Run 10 μL of the labeled probe alongside a DNA size marker (e.g., λ DNA restriction enzyme[RE] digested with *Hin*dIII) on 1% agarose gel without any ethidium bromide in subdued light to check the quality of the labeling reaction. Observe gel to see if the reaction worked. There should be a smear of visible fluorescence (red for rhodamine and green for fluorescein) on the gel. Stain gel with EtBr for 30 min and observe again. The size range of labeled probe should be from 200 bp to 500 bp (compared to λ/*Hin*dIII) and this should correspond to the fluorescent smear previously observed on the gel. *See* **Note 6** for changing the conditions to obtain the appropriate DNA size range.

### 3.4.2. Chromosome Painting Probes

Directly labeled chromosome-specific paints are ready to use in hybridization. These paints are obtained commercially from Cambio Laboratories or others. Go to **Subheading 3.5.2.** for hybridization protocol.

## 3.5. FISH Hybridization

### 3.5.1. FISH

1. Add the appropriate amount of probe per hybridization to a 1.5-mL microcentrifuge tube as per the list that follows. Prepare highly repetitive probes in a different tube than single copy sequence probes.

   200–500 ng for YACs
   100–200 ng for plasmids, PACs, and BACs (except highly repetitive sequence probes, e.g., centromeric, telomeric probes)
   80–100 ng for cosmids
   20–50 ng for highly repetitive sequence probes (e.g., centromeric, telomeric probes)

2. Add 5 µg (5 µL of 1 mg/mL stock) of Cot-1 DNA into each tube except highly repetitive sequence probes. Mix.
3. Add 2 vol of 100% ethanol. Mix.
4. Dry probes in a rotary vacuum dryer (spin vac).
5. The total volume of probe put on per half of the slide (per hybridization) is 10 µL. For two probes to be used in one hybridization, resuspend the probes accordingly, for example, for a single hybridization of centromeric probe and the single copy sequence, resuspend single copy sequence probe in 8 µL of hybridization buffer per hybridization and centromeric probe in 2 µL of hybridization buffer per hybridization.
6. Denature the probe DNA at 75–80°C for 5–6 min.
7. Chill the DNA on ice and centrifuge quickly to get all the liquid down to the bottom of the tube.
8. Preanneal repetitive sequences by incubation at 37°C for 20 min.
9. Mix appropriate amounts of the two probes and apply onto one half of the previously denatured slide. Cover with a 22 mm × 22 mm no. 1.5 coverslip (take care not to trap any air bubbles underneath; if so, ease them out using a Gilson pipet tip) and seal with rubber cement.
10. Place slides in a humid chamber and incubate at 37°C for 24 h (can be incubated more but there is a risk of increased background signal).
11. Proceed to posthybridization washes.

### 3.5.2. Chromosome Painting

For Cambio Laboratories chromosome-specific paints:

1. Warm paints to 42°C and mix well before use.
2. Aliquot 10 µL of paint for each half slide into a microcentrifuge tube and follow **steps 6–11** in **Subheading 3.5.1.**

### *3.5.3. Paraffin FISH*

1. Prepare probes as for one or more color FISH.
2. After resuspending probe pellet in hybridization buffer, add 10 µL of probe on each section. Cover with a 22 mm × 22 mm no. 1.5 coverslip and seal with rubber cement.
3. Place slides on the filter paper in a chamber directly above a box containing 250 mL of water at 4°C (described in **Subheading 2.5., step 3**) and treat in a microwave (Sanyo Super showerwave, 800 W/E) for 3.5 min at power level "simmer" and then 4 min at power level "low." The temperature of the water during microwave treatment is critical and is around 65°C after the first 3.5-min treatment and 75°C after the final treatment.
4. Leave slide in the microwave oven for a few minutes and then transfer them into a 37°C incubator for 24 h.
5. Proceed to posthybridization washes.

## 3.6. Posthybridization Washes

1. Remove coverslip from one slide and immerse in 50% v/v formamide–2X SSC solution at 42°C. Repeat for other slides.
2. Leave slides in solution for 5 min. Repeat two more times.
3. Change solution to 2X SSC at 42°C and incubate slides for 5 min. Repeat two more times.
4. Discard solution and add PBS–Igepal solution at RT. Agitate slides for 10–30 s.
5. Rinse slides with distilled water for 3X.
6. Dehydrate slides by immersing them in 70%, 90%, and 100% ethanol for 2 min each.
7. Air-dry slides in darkness.
8. Apply 10 µL of DAPI solution and cover with no. 0 coverslip. Take care not to trap any air bubbles. Carefully ease them out using a Gilson pipet tip.
9. Slides are ready for image capture and analysis.

## 3.7. Image Capture and Analysis

The signals should be visible under the appropriate filter through the microscope (*see* **Notes 7** and **8**). After focal adjustments are done, 10–20 metaphases should be captured per experiment for metaphase FISH. This should provide enough data for the conclusion of the experiment.

Depending on the experiment adequate number of images of interphase nuclei should be captured per hybridization for statistical analysis to be carried out. Take care not to mistake normal nuclei with the tumor nuclei which usually have a different morphology. Try to observe the signals through the microscope before capturing them. Some signals may be in a different focal plane, which should be observed by adjusting the focus of the microscope. Capturing cells individually is a better approach than capturing them as a group. Some signals can be brighter than others. When captured as a group, the brighter signals would shorten the exposure time, thus rendering the weak signals invisible.

For released chromatin FISH, images are captured in the same way. In the region of 25–30 images should be captured for analysis. For each pair of signals, the size of overlap or gap in kilobase pairs is calculated by using the mean value of the known kilobase length of each of the two probes divided by its measured signal length as a multiplication factor for the measured length of the overlap or gap ***(18)***.

For FISH on paraffin wax-embedded sections at least 50 nuclei should be captured for each experiment (*see* **Note 9**). Statistical analysis is needed for a final result.

## 4. Notes

1. An alternative to using NaOH-based chromatin releasing solution is 70% formamide in 2X SSC, pH 7.0, solution. Longer signals are obtained in DNA released with sodium hydroxide than with formamide ***(18)***. Sodium hydroxide treatment results in complete disruption of nuclei, with the slides being covered by an irregular network of chromatin fibres. With formamide, the borders of most of the disrupted nuclei can still be defined, allowing hybridization signals from the same nucleus to be identified ***(26)***. Hybridization signals are seen as either dotted linear arrays or as continuous lines. The reason for this is not yet clear.
2. Although colcemid time can be varied between 30 and 90 min to obtain more cells in metaphase, the longer incubation times will give more condensed chromosomes. Therefore, 30 min of colcemid incubation should be ideal for adequate numbers of metaphase spreads.
3. The quality of the metaphase spreads is critical and depends on the humidity and temperature of the laboratory, how the material was processed, and the way the slides are made. Extremely dry conditions cause the chromosomes not to spread and very humid conditions give rise to what is known as "chromosome soup." The following conditions can be tried. Slides can be humidified by placing them into ethanol and then iced water, or frosting them in freezer, or holding them over a water bath at 80°C before dropping the cell suspension. After the cells are dropped, the slides can be dried by putting them onto a hotplate at 54°C, or inside a 37°C incubator in a humid chamber, or leaving them at RT. These conditions can be tried alternatively. We find that holding the slide over the water bath and then drying it at RT provides good quality metaphases.
4. Presence of excess cytoplasm covering metaphase chromosomes or nuclei reduces the effect of denaturation and signal hybridization, and may increase background signal. The fixative can be modified by increasing the ratio of glacial acetic acid up to 1.5 and reducing the ratio of methanol to 2.5. This should reduce the amount of cytoplasm on the slide. Also pretreatments with Proteinase K and pepsin may help to reduce the effect of cytoplasm.
5. The denaturation of the target DNA on the slide should be optimal. Under- or overdenaturation will lead to aberrant signal visualization and chromosome morphology, especially on metaphase chromosomes. Swollen chromosomes that

do not take up DAPI counterstain or nuclei with wrinkled appearance indicate overdenaturation. Bright DAPI with very good chromosome morphology but poor signal indicates underdenaturation.

6. If the labeled DNA is not in the required size range, adjust the amount of DNase enzyme mix (0.5 µL, 1.0 µL, 1.5 µL, or 2.0 µL) and DNA polymerase (0.7 µL, 1 µL) in the labeling reaction accordingly. When a labeling problem arises, it is necessary to run a 1% agarose gel without EtBr in subdued light, after Sephadex purification, to pinpoint the problem.
7. No signal observed on the slide usually indicates that the probe was not labeled well. The labeling can be checked as described in **Note 6**. The purity and amount of DNA used in labeling is critical and is a common cause of problems in FISH experiments.
8. Background or nonspecific binding of probe may suggest that the stringency of washing should be increased or that Cot-1 suppression is inadequate.
9. Auto fluorescence of paraffin after FISH indicates that pretreatment of slides needs to be optimized.

For further troubleshooting, please refer to Notes 1–3, 5, 7, and 13 in Chapter 15.

## References


1. Pardue, M. L. and Gall, J. G. (1969) Molecular hybridization of radioactive DNA to the DNA of cytological preparations. *Proc. Natl. Acad. Sci. USA* **64,** 600–604.
2. John, H., Birnstiel, M., and Jones, K. (1969) RNA-DNA hybrids at the cytological level. *Nature* **223,** 582–587.
3. Bauman, J. G. J., Wiegant, J., Borst, P., and van Duijn, P. (1980) A new method for fluorescence microscopical localization of specific DNA sequences by *in situ* hybridization of fluorochrome labeled RNA. *Exp. Cell Res.* **128,** 485–490.
4. Langer, P. R., Waldrop, A. A., and Ward, D. C. (1981) Enzymatic synthesis of biotin-labeled polynucleotides: novel nucleic acid affinity probes. *Proc. Natl. Acad. USA* **78,** 6633–6637.
5. van Proijen-Knegt, A. C., van Hoek, J. F., Bauman, J. G., van Duijn, P., Wool, I. G., and van der Ploeg, M. (1982) *In situ* hybridization of DNA sequences in human metaphase chromosomes visualized by an indirect fluorescent immunocytochemical procedure. *Exp. Cell Res.* **141,** 397–407.
6. Landegent, J. E., Jansen in de Wal, N., Baan, R. A., Hoeijmakers, J. H., and van der Ploeg, M. (1984) 2-Acetylaminofluorene-modified probes for the indirect hybridocytochemical detection of specific nucleic acid sequences. *Exp. Cell Res.* **153,** 61–72.
7. Tchen, P., Fuchs, R. P. P., Sage, E., and Leng, M. (1984) Chemically modified nucleic acids as immuno-detectable probes in hybridization experiments. *Proc. Natl. Acad. Sci. USA* **81,** 3466–3470.
8. Summersgill, B., Goker, H., Weber-Hall, S., Huddart, R., Horwich, A., and Shipley, J. (1997) Molecular cytogenetic analysis of adult testicular germ cell tumors and identification of regions of consensus copy number change. *Br. J. Cancer* **77,** 305–313.

9. Speicher, M. R., Stephen, G. B., and Ward, D. C. (1996) Karyotyping human chromosomes by combinatorial multi-fluor FISH. *Nature Genet.* **12,** 368–375.
10. Schrock, E., du Manoir, S., Veldman, T., Schoell, B., Wienberg, J., Ferguson-Smith, M. A., et al. (1996) Multicolor spectral karyotyping of human chromosomes. *Science* **273,** 494–497.
11. Landegent, J. E., Jansen in de Wal, N., Dirks, R. W., Baao, F., and van der Ploeg, M. (1987) Use of whole cosmid cloned genomic sequences for chromosomal localization by non-radioactive *in situ* hybridization. *Hum. Genet.* **77,** 366–370.
12. Lichter, P., Tang, C. J., Call, K., Hermanson, G., Evans, G. A., Housman, D., and Ward, D. C. (1990) High-resolution mapping of human chromosome 11 by *in situ* hybridization with cosmid clones. *Science* **247,** 64–69.
13. Ioannou, P. A., Amemiya, C. T., Garnes, J., Kroisel, P. M., Shizuya, H., Chen, C., et al. (1994) A new bacteriophage P1-derived vector for the propagation of large human DNA fragments. *Nat. Genet.* **6,** 84–89.
14. Shizuya, H., Birren, B., Kim, U., Mancino, V., Slepak, T., Tachiiri, Y., and Simon, M. (1992) Cloning and stable maintenance of 300-kilobase-pair fragments of human DNA in *Escherichia coli* using and F-factor-based vector. *Proc. Natl. Acad. Sci. USA* **89,** 8794–8797.
15. Burke, D. T., Carle, G. F., and Olson, M. V. (1987) Cloning of large segments of exogenous DNA into yeast by means of artificial chromosome vectors. *Science* **236,** 806–812.
16. Trask, B. J., Massa, H., Kenwrick, S., and Gitschier, J. (1991) Mapping of human chromosome Xq28 by two-color fluorescence *in situ* hybridization of DNA sequences to interphase nuclei. *Am. J. Hum. Genet.* **48,** 1–15.
17. Senger, G., Ragoussis, J., Trowsdale, J., and Sheer, D. (1993) Fine mapping of the MHC class II region within 6p21 and evaluation of probe ordering using interphase fluorescence *in situ* hybridization. *Cytogenet. Cell Genet.* **64,** 49–53.
18. Senger, G., Jones, T. A., Fidlerova, H., Sanseau, P., Trowsdale, J., Duff, M., and Sheer, D. (1994) Released chromatin: linearized DNA for high resolution fluorescence *in situ* hybridization. *Hum. Mol. Genet.* **3,** 1275–1280.
19. Weber-Hall, S., McManus, A., Anderson, J., Nojima, T., Abe, S., Pritchard-Jones, K., and Shipley, J. (1996) Novel formation and amplification of the *PAX7–FKHR* fusion gene in a case of alveolar rhabdomyosarcoma. *Genes Chromosom. Cancer* **17,** 7–13.
20. Parra, I. and Windle, B. (1993) High resolution visual mapping of stretched DNA by fluorescent hybridization. *Nat. Genet.* **5,** 17–21.
21. Nederlof, P. M., van der Flier, S., Wiegant, J., Raap, A. K., Tanke, H. J., Ploem, J. S., and van der Ploeg, M. (1990) Multiple fluorescence *in situ* hybridization. *Cytometry* **11,** 126–131.
22. Nederlof, P. M., Robinson, D., Abuknesha, R., Hopman, A. H., Tanke, H. J., and Raap, A. K. (1989) Three-color fluorescence *in situ* hybridization for the simultaneous detection of multiple nucleid acid sequences. *Cytometry* **10,** 20–27.
23. Ried, T., Baldini, A., Rand, T. C., and Ward, D. C. (1992) Simultaneous visualization of seven different DNA probes by *in situ* hybridization using combinatorial fluorescence and digital imaging microscopy. *Proc. Natl. Acad. Sci. USA* **89,** 1388–1392.

24. Dauwerse, J. G., Wiegant, J., Raap, A. K., Breuning, M. H., and van Ommen, G. J. (1992) Multiple colors by fluorescence *in situ* hybridization using ratio labeled DNA probes create a molecular karyotype. *Hum. Mol. Genet.* **1,** 593–598.
25. Lu, Y.-J., Birdsall, S., Summersgill, B., Smedley, D., Osin, P., Fisher, C., and Shipley, J. (1999) Dual color fluorescence *in situ* hybridization to paraffin-embedded samples to deduce the presence of the der(X)t(X;18)(p11.2;q11.2) and involvement of either the *SSX1* or *SSX2* gene: a diagnostic and prognostic aid for synovial sarcoma. *J. Pathol.* **187,** 490–496.
26. Fidlerova, H., Senger, G., Kost, M., Sanseau, P., and Sheer, D. (1994) Two simple procedures for releasing chromatin from routinely fixed cells for fluorescence *in situ* hybridization. *Cytogenet. Cell Genet.* **65,** 203–205.

# 15

# Comparative Genomic Hybridization

**Rebecca Roylance**

## 1. Introduction

Comparative genomic hybridization (CGH) is a molecular cytogenetic technique used to screen the entire genome for gains and losses of genetic material *(1)*. It is being used increasingly in the study of cancer genetics to identify genes important in the initiation, progression, and, of particular relevance here, metastasis of tumors *(2–4)*. One of the advantages of the technique is that the entire genome is examined in a single experiment, so there is no necessity to know the genetic region of interest prior to investigation. Once regions of gain or loss have been identified, these regions can be defined further using fluorescence *in situ* hybridization (FISH) (described in Chapter 14 by Goker and Shipley) or molecular genetic techniques.

CGH is essentially a modified *in situ* hybridization. Differentially labeled test or tumor DNA (green) and reference or normal DNA (red) are cohybridized to normal metaphase spreads. Differences in the copy number between test and reference DNA are seen as differences in the ratio of green to red fluorescence intensity on the metaphase chromosomes. Images of the metaphases are captured and quantification of the fluorescence ratios performed using a digital image analysis system. Regions of chromosomal gain are seen as an increased fluorescence ratio whereas regions of loss are seen as a decrease in the fluorescence ratio. Losses are detectable when the region affected exceeds 10 Mb *(5)*, smaller regions of gain are detected if there is high level amplification, for example a 2-Mb region that is amplified five times will be visualized *(6)*. For each tumor, 5–10 metaphases are analyzed and an average fluorescence ratio for each chromosome obtained. For the investigation of genes involved in metastasis, the experiments can be approached in one of two ways. The primary tumor DNA and metastasis DNA can be compared

From: *Methods in Molecular Medicine, vol. 57:*
*Metastasis Research Protocols, Vol. 1: Analysis of Cells and Tissues*
Edited by: S. A. Brooks and U. Schumacher © Humana Press Inc., Totowa, NJ

in separate experiments, each hybridized with normal DNA. Alternatively, a modified CGH can be used whereby the primary tumor is the test DNA and the metastasis is the reference DNA *(7)*.

Setting up CGH experiments is a multistep process, all of which are important for a successful result. The steps involved are described in the following subheadings.

### 1.1. Preparation of Metaphase Spreads

These are prepared using blood from a normal healthy volunteer. Blood is cultured in the presence of a mitogen (e.g., phytohemagglutinin [PHA]) which specifically stimulates cell division of the T-lymphocytes. After a period of incubation, usually 72 h, the cells are harvested. Just prior to harvesting, cell division is arrested in metaphase by the addition of colcemid. The cells are treated with a hypotonic solution to make them swell and then fixed using methanol–acetic acid. The metaphase "spreads" are prepared by dropping the fixed cells onto cleaned microscope slides.

### 1.2. Preparation of Probes: DNA Extraction and Labeling

DNA is extracted from the source material using standard methods. Providing there is sufficient DNA (at least 1 μg), it is labeled directly by nick translation. Fluorescently labeled nucleotides are incorporated directly into the DNA; conventionally the test DNA is labeled green and the reference DNA red. However, if only a small amount of DNA is available, it must first be amplified using degenerate oligonucleotide primed-polymerase chain reaction (DOP-PCR) *(8)*. This uses a degenerate primer and has two stages, initial low stringency cycles, where the specific bases at the 3' end of the oligonucleotide theoretically primes every 4 kb along the template DNA and then an increased number of cycles with high stringency, whereby the oligonucleotide "tailed" DNA from the initial cycles is then amplified. DNA amplified in this way can either be labeled by nick translation or by further DOP-PCR, with fluorescent nucleotides incorporated into the PCR reaction.

Once labeled with their respective fluorochromes, the probes are mixed and precipitated in the presence of an excess of unlabeled human Cot-1 DNA. They are resuspended in hybridization mix, denatured and preannealed, before being applied to the denatured metaphase chromosomes on the slides. The presence of Cot-1 DNA and the preannealing step are to suppress hybridization of the interspersed repetitive DNA sequences present in the DNA probes.

### 1.3. Hybridization

The slides are left to hybridize for 72 h.

### *1.4. Posthybridization Washes*

The slides are washed using solutions of differing stringency to remove unbound probe and any nonspecific hybridization.

### *1.5. Image Acquisition and Analysis*

Images are captured using an epifluorescence microscope, equipped with a triple color filter set and a cooled CCD (charge-coupled device) camera. Although color changes should be visible by eye when viewed using a microscope, they need to be quantified. This is done by digital image analysis using commercially available software.

## 2. Materials

### *2.1. Blood Cultures and Preparation of Metaphase Slides*

1. Sodium heparin: 50 mg/mL stored at 4°C.
2. PHA lyophilized stored at 4°C. It is made up in sterile water to 9 mg/mL and stored in aliquots at –20°C.
3. Culture medium: RPMI 1640, fetal calf serum (10%), L-glutamine (4 m*M*).
4. Colcemid solution: 10 µg/mL, stored at 4°C.
5. Hypotonic solution: 0.075 *M* potassium chloride warmed to 37°C prior to use.
6. Fixative solution: 3 vol:1 vol methanol–acetic acid (glacial), prepared immediately prior to use and kept at 4°C.
7. Clean 76 mm × 26 mm microscope slides stored in 500 mL of 70% ethanol with 0.05% v/v concentrated hydrochloric acid and dried immediately with lint-free tissues (e.g., Kimwipes®) before cells are dropped.
8. Ethanol series: 70%, 95%, and 100%.

### *2.2. DNA Extraction*

#### *2.2.1. DNA Extraction from Fresh/Frozen Tissue*

1. Digestion buffer: 100 m*M* NaCl, 10 m*M* Tris-HCl, pH 8.3, 25 m*M* EDTA, 0.5% w/v sodium dodecyl sulfate (SDS), 0.1 mg/mL of Proteinase K. The buffer should be prepared freshly prior to use. All stock solutions, except the Proteinase K, should be autoclaved. Stock Proteinase K should be 20 mg/mL and stored in aliquots at –20°C.
2. Phenol–chloroform: 50:50 buffered phenol and chloroform, stored at 4°C.
3. Chloroform–isoamylalcohol (24:1), stored at 4°C.
4. Distilled water.
5. 3 *M* sodium acetate, pH 5.2: 246.12 g of anhydrous sodium acetate dissolved in 800 mL of water, and pH adjusted to 5.2 with glacial acetic acid, before making up to a total volume of 1 L. Stored in autoclaved aliquots prior to use.
6. 70% and 100% ethanol.
7. 10 mg/mL RNase A solution. Made up in 10 m*M* Tris-HCl, pH 7.5, and 15 m*M* NaCl, and heated to 100°C for 15 min before cooling slowly to room temperature. Stored in aliquots at –20°C.

### 2.2.2. DNA Extraction from Blood

(This is for the reference DNA, although normal tissue adjacent to the tumor can also be used. It is preferable for the reference DNA to be from the same patient as the tumor DNA but if necessary it can be obtained from a healthy volunteer.)

1. Distilled water, to be used ice cold.
2. Nuclei lysis buffer: 10 m*M* Tris-HCl, 400 m*M* NaCl, 2 m*M* EDTA (0.6 g of Tris, 11.70 g of NaCl, and 0.37 g of Na-EDTA dissolved in 500 mL of water). Autoclaved and stored at 4°C, prior to use.
3. 0.1% Nonidet P-40 (NP-40).
4. Proteinase K buffer: 2 m*M* EDTA, 1% v/v SDS, autoclaved prior to use and stored at 4°C.
5. Proteinase K solution: 2 mg of Proteinase K dissolved in 1 mL of Proteinase K buffer, and prepared freshly prior to use.
6. Saturated ammonium acetate: 148 g of $NH_4Ac$ in 100 mL of water.
7. 10% w/v SDS solution.
8. 70% and 100% ethanol.

### 2.2.3. DNA Extraction from Formalin-Fixed Paraffin-Embedded Tissue

1. Xylene (low in sulfur).
2. 30%, 60%, 80%, and 100% ethanol.
3. Digestion buffer: 75 m*M* NaCl; 2.5 m*M* EDTA; 100 m*M* Tris-HCl, pH 8.0 with 0.5% v/v Tween-20 and 0.5 mg/mL of Proteinase K.
4. Then the same materials as **items 2–7** of **Subheading 2.2.1.**

## 2.3. Labeling of Probes

### 2.3.1. Nick Translation

1. DNA solution to be labeled (at least 1 µg is required).
2. 10X Nick translation buffer: 0.2 m*M* dATP, 0.2 m*M* dCTP, 0.2 m*M* dGTP in 500 m*M* Tris-HCl, pH 7.8, 50 m*M* $MgCl_2$, 100 µ*M* β-mercaptoethanol, 100 µg/mL of (BSA) (nuclease free), stored in ready-to-use aliquots at –20°C.
3. Fluorochromes: Fluorescein isothiocyanate (FITC)–12-dUTP and Texas red–5-dUTP; these are light sensitive and so should be handled under low lighting conditions.
4. DNA polymerase I (0.5 U/µL)/DNase I (0.4 mU/µL) enzyme mix (Gibco BRL).
5. DNA polymerase I (10U/µL) (Promega).
6. Distilled water.
7. 0.5 *M* EDTA.

### 2.3.2. DOP-PCR

1. DNA solution (100 pg–100 ng).
2. 10X PCR buffer: 30 m*M* $MgCl_2$, 500 m*M* KCl, 100 m*M* Tris-HCl, pH 8.4, 1 mg/mL of gelatin. All stock solutions should be autoclaved prior to preparation of the buffer.

3. Nucleotides for primary reaction: 2.5 m*M* of each dATP, dGTP, dCTP, and dTTP made up in sterile distilled water.
4. Nucleotides for labeling reaction: 2.5 m*M* of dATP, dGTP, and dCTP and 1.25 m*M* dTTP made up in sterile distilled water.
5. 6-MW primer: 50 µm 5'CCG ACT CGA GNN NNN NAT GTG G3' where N = A,C,G or T in roughly equal proportions.
   The materials in 2–5 of the above should be stored in ready-to-use aliquots at –20°C.
6. Distilled water.
7. *Taq* polymerase (5 U/µL).
   If this does not work, or before attempting this, a DOP-PCR kit is available (Boehringer cat. no. 1644 963).

### 2.4. Mixing, Precipitating, and Denaturing of Probes

1. Labeled DNA.
2. 1 mg/mL Human Cot-1 DNA (Gibco BRL).
3. 3 *M* Sodium acetate, pH 5.2.
4. 100% Ethanol stored at –20°C.
5. Deionized formamide (molecular biology grade) stored at –20°C, in ready-to-use aliquots.
6. 20X Saline sodium citrate (20X SSC): 175.3 g of NaCl and 88.2 g of sodium citrate dissolved in 800 mL of water, pH adjusted to 7.0 with a few drops of a 10 *N* solution of NaOH, before volume adjusted to 1 L. Stored in autoclaved aliquots.
7. 2X hybridization buffer: 20% Dextran sulfate, 4X SSC, pH 7.0, stored at –20°C in ready-to-use aliquots.

### 2.5. Denaturing of Chromosomal DNA on Slides

1. Denaturation solution for slides: 70% v/v formamide (deionized as described in **Subheading 2.4.5.**), 2X SSC, pH 7.0.
2. 22 mm × 50 mm coverslips.
3. Ethanol series: Ice-cold 70%, 95%, and 100%.

### 2.6. Hybridization

1. 22 mm × 22 mm coverslips.
2. Rubber sealant.
3. Humid chamber, for example, plastic slide box with a lid and a layer of moistened paper in the bottom. The slides can then be placed horizontally in the boxes resting on the side "ledges" above the moistened paper. Boxes should be sealed with tape.

### 2.7. Posthybridization Washes

1. Formamide solution: 50% v/v Formamide, 2X SSC, pH 7.0. **Note:** The formamide used here does not need to be deionized.
2. SSC solution: 2X SSC, pH 7.0.
   Solutions 1 and 2 should both be prepared freshly prior to use and then warmed to 42°C. It is important to check they are at the correct temperature prior to use.

3. SSCT: 4X SSC, 0.05% v/v Tween-20, pH 7.0.
4. Ethanol series: 70%, 95%, and 100%.
5. DAPI solution: 4,6-Diamidino-2-phenylindole (DAPI), which is a counterstain, made up in Citifluor; a mounting medium containing antifade, 0.15 μg/mL, stored at 4°C.

### 2.8. Image Acquisition and Analysis

1. Epifluorescence microscope equipped with a triple-color epifluorescence filter set (selective for the fluorochromes DAPI, FITC, and rhodamine) in combination with a cooled CCD camera and suitable computer hardware.
2. Software for analysis is commercially available (e.g., QUIPS™, Vysis Inc., Downers Grove, IL; CytoVision CGH system, Applied Imaging International).

## 3. Methods

### 3.1. Blood Cultures and Harvesting of Cells

Obviously care needs to be taken when working with biological material. A laboratory coat and gloves should be worn and all steps should be carried out in a tissue culture hood, until the cells are fixed. Tubes containing blood should be disposed of carefully prior to incineration. Discarding of supernatant up to the fixative steps should be into chloros. (Do not put fixative into chloros, as this reacts and leads to an offensive odor!)

1. Take 4 mL of blood from a volunteer and place immediately into a tube containing 40 μL of sodium heparin (10 μL of heparin/1 mL of blood), then gently agitate the tube to ensure thorough mixing. (The amount of blood taken is arbitrary–4 mL gives eight tubes which can be reasonably harvested by one person and can yield approx 160 slides, if good metaphases are obtained from each tube.)
2. In separate universal tubes add 0.5 mL of heparinized blood to 9.5 mL of culture medium and 0.1 mL of PHA and mix.
3. Incubate the tubes at 37°C with 5% $CO_2$ for 72 h, with the lids left loose. Gently agitate the tubes each day.
4. Prior to harvesting add 100 μL of colcemid (final concentration, 0.1 μg/mL) and incubate for 15–20 min at 37°C (*see* **Note 1a**).
5. Transfer blood to 15-mL Falcon tubes and centrifuge at 500*g* for 5 min.
6. Remove all of the supernatant with a pipet and resuspend the pellet in the very small amount of remaining medium by flicking the tube gently.
7. Add 10 mL of prewarmed hypotonic KCl. The first few drops should be added carefully, flicking the tube gently all the time. Make up to 10 mL and incubate at 37°C for 15–20 min.
8. Add 0.6 mL of ice-cold fixative solution (as a top layer to the hypotonic solution), close the lid, and gently invert the tube twice to mix the solutions.
9. Centrifuge at 500*g* for 5 min.
10. Remove the supernatant with a pipet and resuspend the pellet completely by flicking the tube gently.

11. Add a few drops of fixative solution *very carefully* while flicking the tube gently (*see* **Note 1b**), then make up to 10 mL, gently agitating the tube all the time.
12. Centrifuge at 500*g* for 5 min.
13. Remove the supernatant with a pipet.
14. Resuspend the pellet by flicking the tube gently and again add 10 mL of fixative solution carefully, agitating the tube all the time.
15. Repeat **steps 12–14** at least five times. After the second fixative step the tubes should be left overnight at 4°C. Thereafter fixative solution does not need to be added slowly, but fresh fixative must be prepared before proceeding.
16. Repeat **steps 12** and **13**, then resuspend the pellet in 0.5–1 mL of fixative solution.
17. Using a siliconized glass pipet, drop from a distance of about 40 cm one drop of fixed material onto each end of a clean microscope slide (*see* **Subheading 2.1., item 7** and **Note 1c**), which has been moistened by breathing on it (the slide should still be moist when the fixed material is placed on the slide).
    **Note:** Two separate drops on each slide means two hybridizations can be done on the same slide.
18. Using a pipet, "wash" the slides with fixative (before they dry) and then place in a warm humid environment to dry (e.g., over a water bath set at 55–60°C) (*see* **Note 1d**).
19. Slides should be examined using a microscope to ensure there are sufficient metaphases. If there are too many then more fixative can be added to the pellet, if there are too few then centrifuge again and resuspend the pellet in less fixative. The amount of cytoplasm should also be noted (*see* **Note 1e**). Even if the slides look satisfactory when examined under phase-contrast microscopy they may not hybridize well and so need to be tested prior to hybridization (*see* **Note 1f**).
20. Once the slides have dried they should be dehydrated in an ethanol series—70%, 95%, and 100%—for 3 min each and then left at room temperature overnight before storage.
21. Slides should be stored at –20°C, in a box containing a desiccant, for example, a small amount of silica gel wrapped in a piece of perforated parafilm, until use.
22. Fixed material can be stored temporarily (days) at 4°C, or for longer periods (weeks) at –20°C.
    **Note:** Prepared slides with metaphase spreads, especially for CGH, can be purchased from Vysis UK.

### 3.2. DNA Extraction

It is very important that the tissue samples are examined first by a pathologist to confirm that there is at least 70% of tumor present in the samples. The presence of contaminating normal tissue means that genetic changes are much more difficult to detect reliably *(6)*. For the following methods the volumes used will depend on the amount of tissue available and should therefore be adjusted accordingly. Tissue handling steps should be carried out in laminar flow cabinet. Phenol–chloroform should be used in a fume hood and disposed of as appropriate.

### 3.2.1. DNA Extraction from Frozen/Fresh Tissue

1. Tissue chunks should be minced in a Petri dish using a scalpel, and then placed in approx 1 mL of extraction buffer. Tissue should be scraped from slides using a drawn-out sterile glass capillary tube and then placed into a small Eppendorf tube with 50–300 µL of digestion buffer (depending on size of sections).
2. Leave in buffer at 37°C for 1 h if very small, or overnight for larger samples.
3. Add an equal volume of phenol–chloroform, mix by inverting the tube, and centrifuge at 2000*g* at 4°C for 15 min.
4. Remove the top layer and place in fresh tube, carefully avoiding the white interface and discarding the bottom layer.
5. Repeat **steps 3** and **4**.
6. Add an equal volume of chloroform–isoamylalcohol, mix by inverting the tube several times and centrifuge at 2000*g* at 4°C for 15 min.
7. Remove the top layer and put in a fresh tube, again avoiding the interface and discarding the bottom layer.
8. Add 0.1 vol of 3 *M* sodium acetate, pH 5.2, and add 2–3 vol of ice-cold 100% ethanol.
9. Gently invert the tube. If there is a large amount of DNA it should now be visible as a precipitate which can be hooked out using a sealed glass Pasteur pipet. Wash briefly in 70% ethanol, and then allow to dry before resuspending in distilled water and go to **step 12**.
10. If, however, no DNA is visible and very small amounts of DNA may not be, the tube should be put at –70°C for 2h to overnight and then centrifuged at 2000*g* at 4°C for 15–30 min. A pellet should now be visible.
11. Remove as much as possible of the supernatant, wash with 70% ethanol and centrifuge again at 2000*g* at 4°C for 15 min. Remove the ethanol and air-dry the pellet at room temperature before resuspending in distilled water.
12. Add 100 µg/mL of RNase A, then leave at 37°C for 30 min–1 h.
13. Determine the quantity of DNA spectrophotometrically and the quality by running on a 1% agarose gel.
14. DNA should be stored at 4°C short term but at –20°C for the longer term.
    **Note:** Alternatively a kit for extracting DNA from tissue can be used, which also gives good results (Qiagen cat. no. 29304).

### 3.2.2. DNA Extraction from Blood

1. Take 50-mL Falcon tubes and add 5–10 mL of blood (it can be fresh or frozen; if frozen allow to thaw first) into each tube.
2. Add ice-cold water to each tube to make a final volume of 50 mL, then invert tubes to mix well and lyse the red blood cells.
3. Centrifuge for 20 min at 4°C and 1300*g*.
4. Discard supernatant by pouring off into chloros and then carefully invert over a paper towel to remove the remaining lysed red blood cells, leaving a white nuclear pellet.
5. Add 25 mL of 0.1% v/v NP-40 to the nuclear pellet and centrifuge at 4°C and 1300*g*.
6. Discard the supernatant and carefully invert over a paper towel to remove remaining liquid, then add 3 mL of nuclei lysis buffer and vortex-mix to resuspend the pellet completely.

7. Add 200 μL of 10% w/v SDS and 600 μL of proteinase K solution, mix by inversion, and incubate at 60°C for 1.5–2 h (or overnight at 37°C).
8. Add 1 mL of saturated ammonium acetate and shake vigorously for 15 s. Allow to stand at room temperature for 10–15 min and then centrifuge for 20 min at 1300*g*.
9. Pour the supernatant into a separate tube. Precipitate the DNA by adding 2 vol of cold 100% ethanol to the supernatant and mix gently by inversion. The DNA should become visible as a precipitate and can be hooked out on the tip of a sealed glass Pasteur pipet.
10. Wash in 70% ethanol and then resuspend in distilled water.
11. Then go to **step 12** of **Subheading 3.2.1.**

### 3.2.3. DNA Extraction from Formalin-Fixed Paraffin Wax-Embedded Tissue

The method described below is for tissue sections on slides but it can be easily adapted to tissue sections in Eppendorf tubes. 5 × 10 μm sections should yield sufficient DNA to nick translate, but obviously it does depend on the area of the sections, generally 1 $cm^2$ should give a reasonable yield.

1. Dewaxing the slides: Take five 10-μm sections that are not hematoxylin and eosin (H&E) stained (as DNA extracted from H&E sections will not PCR), but one H&E slide is needed for reference. Place slides in xylene for 15 min, then 10 min in 100% ethanol and 1 min each in 80%, 60%, and 30% ethanol, taking care not to let the slides dry out at all between steps. Slides can be left at this stage in sterile water, but if they are to be left for a few weeks they should be stored at 4°C and the water changed regularly.
2. By comparing the slides to the H&E slide, scrape off the regions of tumor using a drawn-out sterile glass capillary tube and place into an Eppendorf tube containing 300 μL of digestion buffer.
3. Leave at 55°C overnight and then add more Proteinase K to a final concentration of 1 mg/mL, then leave at 55°C for at least 3 d.
4. Then proceed to **step 3** in **Subheading 3.2.1.** as for DNA extraction from frozen tissue

   **Note:** The Qiagen kit can also be used for formalin-fixed paraffin wax-embedded tissue.

## 3.3. Probe Labeling

### 3.3.1. Nick Translation

Remember that once the fluorochromes are added, all the steps should be done under low lighting conditions.

1. Combine 5 μL of the nick translation buffer, 1 μL of FITC–12-dUTP to label the test DNA or 1 μL of Texas red–5-dUTP to label the reference DNA (*see* **Note 2**), 1 μg of DNA, 10 μL of DNA polymerase I/DNase mix, and 1 μL of DNA polymerase I and distilled water to make a total volume of 50 μL.

   **Note**. The enzymes should be added last and the tubes kept on ice.

2. Incubate at 15°C for 2 h, then leave on ice while 5 µL is run on a 1% agarose gel.
3. Probe fragments should form a smear ranging in size between 500 and 2000 bp; this is particularly important (*see* **Note 3**).
4. If the fragments are the correct size, the reaction should be stopped by adding 2 µL of EDTA and heating to 70°C for 5 min to denature the enzymes.
5. Probes can be stored until use at –20°C.

### *3.3.2. DOP-PCR Labeling (*see ***Note 4***)*

Small amounts of DNA cannot be labeled by nick translation. The DNA is amplified using DOP-PCR and as little as 100 pg has been reported to give reproducible amplification ***(9)***. If sufficient DNA is produced some authors label the products using nick translation (*see* **Note 3**). However, in our experience a satisfactory sized probe, particularly from formalin-fixed paraffin-embedded tissue, is not achieved in this way, so the DNA should be labeled also by DOP-PCR.

#### 3.3.2.1. Primary Reaction

1. Mix 5 µL of 10X PCR buffer, 5 µL of primary dNTPs, 2 µL of 6 MW, 100 pg–100 ng DNA (optimum 20–100 ng), 0.5 µL of *Taq* polymerase, and distilled water to make a total of 50 µL. DNA should be added last.
   **Note:** It is very important to have a negative control in each experiment; a tube containing all the above but with no DNA.
2. Primary reaction conditions: 96°C for 10 min, then 9 cycles of 94°C for 1 min, 30°C for 1.5 min, 72°C for 3 min with ramp to 72°C 1°C/4.2 s, then 35 cycles of 94°C for 1 min, 62°C for 1 min, and 72°C for 3 min, then 72°C for a further 10 min.
3. Run 10 µL on a 1% agarose gel (*see* **Fig. 1**). The PCR product varies in size, particularly if the source of DNA is formalin-fixed paraffin wax-embedded tissue, when the fragments may be in the range of 50–1000 bp. The negative control should not contain a smear; if it does then the products should be discarded and not used (*see* **Note 4**). Unused product should be stored at –20°C.

#### 3.3.2.2. Labeling Reaction

1. Mix 1–5 µL (depending on concentration as determined by gel) DNA from the primary reaction, 5 µL of PCR buffer, 4.4 µL of labeling nucleotides, 2 µL of 6-MW, 1 µL of FITC–12-dUTP (for the test) or 1 µL of Texas red–5-dUTP (for the reference), 1 µL of *Taq* polymerase, and distilled water to make a total of 50 µL. Again set up a negative control.
2. Labeling reaction conditions: 96°C for 10 min, then 40 cycles of 94°C for 1 min, 55°C for 1 min, 72°C for 3 min, and then 72°C for 10 min.
3. Run 10 µL of on a 1% agarose gel. If not used immediately products should be stored at –20°C.

### *3.4. Mixing, Precipitating, and Denaturing of Probes*

1. Mix 1 µg of labeled test DNA and 1 µg of labeled normal DNA in an Eppendorf tube (*see* **Note 5**), 50 µL of human Cot-1 DNA (*see* **Note 6**) and 0.1 vol of 3 *M* sodium acetate, and 2 vol of cold absolute ethanol.

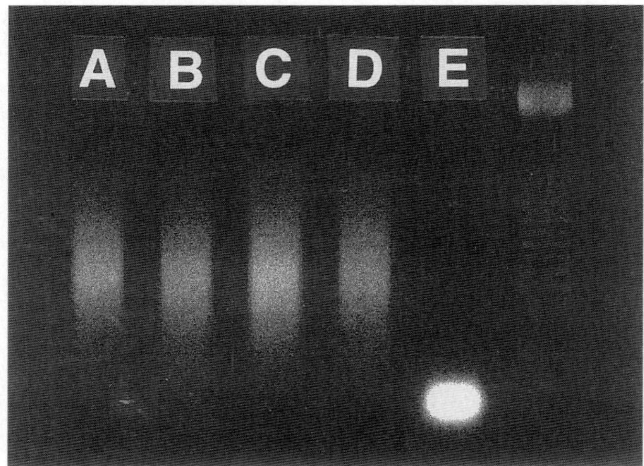


Fig. 1. A 1% agarose gel showing DOP-PCR products. **(A,B)** DNA amplified from frozen tumor DNA. **(C,D)** DNA amplified from normal genomic DNA. **(E)** Negative control only unused primers are seen. The marker lane contains 100-bp ladder.

2. Place on dry ice for 30 min or leave at –20°C overnight.
3. Centrifuge at 15000*g* at 4°C for 15 min.
4. Aspirate supernatant carefully and leave pellet to air-dry, taking care not to overdry the pellet, otherwise it will be very difficult to resuspend.
5. Resuspend pellet in 5 µL of deionized formamide and leave at 37°C for 30 min.
6. Add 5 µL of 2X hybridization mix and mix carefully by pipeting to ensure that the pellet has been thoroughly resuspended.
7. Denature at 75°C for 5 min.
8. Preanneal at 37°C for 30 min–1 h (while the probes are preannealing the slides can be prepared; *see* **Subheading 3.5.**) (*see* **Note 6**).
9. Place the 10 µL of the denatured probe on half of the denatured slide, which has been placed on a hotplate at 37°C. Cover each probe with a 22 mm × 22 mm coverslip, taking care to ensure there are no air bubbles.
10. Seal coverslip with rubber sealant and leave on the hotplate until the rubber sealant dries.
11. Leave to hybridize in a humid chamber for 72 h at 37°C.

### 3.5. Denaturation of Chromosomal DNA on Slides

This is a crucial step and each batch of slides must have the optimal denaturation conditions determined. To do this, denature test slides for different times (e.g., range 2–5 min), stain them with DAPI, and then examine them under the microscope (*see* **Note 1f**).

1. Remove slides from storage just prior to use and mark the areas containing metaphases with a diamond pen.

2. Denature slide on a hotplate at 73°C for the predetermined optimum denaturation time (depending on slide batch as discussed previously) with denaturation solution under a 22 mm × 50 mm coverslip.
3. Place immediately into ice cold 70% ethanol for 3 min, then dehydrate through an ethanol series for 3 min each.
4. Air-dry the slides. They are then ready for use.
5. Just before use, place slides on a hotplate at 37°C.

### 3.6. Posthybridization Washes (see Note 7)

**Note:** These steps should be done under low lighting conditions and formamide waste should be carefully disposed of.

1. The rubber cement should be gently removed and the slide shaken to flick off the coverslips; if they do not come off, then they can be encouraged to come off by putting the slide in the Coplin jar containing formamide solution.
2. 3 × 5 min washes in formamide solution in a shaking water bath.
3. 3 × 5 min washes in SSC solution in a shaking water bath.
4. 1 × 5 min washes in SSCT at room temperature gently shaking the Coplin jar. **Note:** during **steps 2–4** the slides should never become completely dry.
5. Dehydrate slides in an ethanol series for 2 min each and air-dry in darkness.
6. Mount in DAPI solution, approx 30 µL per slide and cover with a 22 mm × 50 mm coverslip.
7. Store slides in the dark in a cardboard folder at 4°C; but capture images as soon as possible.

### 3.7. Image Acquisition and Analysis

For analysis of each sample, images of 5–10 metaphases should be captured and the raw data saved (*see* **Fig. 2A–C** and **E**). Metaphases should be captured using the 63× or 100× objective lens with the field diaphragm closed down to the edge of the image acquisition area (this increases the contrast within the image). It is most important that the bulb is well focused and that there is homogeneous illumination of the optical field, to avoid variations of the fluorescence ratio. Metaphases should have smooth intense color that is not granular in appearance and heterochromatin regions should not be brightly stained (*see* **Notes 1**, **3**, and **6**).

Semi-automated karyotyping and CGH analysis is done using commercially available software. First a digitally inverted DAPI image is generated (*see* **Fig. 2D**) that gives a banding pattern resembling classical G-banding, and is used for karyotyping. The software has tools to divide touching and crossed-over chromosomes (alternatively crossovers can be excluded). The position of axes and centromeres for each chromosome are checked. An automated karyotype is then generated, but the accuracy does need to be confirmed, and any corrections made (*see* **Fig. 3A**). Once the karyotype is acceptable the software generates CGH profiles for each chromosome by calculating the red-to-green ratio, pixel by pixel, along the length of each chromosomal axis. After each metaphase has been analyzed separately, a

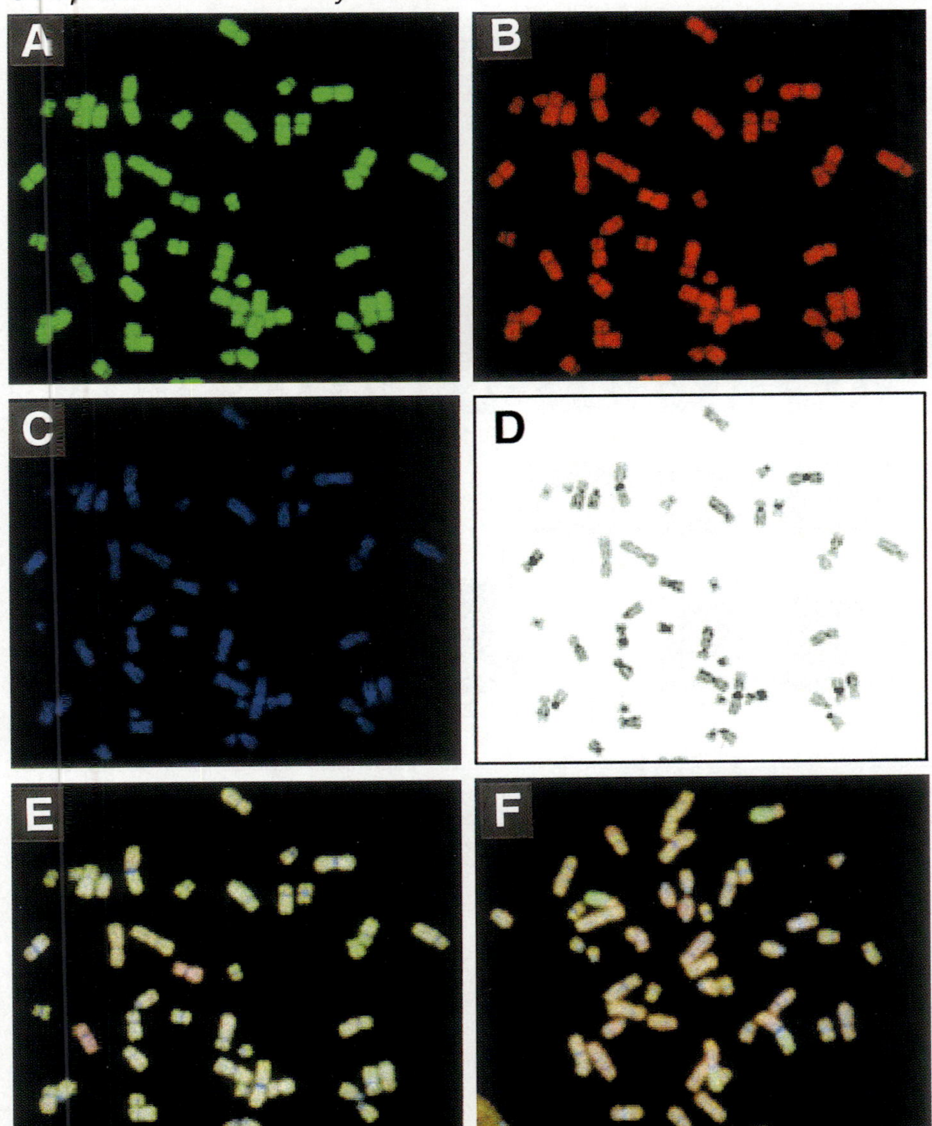


Fig. 2. **(A–E)** Example CGH images of normal male DNA (green) cohybridized with normal female (red) DNA to normal female metaphase spreads. **(A)** FITC image of normal male DNA. **(B)** Texas red image of normal female DNA. **(C)** DAPI image of the metaphase chromosomes. **(D)** The inverted DAPI image, giving a banding pattern resembling G-banding, which is used to karyotype the chromosomes. **(E)** This is the composite image of images **A**, **B**, and **C**. The two X chromosomes appear visibly red, which is as expected from the hybridization, that is the male DNA shows apparent loss of the X chromosome. **(F)** This is an example of a CGH image of breast tumor DNA (green) cohybridized with normal DNA from the same patient (red) to normal female metaphase spreads. Several regions of obvious gain (excessively green) and loss (excessively red) are visible.

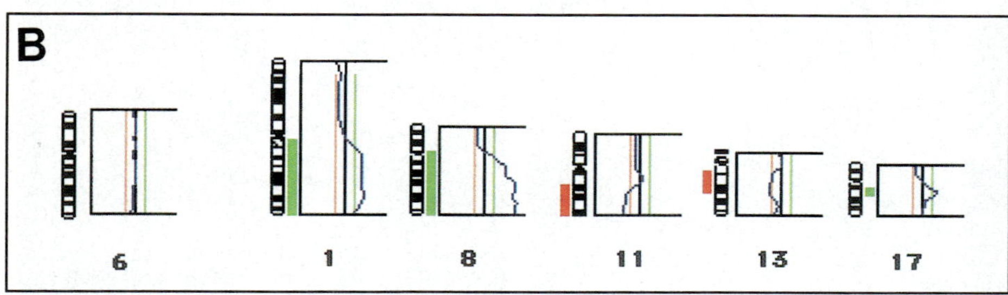

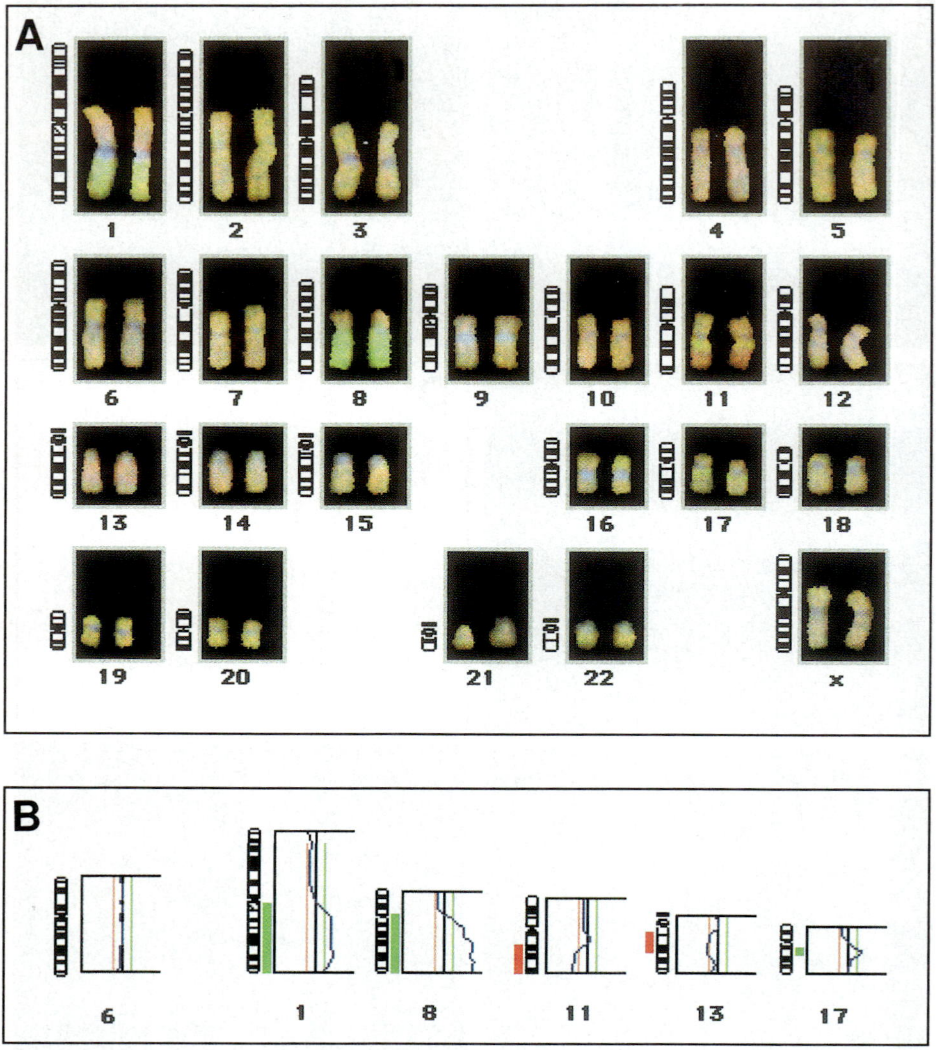

Fig. 3. **(A)** This is the karyogram of the CGH image shown in **Fig. 2F**. **(B)** These are selected average fluorescence ratios following analysis of 10 metaphases from the tumor seen in **Fig. 2F**. Chromosome 6 shows no changes and an average fluorescence ratio of 1.0. The profiles of chromosomes 1, 8, 11, 13, and 17 are shown. The red line to the left is set at 0.8 and the green line to the right is set at 1.2. It can be seen that the regions that appear excessively green or red correspond to regions of the profiles showing gain and loss; note also the two discrete green signals representing amplification on chromosome 17.

collection of 5–10 metaphases are combined and an average ratio for each chromosome obtained (*see* **Fig. 3B**). Conventionally gains are considered significant when the ratio of test to reference DNA is 1.2:1 or greater and losses considered significant when the ratio is 0.8:1 or less, although other limits are used, for example, 1.15 and 0.85. For an improved interpretation of the CGH ratio profiles the analysis software can be extended to calculate statistical confidence intervals, and *t*-statistics can then be applied ***(10)***. High-level amplification has been considered when the ratio exceeds various limits, the minimum used is 1.4. However, some authors infer high-level amplification when by visual inspection areas of discrete intense green signal are noted, with a corresponding profile suggesting gain ***(11)***.

## 4. Notes

1. The quality of the slides is probably the single most important factor in determining the success of a CGH experiment ***(12)***.
   a. The chromosome size and quantity of metaphases is variable. Colcemid times longer than 20 min will give more metaphase spreads but they will be more condensed. Therefore a balance needs to be achieved between chromosome length and number. One way of obtaining numerous chromosomes of a certain size is to synchronise the cell culture. This can be done using methotrexate (MTX), which acts to block mitosis. Prepare bloods up to **Subheading 3.1., step 4**. Then add 0.1 mL of MTX ($1 \times 10^{-5}$ *M* stock stored in aliquots in dark tubes at –20°C) to the cells at 65–72 h. Continue to incubate them at 37°C, but in the dark, for 17 h (i.e., overnight). Centrifuge at 500*g* for 5 min and discard the supernatant containing the MTX. Wash the cells twice in culture medium, centrifuge at 500*g* for 5 min, and discard the medium each time. After the final wash add 10 mL of culture medium and add 0.1 mL of thymidine solution ($1 \times 10^{3}$ *M* stock autoclaved and stored in aliquots at –20°C), which acts to release the mitotic block. Incubate the cultures for between 5.5 and 6 h (they can be left as long as 10 h). Then add 20 µL colcemid and leave for 10–15 min; then proceed to **Subheading 3.1., step 5**.
   b. Fixation of the cells. It is most important to add the first fix slowly, because if it is added too quickly it will result in the pellet clumping and poor fixation. Proper fixation ensures crisp, well spread chromosomes. Placing the cells at 4°C also helps facilitate fixation. The more times the pellets are washed with fixative the cleaner the final preparation will be, with less cytoplasm.
   c. Spreading of the metaphase chromosomes is variable. Ideally metaphases should be well spread with not too many overlapping chromosomes. Metaphases spread well when they are well fixed (*see* **Note 1b**) and when the microscope slides are clean, hence the importance of storing them in ethanol with a small amount of hydrochloric acid and drying them, just prior to dropping the spreads, with lint-free tissue. The other important factor is the atmospheric conditions (*see* **Note 1d**).
   d. Slide making conditions are important. However, conditions can vary considerably (temperature and humidity) and so if the metaphase spreads do not look optimal on one day, store the fixed material and try again another

day. The conditions noted here work well in our laboratory but they are subject to great variability. If the conditions are too cold/dry the metaphases will not spread. Conversely, if the conditions are too hot/humid the metaphases overspread and indeed in extreme cases appear as a "chromosome soup."

e. The amount of cytoplasm is critical. If the metaphases appear to be embedded in too much cytoplasm then washing the pellet and/or the slides with more fixative can be tried. The amount of cytoplasm is critical, as too much prevents both optimal denaturation and probe penetration, with a resulting poor hybridization. However, a small amount of cytoplasm may allow the chromosomes to withstand the necessary denaturation. Some authors pretreat the slides with Proteinase K or pepsin ***(13)*** to remove excess cytoplasm, but we find that if the slides are of sufficient quality pretreatment is not necessary. If a batch of slides has a large amount of cytoplasm, they will not be useful for CGH and so should be discarded. Although it is time consuming to prepare a new batch, it is ultimately worthwhile in comparison to setting up many CGH experiments which yield doubtful results. If a new batch has to be prepared a longer hypotonic stage may be helpful.

f. It is necessary to test the quality of each batch of slides, prior to use; again, if they are of insufficient quality then they should be discarded. Slides are quality tested by denaturing them (using a range of denaturation times e.g., 2–5 min), staining with DAPI, and then examining them under the microscope. This is necessary as slides that look satisfactory down the microscope may not necessarily withstand denaturation well. Optimal metaphase spreads are ones that tolerate enough denaturation to achieve uniform hybridization along each chromosome but do not lose their structure and morphology. By implication from this, slides that are underdenatured will not hybridize, and if overdenatured the chromosomes appear fat without normal morphology. Once an optimal denaturation time has been defined the slides should have a test hybridization with normal male and female DNA; all autosomes should have a ratio between 0.9 and 1.1 and the X chromosome should have a ratio of <0.6.

2. Within each experiment a control should be set up with normal male (green) and normal female DNA (red). The fluorescence ratios, particularly of the X chromosome, should be as in **Note 1f**. If using DOP-PCR to label DNA a useful internal control for an experiment is to pair tumor DNA with DNA from the opposite sex. The profile of the X chromosome will be a guide to the quality of the experiment, that is, a ratio that is less than expected will suggest a poor quality experiment.
3. The size and quality of the probe is an important step in determining the quality of the CGH. If the probe fragments are too large then the incubation time should be increased or more enzyme (DNase–Polymerase mix) added; if the fragment size is too small then repeat with a shorter incubation time or by adding less enzyme. If the probes are too large then the resulting hybridization will be poor with typically a spotty or speckled appearance. DNA contaminated with protein will label inefficiently and give a similar hybridization appearance. If a large amount of contaminating protein is present, the probe will fail to nick at all and in this situation the DNA needs to be repurified. If it is desirable to attempt nick translation of DNA amplified by DOP-PCR then less enzyme and shorter incubation times are needed.

4. DOP-PCR is a difficult technique. It is often not very reproducible and is prone to contamination. Separate pipets should be used for PCR only. If possible it should be done in a dedicated room or area with surfaces, pipets, and tubes subjected to ultraviolet (UV) irradiation. If contamination occurs, this should be dealt with by preparing fresh stock solutions and cleaning the pipets used and taking the usual precautionary methods for any PCR. DNA that has been extracted from H&E sections will not PCR. Although such DNA run on a gel will look satisfactory, it is only when it is amplified that the problem will come to light. Similarly if only small amounts of frozen tissue are available that need to be amplified, problems arise if the tissue has been placed in the embedding medium OCT (optimum cutting temperature) compound. In this situation in our experience the DNA does PCR and reasonable smears are obtained but it results in very poor quality hybridization. If the DOP-PCR still fails to work different magnesium concentrations can be tried, some authors use PCR buffer containing 20 m*M* magnesium. Also, *see* **ref. *14***.
5. Some authors use less than 1 µg of nick-translated labeled probe; however, in our experience this amount gives a strong uniform hybridization. However, if using DOP-PCR labeled products, it may be necessary to use less probe.
6. The amount of Cot-1 and preannealing times are important, as if there is insufficient suppression of the repetitive sequences this will result in strong staining of the heterochromatin regions of chromosomes 1, 9, 16, and 19 and high intensity of nonspecific fluorescence at chromosomal regions that have a high content of repetitive sequences. Both these changes can not be corrected for during image analysis and will result in making it difficult to detect small copy number alterations in the heterochromatin regions and inaccuracies in the fluorescence ratio in regions of repetitive sequences.
7. Posthybridization washes are probably subject to the most variability between different authors. Those outlined here give good results with a very good signal-to-noise ratio.

## Acknowledgments

Many thanks to Patricia Gorman, Emanuela Volpi, Denise Sheer, Richard Poulsom, Kelly Woodford-Richens, and Hanan Lamlum for their helpful comments during the preparation of this manuscript.

## References

1. Kallioniemi, A., Kallioniemi, O.-P., Sudar, D., Rutovitz, D., Gray, J., Waldman, F., and Pinkel, D. (1992) Comparative genomic hybridization for molecular cytogenetic analysis of solid tumours. *Science* **258,** 818–821.
2. Cher, M. L., Bova, G. S., Moore, D. H., Small, E. J., Carroll, P. R., Pin, S. S., et al. (1996) Genetic alterations in untreated metastases and androgen-independent prostate cancer detected by comparative genomic hybridization and allelotyping. *Cancer Res.* **56,** 3091–3102.

3. Gronwald, J., Storkel, S., Holtgreve-Grez, H., Hadaczek, P., Brinkschmidt, C., Jauch, A., et al. (1997) Comparison of DNA gains and losses in primary renal clear cell carcinomas and metastatic sites: importance of 1q and 3p copy number changes in metastatic events. *Cancer Res.* **57,** 481–487.
4. Kuukasjarvi, T., Karhu, R., Tanner, M., Kahkonen, M., Schaffer, A., Nupponen, N., et al. (1997) Genetic heterogeneity and clonal evolution underlying development of asynchronous metastasis in human breast cancer. *Cancer Res.* **57,** 1597–1604.
5. Bentz, M., Plesch, A., Stilgenbauer, S., Dohner, H., and Lichter, P. (1998) Minimal sizes of deletions detected by comparative genomic hybridization. *Genes Chromosomes Cancer* **21,** 172–175.
6. Kallioniemi, O.-P., Kallioniemi, A., Piper, J., Isola, J., Waldman, F. M., Gray, J. W., and Pinkel, D. (1994) Optimizing comparative genomic hybridization for analysis of DNA sequence copy number changes in solid tumors. *Genes Chromosomes Cancer* **10,** 231–243.
7. Nishizaki, T., Devries, S., Chew, K., Goodson, W. H., Ljung, B. M., Thor, A., and Waldman, F. M. (1997) Genetic alterations in primary breast cancers and their metastases—direct comparison using modified comparative genomic hybridization. *Genes Chromosomes Cancer* **19,** 267–272.
8. Telenius, H., Carter, N. P., Bebb, C. E., Nordenskjold, M., Ponder, B. A., and Tunnacliffe, A. (1992) Degenerate oligonucleotide-primed PCR: general amplification of target DNA by a single degenerate primate. *Genomics* **13,** 718–725.
9. Speicher, M. R., du Manoir, S., Schrock, E., Holtgreve, G. H., Schoell, B., Lengauer, C., et al. (1993) Molecular cytogenetic analysis of formalin-fixed, paraffin-embedded, solid tumors by comparative genomic hybridization after universal DNA-amplification. *Hum. Mol. Genet.* **2,** 1907–1914.
10. Moore, D. H., 2nd, Pallavicini, M., Cher, M. L., and Gray, J. W. (1997) A *t*-statistic for objective interpretation of comparative genomic hybridization (CGH) profiles. *Cytometry* **28,** 183–190.
11. Solinas-Toldo, S., Wallrapp, C., Muller-Pillasch, F., Bentz, M., Gress, T., and Lichter, P. (1996) Mapping of chromosomal imbalances in pancreatic carcinoma by comparative genomic hybridization. *Cancer Res.* **56,** 3802–3807.
12. Karhu, R., Kahkonen, M., Kuukasjarvi, T., Pennanen, S., Tirkkonen, M., and Kallioniemi, O. (1997) Quality control of CGH: impact of metaphase chromosomes and the dynamic range of hybridization. *Cytometry* **28,** 198–205.
13. Courjal, F. and Theillet, C. (1997) Comparative genomic hybridization analysis of breast tumors with predetermined profiles of DNA amplification. *Cancer Res.* **57,** 4368–4377.
14. Kuukasjarvi, T., Tanner, M., Pennanen, S., Karhu, R., Visakorpi, T., and Isola, J. (1997) Optimizing DOP-PCR for universal amplification of small DNA samples in comparative genomic hybridization. *Genes Chromosomes Cancer* **18,** 94–101.

16

# RNA Purification and Quantification Methods

**Adam Jones, Chisato Fujiyama, and Kenneth Smith**

## 1. Introduction

Analysis of gene expression within tumors is frequently performed on the messenger RNA (mRNA). This chapter first describes the purification of mRNA from either tissues or cell culture and then describes the quantification by ribonuclease protection assay.

Having initially purified the cellular RNA the technique of ribonuclease protection relies on generation of a radioactive probe of antisense messenger RNA that is homologous to, and therefore capable of hybridizing with, the complementary strand of endogenous sense mRNA. A cocktail of single-strand-specific ribonucleases is then added to degrade any regions of the probe not protected by hybridization. Double-stranded mRNA, which is resistant to ribonuclease degradation, is then separated by polyacrylamide gel electrophoresis (a technique already described in Chapter 11 by Blancher and Jones) and subsequently detected by autoradiography.

Premature contamination with ribonucleases that are both stable and active is one of the main reasons for the failure of these techniques. Contamination with exogenous ribonucleases is avoided by the following (1) Gloves should be worn—skin is a plentiful source of ribonucleases. (2) Plasticware not touched by ungloved hands can be considered ribonuclease free. All glassware should be baked at 250–300°C for at least 4 h or rinsed with chloroform before use. (3) All solutions should be made up with water treated overnight with diethylpyrocarbonate (DEPC) which inactivates ribonucleases by covalent modification.

RNA purification begins with inactivation of endogenous ribonucleases with guanidinium thiocyanate. Ultracentrifugation of the sample through a cushion of caesium chloride utilizes the greater density of RNA to separate it from

From: *Methods in Molecular Medicine, vol. 57:*
*Metastasis Research Protocols, Vol. 1: Analysis of Cells and Tissues*
Edited by: S. A. Brooks and U. Schumacher © Humana Press Inc., Totowa, NJ

DNA, protein, and other cellular contaminants. Extraction with phenol–chloroform further purifies the RNA from protein contaminants. The RNA is then ethanol precipitated, resuspended in DEPC-treated water and stored at –70°C.

After the RNA is purified, a riboprobe specific to the RNA of interest is constructed by cloning a region (100–400 bp) of the cDNA into a plasmid containing bacteriophage RNA promoter(s) such that the cDNA is downstream of the promotor. The construct is linearized with a suitable restriction endonuclease and antisense radiolabeled copies of the mRNA are transcribed from the bacteriophage RNA promoter. These RNA sequences are subsequently purified with deoxyribonucleases to remove template DNA.

The radiolabeled probe is then added to the RNA sample, where it will hybridize with its homologous sense mRNA. Ribonucleases are added to degrade any single-chain RNA—that is, free probe and RNA not hybridized. These protected fragments are recovered with ethanol precipitation. Identification of an appropriate sized probe on subsequent gel electrophoresis will confirm the presence or absence of target mRNA. A second and possibly third probe is also usually made to a known common mRNA molecule, for example, glyceraldehyde-3-phosphate dehydrogenase (GAPDH)—a housekeeping enzyme—to act as a loading control between samples.

This chapter is divided into several parts relating to the various steps in these techniques:

1. RNA purification.
2. Determining RNA concentration and purity.
3. Ribonuclease protection assay—probe generation.
4. Ribonuclease protection assay—hybridization of probe.

## 2. Materials

All materials must be sterile and RNAse free as previously outlined. Where possible, molecular grade reagents should be used.

### *2.1. RNA Purification*

1. DEPC-treated water: Add 1 mL of DEPC to 1 L of double-distilled water, shake vigorously to solubilize the DEPC, and leave overnight at room temperature. Traces of DEPC must then be removed by autoclaving (15 min on liquid cycle); otherwise the RNA may be inactivated by carboxymethylation. Exercise care and use a fume hood when using DEPC, as it is a suspected carcinogen and can also decompose to produce an explosive mixture of ethanol and carbon dioxide.
2. Solution D (denaturing). To 250 g of guanadinium thiocyanate add 303 mL of DEPC-treated water, 17.6 mL of 0.75 *M* sodium citrate and 13.2 mL of 20% *n*-lauroyl sarcosine (sodium salt). Pour the contents into a beaker and stir, with gentle heat, until contents are dissolved. Store in the dark at 4°C for up to 4 mo.
3. 2-Mercaptoethanol.

4. Phosphate buffered saline (PBS).
5. 5.7 *M* Caesium chloride.
6. Water-saturated phenol.
7. Chloroform.
8. 2 *M* Sodium acetate, pH 4.0.
9. 0.75 *M* Sodium citrate.
10. Absolute ethanol.
11. 70% Ethanol.
12. Cell scraper (rubber policeman).
13. Pipets.
14. 50-mL Falcon (Becton–Dickinson) conical tubes.
15. Polypropylene centrifuge tubes to fit appropriate rotor.
16. 10-mL Syringes.
17. 21- or 23-gauge hypodermic needles.
18. RNase-free pipet tips.
19. RNase free 1.5-mL Eppendorf tubes.

It is worth setting aside all reagents, solutions, glassware, and plasticware that are to be used only for RNA work.

### 2.2. Determining RNA Concentration and Purity

1. DEPC-treated water.
2. Agarose gel box.
3. Spectophotometer capable of measuring optical density at 260 and 280 nm.
4. Agarose powder.
5. 1X TBE (10X TBE stock is: 108 g of Tris base, 55 g of boric acid, 40 mL of 0.5 *M* EDTA, pH 8.0; water to 1 L).
6. Ethidium bromide (toxic—**handle with care**).
7. 1 *M* Sodium hydroxide.
8. 70% Ethanol (made with DEPC).
9. Ficoll EDTA (10X stock is: 20 g of Ficoll, 1 g of sodium dodecyl sulfate [SDS], 0.25 g of bromophenol blue, 20 mL of 0.5 *M* EDTA, pH 8.0; water to 100 mL).

### 2.3. Ribonuclease Protection Assay—Probe Generation

1. "Plugged" pipet tips (20 µL, 200 µL, 1000 µL).
2. 10X Transcription buffer (Boehringer Mannheim).
3. 200 m*M* Dithriothreitol (DTT).
4. NTP mix (no CTP—final concentration of 4 m*M* for each NTP).
5. DEPC-treated water.
6. [$\alpha$-$^{32}$P] CTP (Amersham, specific activity 400 Ci/mmol).
7. Placental ribonuclease inhibitor.
8. Linearized DNA probe template.
9. RNA polymerase: either T3, T7, or SP6 depending on vector used—20 U/µL.
10. Absolute ethanol.
11. Silane (dimethyldichlorosilane).

12. Gel diluent (National Diagnostics, Hull, England).
13. Gel concentrate (National Diagnostics, Hull, England).
14. 10X TBE: 0.89 *M* Tris-base, 0.89 *M* boric acid, 0.02 *M* EDTA.
15. 10% Ammonium persulphate (APS).
16. *N,N,N',N'*-Tetramethylethylenediamine (TEMED).
17. RNase free DNase (Boehringer Mannheim).
18. Sephadex G-50 centrifugation columns.
19. 50-mL Falcon (Becton–Dickinson) conical tubes.
20. Loading buffer: 80% w/v deionized formamide; 0.001 *M* EDTA, pH 8.0, 0.1% bromophenol blue, 0.1% xylene cyanol.
21. Whatman 3MM paper.
22. Scintillation vial and scintillation counter.

### 2.4. Ribonuclease Protection Assay—Hybridization of Probe

1. 5X Hybridization buffer: 0.2 *M* PIPES, pH 6.4, 2 *M* NaCl, 0.005 *M* EDTA.
2. Deionized formamide.
3. tRNA.
4. Digestion buffer: 10 m*M* Tris-HCl, pH 7.5, 300 m*M* NaCl, 5 m*M* EDTA.
5. RNase T1.
6. RNase A.
7. 20% SDS (lauryl sulfate).
8. Proteinase K.
9. Phenol.
10. Absolute ethanol.
11. 70% Ethanol (made with DEPC).
12. Electrophoresis plate.
13. Silane (dimethyldichlorosilane).
14. Gel diluent (National Diagnostics, Hull, England).
15. Gel concentrate (National Diagnostics, Hull, England).
16. 10X TBE: 0.89 *M* Tris, 0.89 *M* boric acid, 0.02 *M* EDTA.
17. 10% APS.
18. TEMED.
19. Loading buffer: 80% w/v deionized formamide, 0.001 *M* EDTA, pH 8.0, 0.1% bromophenol blue, 0.1% xylene cyanol.
20. Whatman 3MM paper.

## 3. Methods

### 3.1. RNA Purification

This is a modification of the caesium chloride method, which we have found to reproducibly give fairly high yields of pure RNA from small quantities of tissues or cells. Several commercial kits are available for RNA purification. These are much quicker but are more expensive and generally give substantially lower yields of RNA.

1. Add 360 μL of 2-mercaptoethanol to 50 mL of solution D on the day of the experiment.
2. For cells:
   a. Aspirate medium off cells. Wash with PBS × 1 and aspirate dry. (Note that the degree of cellular confluence alters the level of several mRNAs. Most people would aim for cells in the exponential growth phase—50–70% confluent.)
   b. Add 4 mL of solution D to a 10-cm plate. Scrape cell lysate off with cell scraper and aspirate into a 50-mL falcon tubes. Place in ice.
3. For tumor tissue:
   a. Tissues that have been snap-frozen and stored, preferably in liquid nitrogen, are placed on a Petri dish on ice and briefly thawed so that they can be minced finely with a disposable scalpel.
   b. Transfer the mince to a porcelain mortar, chill with liquid nitrogen, and pulverize with a porcelain pestle to achieve a powder consistency. The powder is scraped into a 50-mL falcon tube using a chilled scalpel and 10 mL of solution D added. Minimize the time taken to thaw and mince the tissue, as endogenous RNases will become activated.
4. Syringe DNA from the Falcon tube on ice through a 21- or 23-gauge hypodermic needle. This shears the genomic DNA.
5. Pipet 1.5 mL of 5.7 *M* caesium chloride into a Beckman centrifuge tube (13 × 51 mm if using SW 55 rotor or 3.0 mL into 14 × 89 mm for SW 41 rotor).
6. Pipet cell or tissue lysate on top of caesium chloride. Do this slowly down the side wall of tube. The caesium chloride and lysate remain as two separate layers with a clearly seen interface. Fill the centrifuge tube to the top with additional solution D if necessary.
7. Centrifuge tubes at 23°C for 17 h at 36,000 rpm or 3 h at 54,000 rpm at 20–25°C.
8. After centrifugation remove tubes. Using a 1-mL pipet tip aspirate all but the last 1 cm of solution. Slice through the tube just above this level using a hot scalpel. This gives greater control to aspirate the last 1 cm or so. Do this carefully; the RNA is found at the bottom of the tube and looks like a 3–5 mm gelatinous clear contact lens.
9. Resuspend this RNA pellet with 300 μL of DEPC-treated water. Leave for 1 min then transfer to a new Eppendorf tube. Repeat resuspension again using 100 μL DEPC-treated water and transfer to the same Eppendorf tube.
10. Add 200 μL of phenol and 200 μL of chloroform to the Eppendorf tube. Vortex-mix for 15 s. Centrifuge at room temperature for 2 min at 14,000 rpm. This results in two phases, a top aqueous phase that contains the RNA and a lower organic phase. Aspirate the top aqueous phase and transfer to a new Eppendorf tube. Discard the rest.
11. To this Eppendorf tube add 400 μL of chloroform, to eliminate residual phenol. Vortex-mix for 15 s. Centrifuge at room temperature for 2 min at 14,000 rpm. This again results in two phases, a top aqueous phase which contains the RNA and a lower organic phase. Aspirate the top aqueous phase and transfer to a new Eppendorf tube. Discard the rest.

12. Add 44 μL of 2 *M* sodium acetate to the Eppendorf tube (making a final concentration of approx 0.2 *M* sodium acetate). Precipitation of RNA is more efficient in the presence of salts.
13. Add 2.5 vol of absolute ethanol, that is, 444 μL × 2.5 = 1100 μL of absolute ethanol. In practice if using 1.5-mL Eppendorf tubes, fill them to the top.
14. Leave at –70°C for a minimum of 30 min. This can be prolonged to several days if more convenient.
15. Next, centrifuge at 4°C for 10 min at 14,000 rpm.
16. A white pellet should be seen at the bottom of the Eppendorf tube, containing the RNA. Aspirate all the overlying fluid.
17. Resuspend the pellet in 1 mL of 70% ethanol. Vortex-mix for 15 s, then centrifuge at 4°C for 10 min at 14,000 rpm.
18. Aspirate all the ethanol. Centrifuge (room temperature) for a few seconds, then aspirate any remaining ethanol. The remaining pellet should be fairly pure RNA. Resuspend this in 50 μL of DEPC. The concentration and then the purity of the RNA now needs calculating.

### 3.2 Determining RNA Concentration and Purity

1. To each of two Eppendorf tubes add 98 μL of DEPC. To these add 2 μL of the RNA–DEPC solution described previously. Use these two Eppendorf tubes to perform the tests in **step 2** in duplicate.
2. Using a spectophotometer, measure the optical density at 260 and 280 nm. For RNA an optical density at 260 nm of 1 equates to a concentration of 40 μg/mL. Therefore,

   $$\text{Optical density at 260 nm} \times 40 \times \text{dilution factor (50 in this case)} \\ = \text{concentration of RNA in } \mu\text{g/mL}.$$

   Ideally, the concentration should be around 1 μg/μL.
3. The optical density at 260 nm divided by the optical density at 280 nm provides an estimate of the purity of the RNA. Ideally, this figure should be ≥1.7.
4. Now run an 1% agarose gel to ensure the RNA is intact.
   a. First soak the gel box in 1 *M* sodium hydroxide for 30 min to denature any RNases. Rinse thoroughly with distilled water before use.
   b. Prepare a 1% agarose gel in TBE (0.6 g of agarose in 60 mL of TBE), adding 0.5 μg ethidium bromide per milliliter of gel. This intercalates with RNA and will allow visualization of the RNA under ultraviolet light. **Caution:** Ethidium bromide is extremely toxic and must be handled with due caution.
   c. Add 2 μg (should be approx 2 μL) of RNA to 20 μL of DEPC. Add 4 μL of Ficoll EDTA loading dye to each sample.
   d. Run at 80 V for approx 2 h. Visualize gel under ultraviolet (UV) light. If the RNA is intact there should be two bands corresponding to the 28S and 18S ribosomal RNAs (*see* **Fig. 1**). The intensity of the 28S RNA should be approximately twice that of the 18S band. Messenger RNA species will be visible as a faint smear running the length of the sample lane. If the RNA is degraded the ribosomal bands will be absent and the ethidium staining will be at the bottom of the gel.

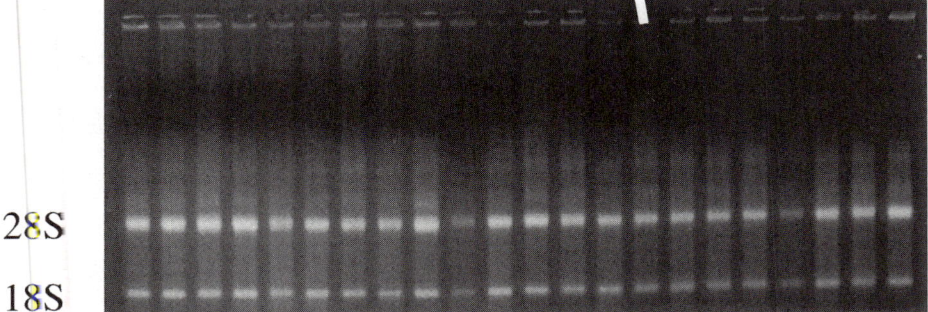


Fig. 1. RNA purification, confirmation of RNA purity (*see* **Subheading 3.2.**). This is a photograph of the agarose gel demonstrating two bands corresponding to the 28S and 18S ribosomes, confirming that the RNA is intact.

### 3.3. Ribonuclease Protection Assay—Probe Generation

Again, at least initially, the fundamental principle of avoiding contact with exogenous ribonucleases applies. Second, this assay utilizes radioactive $^{32}$phosphate and great care must be taken to avoid radioactive contamination. Therefore: (1) The workspace should be separate from other people in the laboratory and the area protected with Perspex screens. (2) Monitor work using a Geiger counter. (3) Dispose of all waste tips, Eppendorf tubes, etc. into a plastic bag within a Perspex box prior to transfer to a radioactivity bin. (4) When transferring samples around the laboratory, carry them behind a Perspex shield.

1. Label one Eppendorf tube for each probe, for example, 1 for probe to mRNA of interest and 1 for the probe to "housekeeper" mRNA (*see* **Notes 1** and **2**).
2. Add all reagents in the following order. Keep reagents on ice once thawed. **Note:** Radioactivity is being introduced at this stage.

   In the following example, we use a riboprobe for the angiogenic factor vascular endothelial growth factor (VEGF) and as loading controls riboprobes for GAPDH sense and GAPDH antisense. GAPDH is an endogenous cytosolic enzyme involved in glycolysis and its mRNA is abundantly expressed in cells. It is frequently used as a control to monitor differences in RNA loading between samples. The GAPDH antisense will hybridize to endogenous GAPDH mRNA and also to the exogenous GAPDH sense. This gives an internal and and external control to assess equal loading.

| | GAPDH antisense | GAPDH sense | VEGF |
|---|---|---|---|
| 10X Transcription buffer (µL) | 1.0 | 1.0 | 1.0 |
| 200 m*M* DTT (µL) | 0.7 | 0.7 | 0.7 |
| NTP mix (no CTP) (µL) | 1.1 | 1.1 | 1.1 |
| [α-$^{32}$P]CTP (µL) | 2.5 | 2.5 | 2.5 |
| DEPC (µL) | 2.3 | 2.3 | 1.8 |
| Placental ribonuclease inhibitor (µL) | 0.7 | 0.7 | 0.7 |

| | | | |
|---|---|---|---|
| Linearized template (μg) | 1.0 | 1.0 | 1.0 |
| RNA polymerase (μL) | 0.7 (T3) | 0.7 (T7) | 0.7 (T7) |
| Total volume (μL) | 10 | 10 | 10 |

3. Mix each Eppendorf tube by aspirating contents up and down a few times.
4. Place Eppendorf tubes in water bath for 15 min at 37°C.
5. During these 15 min start preparing the first gel. This gel is a small gel to check the activity and integrity of the probes. Clean gel apparatus with deionized water, then wipe glass plates with absolute alcohol and finally wipe the back plate (the one with the reservoir) with a few milliliters of silane. This allows the gel to detatch from the back plate and remain on the front plate when the plates are separated at the end of the experiment.
6. When assembling the gel apparatus offset the two plates so that the front plate finishes 2–3 mm above the lower level of the back plate. This allows the gel to be poured in from below to form the initial plug.
7. Make the gel in a sterile 150-mL flask using 49.5 mL of gel diluent, 7.5 mL of 10X TBE, and 18 mL of gel concentrate. This is enough for a 47 × 22 cm gel plate set.
8. Take 10 mL of the above mixture in a universal and add 50 μL of 10% APS (this should be no more than 2 wk old and stored in a refrigerator 4–8°C). Then add 20 μL of TEMED. This causes the gel to set rapidly. Before this happens fill a 1000 μL pipet tip with this solution and run it along the 2–3 mm gap between the plates, injecting as you go. It is best to rest the apparatus at a 5–10° angle and inject enough gel to form a layer about 5 cm up the plates (approx 5 mL). Leave a small amount of gel in the universal, when this has set then so has the plug, and it is then safe to pour in the main part of the gel from the top end.
9. After 15 min in the water bath at 37°C remove samples, centrifuge (14,000 rpm) for a few seconds, and then add 1 μL of DNase 1 to each Eppendorf tube. Return to the water bath for 30 min (37°C).
10. Add 200 μL of APS and 24 μL of TEMED to the gel solution in the 150-mL flask.
11. To pour gel turn apparatus over so that the reservoir (back plate) is on top. Angle the plates at 20° and tilt so that the far corner is the lowest point. The lower (front) plate will overlap the top (back) plate by 2 cm. Along this overlap pour the rest of the gel using a 25-mL pipet (about 1 mL a second). When nearly full, reduce the angle to about $5^0$. Gel should now completely fill the gap between the two plates; when it does, insert the comb and attach a bulldog clip across front and back plate to keep comb fixed in place. This gel should set in 30–60 min.
12. Take one G-50 centrifugation column for each sample. Uncap both ends of the Sephadex column and place it in an Eppendorf and both these into a 50-mL Falcon tube. Centrifuge at 1500 rpm for 2 min. Discard the liquid in the Eppendorf tube and centrifuge again for 1 min at 1500 rpm. Again discard the liquid and replace the Eppendorf tube with a new one.
13. Remove samples from the water bath after 30 min at 37°C. Centrifuge for a few seconds at 14,000 rpm. Load all of the sample into the center of a Sephadex column. Centrifuge loaded Sephadex columns (within a 50-mL Falcon tube) for

4 min at 2500 rpm. The liquid that now appears in the bottom Eppendorf tube is the purified probe. Check its signal on the Geiger counter—it should be "hot."

14. Once the main gel has set fill the gel apparatus with warmed running buffer (1X TBE). Prior to removing the comb, mark on the plate the site of each tooth; this helps with loading if the wells are hard to see. After removing the comb aspirate up and down into each well using the running buffer and a 1-mL Gilson pipet.
15. In new Eppendorf tube, add 9 µL of loading buffer to 1 µL of each probe.
16. Load 10 µL of each sample (9 µL of loading buffer and 1 µL of probe) to separate wells. Avoid any wells with obvious air bubbles beneath them.
17. Run at fixed wattage of 50 W for 30 min. It is sufficient to run the dye for only 5–6 cm.
18. To separate the plates, pour out the running buffer and lay them down with the reservoir (back) plate uppermost. This is then raised up carefully, and because of the silane the gel should detach from this and remain on the lower front plate.
19. Smooth an appropriate sized sheet of Whatman 3MM paper over the gel for a few seconds and then lift the filter paper. The gel should be transferred to the filter paper.
20. Cover the gel with clingfilm and place in an X-ray cassette with X-ray film for 30 min at –70/80°C.
21. During this time quantify the radioactivity of the probes. Take 2 µL of probe and place in a scintillation vial and then read in a scintillation counter. Ensure this is set up for the correct isotope and use a program that reads over 1 min. Work out counts per microliter. Because the sample is measured in the absence of scintillation the value will be in Cerenkov cpm, which depending on the model of scintillation counter will be approx 50% of the actual cpm. The values given below are all in Cerenkov counts. For each mRNA sample to be assayed aim for approx 200,000 cpm of probe to mRNA of interest, 60,000 cpm of GAPDH antisense, and 3000 cpm of GAPDH sense.
22. After 30 min, develop the X-ray film. There should be one band of exposure for VEGF and one band for GAPDH sense and antisense (similar molecular weight). If additional band(s) appear (always of smaller molecular weight, and of variable intensities), three options are possible:
    a. If the additional band intensities are weak, <10 % of the total, then the probe can be used without further purification.
    b. The probe preparation can be separated on a 1% agarose gel, as described in **Subheading 3.1.**; the radioactive band detected by brief autoradiographic exposure; the band cut out of the gel using a scalpel; and the $^{32}$P-probe eluted. This step will require an additional 3 h, however, it may be time well spent as later problems with high backgrounds are often minimized by this probe clean-up step.
    c. Start again and relabel the probe.

### 3.4. Ribonuclease Protection Assay—Hybridization of Probe

1. Make up a hybridization cocktail consisting of the following (*see* **Note 1**):
   8 µL of 5X hybridization buffer
   32 µL of formamide
   2 5 µL of tRNA
   Volume as in **Subheading 3.3., step 21** for VEGF probe

Volume as in **Subheading 3.3., step 21** for GAPDH antisense
Volume as in **Subheading 3.3., step 21** for GAPDH sense
= Total volume *X*.

Multiply $X \times Y$ samples + 2 of mRNA to be tested to give volumes to be used for hybidisation cocktail.

2. For each mRNA sample, take 10 µg of mRNA and make up the volume to 10 µL with DEPC in a new Eppendorf tube. This highlights the importance of using RNA preparations with a concentration of ≥1 µg/µL. More dilute preparations may need to be concentrated by ethanol precipitation (time consuming). Add *X* µL of hybridization cocktail to each sample.
3. One control vial containing 25 µg of tRNA in 10 µL of DEPC water should also receive *X* µL of hybridization cocktail. Keep unused cocktail in a Perspex box at –20°C.
4. Heat/boil these samples to 100°C for 3 min to untangle the RNA and then place in a water bath at 55°C overnight. Hybridization temperatures vary between 45–60°C. We have found 55°C gives good reproducible results.
5. The next morning, make ribonuclease digestion buffer (in excess of 350 µL per sample). This consists of 10 mL of digestion buffer, 20 µL of RNase T1 (500,000 U/mL) and 20 µL of 20 mg/mL RNase A.
6. Add 350 µL of ribonuclease digestion buffer to each sample, vortex-mix briefly, and place in a water bath at 37°C for 30 min. Any unprotected single-stranded RNA is destroyed.
7. During this 30 min, prepare the 6% nondenaturing polyacrylamide gel (29:1 acrylamide–bisacrylamide). For a large (47 cm × 22 cm) gel capable of running up to 32 samples, mix in a 150-mL flask: 29.6 mL of gel concentrate, 12 mL of 10X TBE, and 78.4 mL of gel diluent.
8. Take 20 mL of this solution and add 75 µL of 10% APS and 15 µL of TEMED to use for the plug as before. When this has set, add 325 µL of 10% APS and 39 µL of TEMED to the remaining solution and pour the main part of the gel.
9. Prepare a solution containing 10 µL of 20% SDS and 2.5 µL of Proteinase K for each sample.
10. After 30 min, remove samples from the 37°C water bath and add 12.5 µL of the above solution to each sample. Vortex-mix briefly and return to the 37°C water bath for a further 15 min.
11. After this 15 min, centrifuge samples for a few seconds and add 400 µL of phenol to each sample.
12. Centrifuge at 4°C at 14,000 rpm for 5 min.
13. Using a Gilson pipet, remove the upper aqueous phase, about 375–400 µL, and place each sample into a new Eppendorf tube.
14. To this add 900 µL of absolute ethanol. Vortex-mix for 10 s and then leave to precipitate at –80°C for 30 min (or longer) in a Perspex box.
15. Remove samples from the –80°C refrigerator and centrifuge at 4°C for 10 min.
16. There should be a small pellet at the bottom of each Eppendorf tube. Aspirate the overlying fluid and discard into a universal within a Perspex box as this will contain radioactivity. (Most of the radioactivity should be in the pellet.)

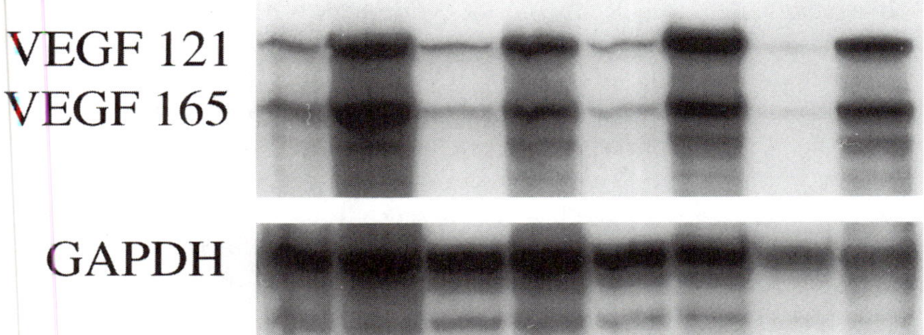


Fig. 2. Ribonuclease protection assay, hybridization of probe (*see* **Subheading 3.4.**). This is an autoradiograph, showing eight RNA samples analyzed for VEGF. The riboprobe detects the two major isoforms of VEGF, VEGF 121, and VEGF 165, and therefore two bands appear. The GAPDH loading control demonstrates underloading of lanes 7 and 8 and therefore the true signal from VEGF will be higher in these lanes when this is taken into account.

17. Wash the pellet by adding 1 mL 70% ethanol, vortex-mix for a few seconds, and then place in a 37°C water bath for 5 min. This step is to remove excess salt from the samples.
18. Remove samples from water bath and centrifuge at 14,000 rpm at 4°C for 10 min.
19. The large gel should be set by now. Add the running buffer to the gel apparatus after warming the buffer in a microwave; set the apparatus to 100 W and start.
20. Remove samples from the cold room centrifuge and aspirate most of the ethanol with a Gilson pipet.
21. Centrifuge for a few seconds and aspirate the rest of the ethanol.
22. Add 10 µL of RNA loading buffer and vortex-mix strongly.
23. Heat Eppendorf tubes at 80°C for 5 min, then place in a boiling bath at 100°C for 4 min to denature the RNA–RNA hybrid.
24. Immediately after this, place samples on ice and begin loading.
25. In addition to samples load one (or two) lanes with 10 µL of the cocktail prepared yesterday (*see* **Note 3**).
26. Run at 100 W for approx 2 h, until the lower colored band reaches the bottom of the gel. After this, pour out the running buffer and lay the apparatus down with the back plate (with reservoir) uppermost. Ease this plate off using a pencil as a lever. The gel should remain attached to the lowermost plate.
27. Place a layer of Whatman 3MM paper on top of the gel, smooth out, and then lift the filter paper. The gel should be transferred to the filter paper; cover this with clingfilm wrap and completely dry the gel for about 1 h on a vacuum-assisted dryer set at 70–80°C.
28. Place in a film cassette with photographic film. Develop after variable times in a 80°C freezer depending on exposure. A significant GAPDH signal should be visible after a 1-h exposure (*see* **Fig. 2**).

29. It is important that the tRNA lane does not contain a signal where the mRNA of interest should be.
30. Quantification of the VEGF and GADPH signals can be done either by phosphoimage analysis or by image analysis of the respective signals. In the case of phosphoimage analysis, the dried-down gel is exposed to a phosphoimage screen overnight at room temperature; the signal is linear over several log orders. Image analysis of photographic film has a much narrower range of linearity and therefore different exposures of GAPDH and VEGF will need to be used for accurate quantification.

## 4. Notes

1. Time considerations. Probe synthesis takes approx 3 h (agarose gel purification adds a further 3 h). Probes are best used the day they are prepared. Hybridization is conveniently carried out overnight, which allows 3–4 h for probe digestion, 2.5 h to load and run the gel, and 30–60 min to dry the gel.
2. Constructing a suitable riboprobe: We find that a length of between 100 and 400 base pairs is optimal. When analyzing hybridizations containing three or more different riboprobes it is important to make sure each riboprobe is of a sufficiently different size length to allow complete resolution on the polyacrylamide gel. Riboprobes can be constructed by either polymerase chain reaction (PCR) or by subcloning a suitable restriction fragment, downstream of an RNA polymerase promoter, into a vector (we have found pBluescript ideal).

   When linearizing the construct prior to labeling with [$\alpha^{32}$P]CTP chose a restriction enzyme that will give either a 5' overhang or a blunt end. 3' Overhangs should be avoided, as they can serve as initiation sites for the polymerase resulting in synthesis of RNA complementary to the riboprobe.

   In cases where riboprobe labeling is incomplete it may be possible to correct this either by adding unlabeled CTP (at a concentration 10% that of the three other NTPs), by changing the RNA polymerase used, or by using a different $^{32}$P-labeled ribonucleotide.
3. If using DNA markers to calculate the size of the VEGF and GAPDH signals, it is important to note that RNA has a 10% lower mobility in the gel than DNA of the same size.

## Suggested Further Reading


1. Chomczynski, P. and Saachi, N. (1987) Single-step method of RNA isolation by acid guanidium thiocyanate–phenol–chloroform extraction. *Anal. Biochem.* **162,** 156–159.
2. Zinn, K., DiMaio, D., and Maniatis, T. (1983) Identification of 2 distinct regulatory regions adjacent to the human β-interferon gene. *Cell* **34,** 865–879.
3. Melton, D. A., Krieg, P. A., Rebagliati, M. R., Maniatis, T., Zinn, K., and Green, M. R. (1984) Efficient in vitro synthesis of biologically active RNA and RNA

hybridization probes from plasmids containing a bacteriophage SP6 promotor. *Nucleic Acids Res.* **12,** 7035–7056.

4. Auselbel, F. M., Brent, R., Kingston, R. E., Moore, D. D., Seidman, J. G., Smith, L. A., and Struhl, K., eds. (1997) RNA Preparation; Ribonuclease Protection Assay, in *Current Protocols in Molecular Biology.* John Wiley and Sons, New York, pp. 4.0.3–4.2.7 and 4.7.1–4.7.6.

# 17

# Northern and Southern Blotting and the Polymerase Chain Reaction to Detect Gene Expression

**Thelma Tennant, Marina Chekmareva, and Carrie W. Rinker-Schaeffer**

## 1. Introduction

For cancer cells to form a metastasis, cells from the primary tumor must overcome the local adhesive forces, migrate and invade the microcirculation, arrest at a secondary site, and then finally proliferate *(1)*. As implied by its multistep nature, cancer metastasis is a complex and dynamic process that is likely to be regulated by a series of genes at each step *(2)*. A variety of approaches have been used to discern the molecular events that regulate this process. It is likely that the ability of a cancer cell to form clinically detectable metastases is influenced by a variety of factors, including alterations in the pattern of gene expression within the cancer cell. Such changes could be the result of genetic or epigenetic modifications *(3)*.

Although there has been a growing emphasis on array-based techniques for high-throughput screening of gene expression patterns, there are several well established protocols that can be used to identify such molecular changes. This chapter describes two of these techniques: Northern and Southern blotting.

### *1.1. Southern Blotting*

E. M. Southern first described a method for immobilizing size-fractionated DNA fragments on a nitrocellulose membrane in 1975 *(1)*. Since then, a number of different variations of this blotting method have been developed, as well as a variety of ways by which scientists can generate and hybridize probes to detect specifically the sequences thus immobilized. Southern blotting is now a general term for a number of different methods by which DNA is transferred from a gel to a membrane, and because nitrocellulose is relatively fragile,

From: *Methods in Molecular Medicine, vol. 57: Metastasis Research Protocols, Vol. 1: Analysis of Cells and Tissues*
Edited by: S. A. Brooks and U. Schumacher © Humana Press Inc., Totowa, NJ

improved membranes have been developed that are more durable and that have been optimized for allowing binding of nucleic acids. Southern blotting remains a widely used and extremely useful technique for the detection of DNA fragments.

One specific application of this method is the determination of the methylation status of a particular gene. There is a body of evidence suggesting that while neoplastic cells generally show widespread genomic hypomethylation, regions of hypermethylation may play a critical role in regulation of gene expression *(2,3)*. A main target of the regional hypermethylation are normally unmethylated CpG sites located in gene promoter regions *(2)*. Hypermethylation of CpG sites is an epigenetic mechanism of gene inactivation *(2)*. This is explored specifically in Chapter 18 by Lilischkis et al., which gives an alternative methodology for analysis of methylation of CpG islands. A role for methylation in the inactivation of tumor suppressor genes, including *p16*, *p53*, and *E-cadherin*, has been demonstrated *(4–6)*. Thus, molecular mechanisms underlying tumorigenesis may include gene inactivation by both genetic changes, such as mutation, and epigenetic modifications, such as promoter methylation *(7)*. Interestingly, there is evidence that genes involved in metastasis, such as tissue inhibitor of metalloproteinase (*TIMP*), harbor several CpG sites in their promoter regions which may be targets for methylation *(8)*.

A classic approach to the evaluation of the methylation status of such promoter region regions is to use restriction enzymes to (1) excise the 5' flanking (promoter) region of the candidate gene and (2) determine if the CpG sites are methylated using methylation-sensitive and -insensitive restriction enzymes that recognize the same restriction site *(5)*. Enzymes commonly used for step 2 include *Msp*I and *Hpa*II *(6)*. An example is shown in **Fig. 1**. Analysis of the 5' flanking region of the gene of interest shows the presence of a putative CpG site (e.g., a cluster of five cctgg sites). To determine the methylation status of these sites, the DNA is digested with *Hin*dIII and *Hae*II, restriction enzymes that recognize palindromic sequences flanking the CpG sites. To determine whether the CpG sites within the 1200-bp fragment are methylated, the DNA is subsequently digested with *Msp*I, a methylation-insensitive restriction enzyme, or *Hpa*II, a methylation-sensitive enzyme. *Msp*I and *Hpa*II are isoschizomers (e.g., recognize the same DNA sequence, …C/CGG…), which vary only in their ability to digest DNA that is methylated. Thus, digestion of the 1200-bp fragment with *Msp*I will yield two major fragments of 700 and 400 bp. Methylation of the CpG site will block the ability of *Hpa*II to cut at this site, resulting in a 1200-bp product. The restriction patterns obtained after digestion are determined by Southern blotting. As detailed in the protocol that follows, the digested DNA is separated electrophoretically, transferred to the desired membrane, and detected by hybridization with an appropriate probe. In

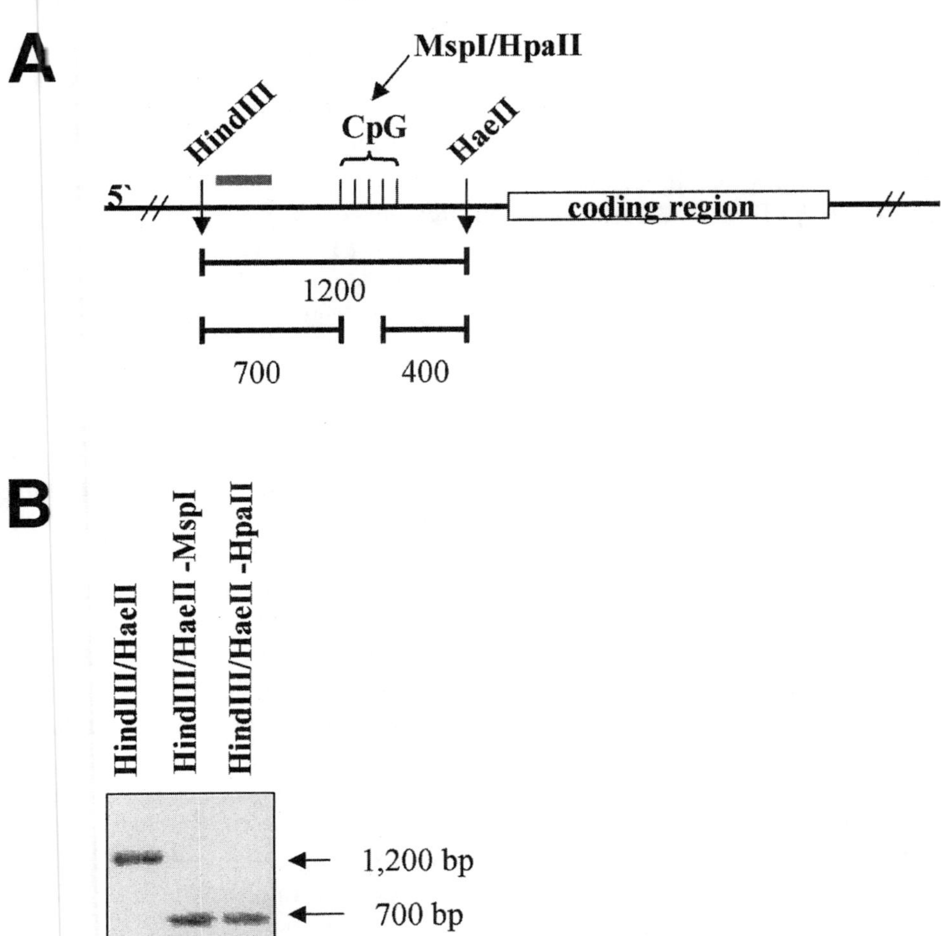


Fig. 1. **(A)** Schematic presentation of promoter hypermethylation study. The probe (shown in gray) was designed to recognize a 302-bp sequence located within the MMP2 promoter region. DNA is digested with *Hin*dIII and *Hae*II restriction enzymes that recognize palindromic sequences flanking the CpG sites. To determine whether the CpG sites within the 1200-bp fragment are methylated, the DNA is subsequently digested with *Msp*I, a methylation-insensitive restriction enzyme, or *Hpa*II, a methylation-sensitive enzyme. Digestion of the 1200-bp fragment with *Msp*I will yield two major fragments of 700 and 400 bp. Methylation of CpG site will block the ability of *Hpa*II to cut at this site, resulting in a 1200-bp product. **(B)** An example of the restriction pattern determined by Southern blotting. The probe recognizes a sequence at the 5' end of the 1200-bp *Hin*dIII–*Hae*II fragment. The digestion of the *Hin*dIII–*Hae*II fragment with either *Msp*I or *Hpa*II results in the production and subsequent detection of a 700-bp fragment, demonstrating that, in this case, the CpG site is not methylated.

our example the (shown in gray) probe recognizes a sequence at the 5' end of the 1200-bp *Hin*dIII–*Hae*II fragment. This probe will detect the 700-bp product generated by *Msp*I digestion of the 1200-bp fragment. If the CpG site is not methylated, the probe will similarly recognize the 700-bp product generated by *Hpa*II digestion. In contrast, methylation of the CpG site will block the ability of *Hpa*II to cut at this restriction site; thus, the probe will detect the full-length form (1200 bp) *Hin*dIII–*Hae*II fragment. **Figure 1B** shows how digestion of the –III–*Hae*II fragment with either *Msp*I or *Hpa*II results in the production and subsequent detection of a 700-bp fragment, demonstrating that, in this case, the CpG site is not methylated.

### 1.2. Northern Blotting

Following the description of Southern blotting in 1975, this method for immobilizing gel fractionated RNA quickly followed, and was first described in publication in 1977 ***(4,9)***. It was initially called "Northern blotting" jokingly, as the opposite of "Southern blotting," but this jargon quickly became standard in the scientific community, and is now an accepted name for this method. Like Southern blotting methods, the methods for fractionating and preserving RNA on a membrane, as well as the materials have been improved, and there are a number of variations upon the same theme. The protocol described here is a basic protocol for Northern blotting and can be used in any kind of molecular research.

An example of Northern blotting in metastasis research is shown in **Fig. 2**. The blot shown contains a number of cell line variants and has been probed for a member of the collagenase family, MMP-2. The purpose of the blot was to determine if these variants of the same cell line had differential levels of MMP-2 expression. Previously, a Southern blot had been performed on these same lines to show that all of these lines contained the gene in question.

## 2. Materials

### 2.1. Southern Blotting

#### 2.1.1. Genomic DNA Extraction

1. TEN9 solution: 50 m*M* Tris-HCl, (pH 9.0), 100 m*M* EDTA, 200 m*M* NaCl.
2. RNase A (10 mg/mL).
3. 20% Sodium dodecyl sulfate (SDS).
4. Proteinase K.
5. PC8 solution (phenol–chloroform, pH 8.0): It should be noted that phenol is a caustic substance and must be handled with extreme care. Appropriate safety procedures should be reviewed before proceeding with the preparation of this solution. Melt 480 mL (500 g) of molecular biology grade phenol by placing the jar in a water bath set at 50°C. Ensure that the lid is securely placed on the jar. Once the phenol has melted, add 0.5 g of 8-hydroxyquinoline and stir using a

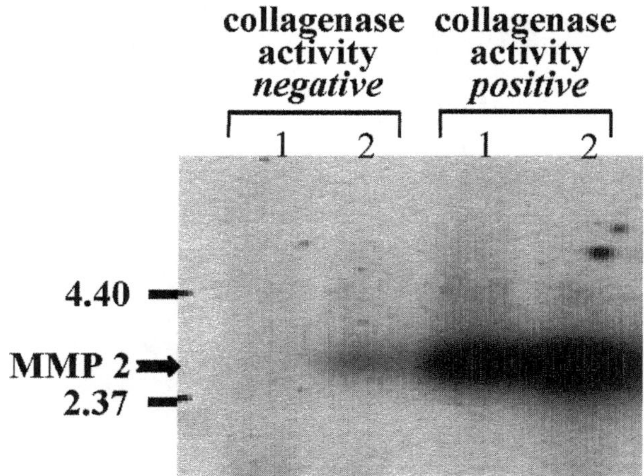


Fig. 2. Expression of *MMP2* mRNA in cell line variants. PolyA RNA was isolated from each of the sources and probed for *MMP2* expression as previously described. The light band seen in *lane 2* is the result of spill over when loading the gel.

magnetic stir bar and plate. Add 500 mL of 50 m*M* Tris, pH 8.0. Securely replace the lid on the jar and stir gently for 10 min. Allow the phases to separate. Remove the top layer (Tris) and discard it into a phenol waste receptacle. Add another 500 mL of 50 m*M* Tris (pH 8.0), and stir again for 10 min, then allow the phases to separate. Check the pH of the Tris after the separation has occurred to make sure that the pH is 8.0. If not, repeat this buffering process until the pH of the Tris layer reaches 8.0. Add 640 mL of molecular biology grade chloroform, and mix well. Allow the phases to separate, and use the bottom phase in DNA extraction.

6. Chloroform/isoamylalcohol (24:1 v/v).
7. 5 *M* NaCl.
8. Absolute ethanol.
9. 70% Ethanol (v/v).
10. 15-mL Polypropylene tubes.
11. Serum separation tubes (Vacutainer Brand SST Tubes).

### 2.1.2. Restriction Enzyme Digestion

1. Restriction enzyme of choice.
2. Restriction enzyme 10X buffer.
3. Glycogen—20 mg/mL.
4. 5 *M* NaCl.

### 2.1.3. Agarose Gel Electrophoresis

1. Agarose (DNA separation grade, low EEO).
2. 1 mg/mL ethidium bromide.

3. Glycerol loading buffer: 50% glycerol in $dH_2O$, with enough bromphenol blue to create a dark blue loading buffer, which will be detectable when added to sample and loaded on a gel.
4. TBE running buffer: 0.89 m*M* Trizma base; 0.89 m*M* boric acid, 0.2 m*M* EDTA, pH 8.0.

#### *2.1.4. Gel Treatment and Capillary Transfer* (10)

1. Depurination solution: 0.25 *M* HCl.
2. Denaturing solution: 0.5 *N* NaOH, 1 *M* NaCl.
3. Neutralizing solution: 1.5 *M* Tris-HCl, pH 7.4, 3 *M* NaCl.
4. Zeta-Probe Blotting Membranes (Bio-Rad).
5. 20X Saline sodium citrate (SSC): 3 *M* NaCl, 0.3 *M* trisodium citrate.
6. Whatman 3MM paper.
7. Absorbent paper towels, cut to the size of the Zeta-Probe membrane.
8. Diethylcarbonate (DEPC)-treated $dH_2O$.
9. Plastic wrap.

#### *2.1.5. Probe Preparation*

1. 10X Polymerase chain reaction (PCR) reaction buffer.
2. 10 m*M* primer pair.
3. 10 m*M* dATP.
4. 10 m*M* dGTP.
5. 10 m*M* dTTP.
6. 2 m*M* [$^{32}$P]dCTP.
7. *Taq* polymerase.
8. $dH_2O$.
9. 25–50 ng of DNA template.
10. Quick Spin Sephadex Columns G-25, Fine (Boehringer Mannheim).

#### *2.1.6. For Hybridization and Detection* (10)

1. Express-Hyb hybridization solution (Clontech).
2. Salmon Testes DNA (Sigma) (10 mg/mL).
3. Wash solution 1: 1 m*M* EDTA, 40 m*M* $Na_2HPO_4$, pH 7.2; 5% SDS.
4. Wash solution 2: 1 m*M* EDTA, 40 m*M* $Na_2HPO_4$, pH 7.2; 1%SDS.
5. Kodak XOMAT AR film.

### *2.2. Northern Blotting*

#### *2.2.1. Total RNA Preparation*

1. DEPC-treated glass 30-mL Corex tubes (*see* **Note 1**).
2. TriZol (GibCo BRL).
3. Chloroform.
4. Isopropanol.
5. 70% DEPC ethanol (70% ethanol made with DEPC-treated $dH_2O$).
6. DEPC-treated $dH_2O$.

#### 2.2.2. mRNA Preparation: (11) (*see Note 2)

1. DEPC-treated glass 30-mL Corex tubes.
2. Large OligodT columns*.
3. High-salt buffer*: 10 m*M* Tris-HCl, pH 7.4, 1 m*M* EDTA, 0.5 *M* NaCl.
4. Low-salt buffer*: 10 m*M* Tris-HCl, pH 7.4, 1 m*M* EDTA, 0.1 *M* NaCl.
5. Elution buffer*: 10 m*M* Tris-HCl, pH 7.4, 1 m*M* EDTA.
6. Sample buffer*: 10 m*M* Tris-HCl, pH 7.4, 1 m*M* EDTA, 3.0 *M* NaCl.
7. Glycogen* (5–10 mg/mL).
8. Absolute ethanol.
9. 80% Ethanol (made with DEPC $dH_2O$).
10. DEPC-treated $dH_2O$.
11. DEPC-treated Corex tubes.

#### 2.2.3. Agarose Gel Electrophoresis

1. 10X MOPS: 0.4 *M* Morpholinopropanesulfonic acid (MOPS), pH 7.0, 100 m*M* sodium acetate, 10 m*M* EDTA, pH 8.0.
2. MOPS gel: (for 200 mL): 2.4 g of agarose, 20 mL of 10X MOPS, 6 mL of 37% formaldehyde, 174 mL of DEPC $dH_2O$. Melt the agarose in $dH_2O$. Allow the flask to cool to handling temperature, then add MOPS and formaldehyde, mix well by swirling the flask, and pour into the gel mold.
3. RNA loading buffer: 100 µL of 10X MOPS, 175 µL of 37% formaldehyde, 500 µL of formamide, 20 µL of 2 mg/mL ethidium bromide. Use 40 µL of loading buffer/ 10 µL of sample.
4. DEPC-treated $dH_2O$.

#### 2.2.4. Gel Capillary Transfer

1. Zeta-Probe Blotting Membrane (Bio-Rad).
2. 20X SSC: 175 g/L of 3 *M* NaCl, 88.2 g/L of 0.3 *M* trisodium citrate.
3. Whatman 3MM paper.
4. Absorbent paper towels, cut to the size of the Zeta-Probe membrane.
5. DEPC-treated $dH_2O$.
6. Plastic wrap.

#### 2.2.5. Random Primer Probe Preparation (12)

1. Megaprime DNA Labeling System (Amersham).
2. [$^{32}$P]dCTP.
3. Template DNA.
4. Quick-Spin Sephadex columns G-25, Fine (Boehringer Mannheim).

#### 2.2.6. Hybridization and Detection (13)

1. Express-Hyb hybridization solution (Clontech).
2. Salmon testes DNA (Sigma).
3. Wash solution 1: 1 m*M* EDTA, 40 m*M* $Na_2HPO_4$, pH 7.2, 5% SDS.
4. Wash solution 2: 1 m*M* EDTA, 40 m*M* $Na_2HPO_4$, pH 7.2, 1% SDS.
5. Kodak XOMAT AR film.

# 3. Methods

## 3.1. Southern Blotting

### 3.1.1. Genomic DNA Extraction

1. A frozen pellet of cells can be used for genomic DNA extraction. Cell pellets may be stored for extended periods of time at –20°C, and still yield good quality DNA by this method.
2. To each pellet of cells, add 5 mL of TEN9 solution, containing 100 μg/mL of RNase A. Resuspend the cell pellet in this solution by pipetting up and down.
3. Add 250 μL of 20% SDS, and mix well. (**Note:** Mix by inversion.)
4. Add 125 μL of Proteinase K. Incubate the sample overnight at 50°C.
5. Following overnight digestion, the solution should appear viscous and be stringy when pipetted.
6. Transfer the material from the overnight digest into a serum separation tube. Add 5 mL of PC8 solution, replace the cap firmly on the tube, and shake vigorously. (**Caution:** Phenol and chloroform are extremely caustic substances. When performing this step, be sure to use gloves and work in a chemical hood.)
7. Centrifuge the serum separation tube at 2000*g* for 5 min.
8. The organic layer of the mixture will be trapped in the bottom of the serum separation tube. Pour the clear aqueous layer into a new serum separation tube, and repeat the PC8 extraction step.
9. Pour the aqueous layer into a new serum separation tube and add 5 mL of chloroform–isoamyl alcohol solution. Recap the tube tightly and shake vigorously to mix.
10. Centrifuge the tube for 5 min at 2000*g*, then carefully remove the aqueous layer and transfer it in to a sterile polypropylene tube.
11. Add 0.33 mL of 5 *M* NaCl, and 10 mL (2 vol) of absolute ethanol to the solution. Mix gently by inverting three or four times. The solution should become cloudy when the NaCl is added, but mixing should cause it to clear. Place the sample at –80°C for 30 min.
12. After the –80°C incubation, collect the DNA by centrifuging at 12,000*g* at 4°C for 10 min.
13. Identify the position of the pellet at the bottom of the tube, and then carefully remove the supernatant, trying not to disturb the pellet. Wash the pellet with 1 mL of 70% ethanol, and spin in a microcentrifuge at top speed for 10 min.
14. Remove the ethanol, and allow the pellet to air-dry. (A speed-vacuum can be used to dry the DNA pellets, but be careful to avoid overdrying, as it can make resuspension difficult.)
15. Resuspend the pellet in 1 mL of $dH_2O$.
16. Quantitate the sample with a spectrophotometer.

### 3.1.2. Restriction Enzyme Digestion

1. This reaction will be carried out in a relatively large volume, and then precipitated in preparation for gel electrophoresis. Prepare a mix in a microcentrifuge tube of the following elements, to a total volume of 180 μL:

15 μg of DNA
18 μL of 10X restriction enzyme buffer
120 U of restriction enzyme
$dH_2O$ to 180 μL

2. Incubate the mixture overnight at 37°C.
3. Because of the large volume of this mixture, it is necessary to precipitate the digested sample before proceeding with electrophoresis. Add 20 μL of glycogen, 33 μL of 5 *M* NaCl, and absolute ethanol to the sample; mix it thoroughly by inverting several times, and store it at –80°C for 1 h.
4. Centrifuge the sample at 14,000*g* in a microcentrifuge, at 4°C for 15 min.
5. Identify the location of the pellet and carefully drain of the supernatant. Rinse with 1 mL of 70% ethanol and centrifuge at 14,000*g* for 15 min.
6. Again, identify the pellet (it tends to be less adherent to the walls of the microcentrifuge tube (after the rinse) and carefully drain off the ethanol. Allow the pellet to air-dry.
7. Resuspend the pellet in 20 μL of $dH_2O$.

### 3.1.3. Agarose Gel Electrophoresis

1. Add 5 μL of glycerol loading buffer to the sample.
2. Load the samples into the gel and run at 88 mV for 1–2 h (watch the dye band carefully).
3. When the dye band nears the bottom of the gel, remove the gel, and visualize using an ultraviolet (UV) box and camera. Make sure to take the picture with a ruler aligned along the edge of the gel—this helps to identify a band size after probing the membrane.

### 3.1.4. Gel Treatment and Capillary Transfer

1. Depurinate the DNA in the gel by soaking it in the depurination solution for 15 min, agitating slowly. Make sure that the gel is immersed in the solution. The blue loading marker should turn yellow when the gel is sufficiently treated.
2. Denature the DNA by soaking it in the denaturing solution for 30 min, slowly agitating.
3. Neutralize the gel in a bath of neutralizing solution for 30 min, again with slow agitation. The blue marker should turn from yellow to blue when this step has been completed.
4. The gel transfer apparatus should have a shallow tray in which to hold the transfer buffer (20X SSC). Cut a strip of Whatman 3MM paper to use as a wick for the capillary transfer. It should be long enough to wrap around the transfer platform, and slightly wider than the gel.
5. Soak the wick in 20X SSC, and wrap it carefully around the transfer platform. Remove any air bubbles that remain between the wick and the platform surface by rolling a pipet across the surface of the wick.
6. Carefully place the gel on the wick, and remove any air bubbles.
7. Dampen the Zeta-Probe membrane in DEPC-treated water. Carefully place the membrane on top of the gel and smooth out the air bubbles.

8. Cut three sheets of Whatman 3MM paper to the size of Zeta-Probe membrane. Dampen them in 20X SSC, and place the sheets on top of the membrane one at a time, smoothing out air bubbles.
9. Place a stack of absorbent paper towels on top of the Whatman 3MM paper. The paper towels absorb the 20X SSC as it travels through the gel and the membrane.
10. Carefully wrap the apparatus with plastic wrap. The wrap should come up to the edges of the gel on all sides—this is particularly important, in order to prevent the DNA from diffusing around the membrane and up into the 3MM paper.
11. Place a weight on top of the apparatus—a recommended weight is a 500-mL glass chemical bottle that contains approx 175 mL of water. A weight that is too heavy will crush the gel, while one that is too light will slow the capillary transfer.
12. Allow transfer to occur overnight—18–24 h. Following transfer, disassemble the apparatus, and immediately fix the DNA on the membrane by crosslinking.

### 3.1.5. Probe Preparation

1. Prepare a mixture of the following (the final volume of the reaction should be 50 μL):
   5 μL of 10X PCR buffer
   1 μL of 10 m*M* primer pair
   1 μL of dATP
   1 μL of dGTP
   1 μL of dTTP
   25–50 ng of PCR DNA template
   $dH_2O$ up to 50 μL
2. Behind a radioactivity shield, add 5 μL of [$^{32}$P]dCTP to the reaction.
3. Add 0.5 μL of *Taq* polymerase, cap the tube, and immediately place the PCR reaction in a PCR thermocycler.
4. Program the PCR thermocycler to only 10 cycles.
5. Resuspend the beads in a Sephadex column, remove the top and bottom plugs, and insert the column into a collection tube. Centrifuge the column at 1800*g* for 5 min. Discard the effluent.
6. Add the probe to the center of the Sephadex column cellulose bed. Insert a new collection tube, and centrifuge the column at 1800*g* for 5 min. The effluent in the collection tube is the probe.
7. Take 1 μL and measure in a scintillation counter to determine the activity of the probe.

### 3.1.6. Hybridization and Detection

1. Wet the membrane by soaking in $dH_2O$ for 5–10 min.
2. Preheat the hybridization oven to 65°C. Place the bottle of Express-Hyb in the oven to heat, as well. This is important, because the Express-Hyb contains elements that come out of solution at room temperature.
3. Prepare 10 μg/mL Express-Hyb of salmon testes DNA by heating it at 95–100°C for 10 min, and then chilling on ice for 5 min.
4. Add 0.15 mL of Express-Hyb per 1 $cm^2$ of membrane to a glass hybridization tube, and add the denatured salmon testes DNA to the Express-Hyb.

5. Carefully place the membrane in the hybridization tube DNA-side facing the inside of the tube, making sure that the edges do not overlap.
6. Allow the membrane to prehybridize for 30 min, rotating slowly at 65°C.
7. Boil the probe at 95–100°C for 10 min, then chill on ice for 5 min.
8. Pour out the prehybridization solution and add fresh Express-Hyb to the vial. Add the boiled probe to this new solution. Allow the probe to hybridize overnight at 65°C, rotating slowly.
9. Following the overnight hybridization, drain the probe and Express-Hyb into a radioactive waste container, and add 10 mL of wash solution 1. Allow this to wash for 30 min in the hybridization oven (65°C and rotating slowly). Repeat this wash twice. (Be sure to perform these steps behind a radioactivity shield, as the membrane and the washes will be radioactive.)
10. Add 10 mL of wash solution 2, and wash for 30 min in the hybridization oven, again at 65°C and 30 min. Repeat this wash twice.
11. Remove the membrane and carefully wrap it in plastic wrap, smoothing out the wrinkles. Place the wrapped membrane in a film cassette. Make sure to orient the membrane in such a way as to make analysis of the developed film simple.
12. In a dark room, place a piece of X-ray film over the membrane. Place the cassette at –80°C for 24 h.
13. Develop the film. If the bands on the autoradiogram are faint, put a new piece of film over the membrane, and store at –80°C for a longer exposure.

## 3.2. Northern Blotting

### 3.2.1. Total RNA Extraction

1. Add 1 mL of Triazol to a pellet of approx $10^6$ cells. Make sure that the pellets are in polypropylene tubes.
2. Resuspend the pellet completely. Let the samples stand at room temperature for 5 min.
3. Add 0.2 mL of chloroform/mL of Triazol. Cap the tube and shake vigorously for 15 s, and allow it to stand at room temperature for 5 min.
4. Centrifuge the tube at 1800*g* for 15 min at 4°C.
5. Remove the aqueous (upper) phase and transfer it to a fresh 15-mL tube. Add 0.5 mL of isopropanol/ml aqueous. Mix the sample well by shaking vigorously, and incubate at room temperature for 10 min.
6. Centrifuge at 12,000*g* in 4°C for 10 min. Remove the supernatant and wash pellet with 70% cold ethanol. After centrifuging, remove the ethanol completely, but do not let the pellet dry out.
7. Resuspend the pellet in 50 µL of DEPC $dH_2O$ (depending on the amount of starting material, this amount may vary).
8. Quantitate the sample with a spectrophotometer.
9. Run a formaldehyde-agarose check gel with 4 µL of the sample to confirm the spectrophotometer readings, and to determine the quality of total RNA. Refer to **Subheading 3.1.3.**

### 3.2.2. mRNA Extraction (11)

1. Invert the oligo dT column to resuspend the cellulose beads.
2. Remove top, then bottom closures, allow storage buffer to drain out into an upright centrifuge (Corex) tube.
3. Add 1 mL of high-salt buffer—drain through under gravity. Do this twice.
4. Heat RNA at 65°C for 5 min.
5. Place the sample on ice. Add 0.2 mL of sample buffer and mix gently by inverting the tubes once or twice.
6. Prewarm the elution buffer aliquots at 65°C (four aliquots per sample, 250 µL each)
7. Discard the flow-through and apply the RNA sample. (Increase volume of RNA sample to 1 mL with elution buffer.) Apply the sample to the top of the cellulose bed and allow it to drain through the column.
8. Centrifuge the column at 350*g* for 2 min.
9. Apply 0.25 mL of high-salt buffer to the column and centrifuge at 350*g* for 2 min, twice.
10. Wash with 0.25 mL of low-salt buffer and centrifuge at 350*g* for 2 min. Repeat this wash 3×.
11. Discard the effluent from the washes. Place the column over a sterile RNase-free 1.5-mL microcentrifuge tube and place this apparatus inside a Corex tube.
12. Elute the column with prewarmed aliquots of elution buffer—350*g* for 2 min, 3×.
13. Place the microcentrifuge tube on ice.
14. Transfer the eluate to a sterile Corex tube and add 100 µL of sample buffer, 10 µL of glycogen, and 2.5 mL of ice-cold absolute ethanol. Incubate at –80°C at least 2 h.
15. Centrifuge the tube at 12,000*g* at 4°C for 10 min.
16. Drain off the supernatant. (Watch the pellet carefully, to ensure that it does not dislodge from the tube.) Wash the pellet once with 80% ice-cold ethanol.
17. Centrifuge at 12,000*g* 4°C for 10 min.
18. Drain off ethanol, (again, watching for the pellet) and carefully remove any residual ethanol with a pipet tip. Resuspend mRNA in 25 µL of DEPC $dH_2O$. (Larger pellets may require more volume.)
19. Run a formaldehyde-agarose check gel with 4 µL of the sample to confirm the spectrophotometer readings, and to determine the quality of the mRNA. Refer to **Subheading 3.1.3.**
20. If the check gel shows that the sample still contains large bands of ribosomal RNA, or if very pure mRNA needed, it may be necessary to repeat this method for an additional polyA column purification of this sample.

### 3.2.3. Agarose Gel Electrophoresis

1. Heat loading buffer/sample at 65°C for 15 min.
2. Place on ice for 5 min.
3. Add bromphenol blue suspended in DEPC $dH_2O$ to the sample.
4. Place the gel into a gel box containing 1X MOPS as the buffer (10X MOPS should be diluted in $dH_2O$).

5. Load the samples into the gel and run at 88 mV for 1–2 h (watch the dye band carefully), and when it nears the bottom of the gel, remove the gel, and visualize using a UV box and camera. Make sure to take the picture with a ruler aligned along the edge of the gel—this helps to identify a band size after probing the membrane.

### *3.2.4. Gel Capillary Transfer*

1. The gel transfer apparatus should have a shallow tray in which to hold the transfer buffer (20X SSC). Cut a strip of Whatman 3MM paper to use as a wick for the capillary transfer. It should be long enough to wrap around the transfer platform, and slightly wider than the gel.
2. Soak the wick in 20X SSC and wrap it carefully around the transfer platform. Roll a clean, disposable pipet across the wick to remove any air bubbles that remain between the wick and the platform surface.
3. Carefully place the gel on the wick and remove any air bubbles.
4. Dampen the Zeta-Probe membrane in DEPC-treated $dH_2O$. Carefully place the membrane on top of the gel, and smooth out the air bubbles.
5. Cut three sheets of Whatman 3MM paper to the size of Zeta-Probe membrane. Dampen then in 20X SSC, and place them on top of the membrane one at a time, smoothing out air bubbles.
6. Place a stack of absorbent paper towels on top of the Whatman 3MM paper. The paper towels absorb the 20X SSC as it travels through the gel and the membrane.
7. Carefully wrap the apparatus with plastic wrap. The wrap should come up to the edges of the gel on all sides—this is particularly important, in order to prevent the RNA from diffusing around the membrane and up into the 3MM paper.
8. Place a weight on top of the apparatus—a recommended weight is a 500-mL glass chemical bottle that contains approx 175 mL of water. A weight that is too heavy will crush the gel, while one that is too light will slow the capillary transfer.
9. Allow transfer to occur overnight—18–24 h. Following transfer, carefully disassemble the apparatus, and immediately fix the RNA on the membrane by crosslinking.
10. Store the membrane in a protective plastic sheet in 4°C.

### *3.2.5. Probe Preparation* (12)

1. Place primers, reaction buffer, water, and DNA at room temperature to thaw.
2. Dilute the template DNA to a concentration of 5–25 ng/µL in $dH_2O$.
3. Aliquot water, primers, and DNA into a microcentrifuge tube. Denature the mixture by heating to 95–100°C for 5 min.
4. Centrifuge the tube briefly to bring the contents to the bottom of the tube. Keeping it at room temperature, add labeling buffer, followed by the radiolabeled [$^{32}$P]dCTP and the enzyme. (Be sure to keep the enzyme on ice until the very last step, and immediately return it to the –20°C after assembling the mixture. Also, make sure to perform this step behind a radioactivity shield.)
5. Mix gently by pipetting up and down and then cap the tube and centrifuge briefly.

6. Incubate the mixture at 37°C for 10 min.
7. Resuspend the beads in a Sephadex column, remove the top and bottom plugs, and insert the column into a collection tube. Centrifuge the column at 1800*g* for 5 min. Discard the effluent.
8. Add the probe to the center of the Sephadex column cellulose bed. Insert a new collection tube and centrifuge the column at 1800*g* for 5 min. This time the effluent in the collection tube is the probe.
9. Take 1 µL and measure in a scintillation counter to determine the activity of the probe.

### *3.2.6. Hybridization and Detection* (13)

1. Rehydrate the membrane by soaking it in $dH_2O$ for 5–10 min.
2. Preheat the hybridization oven to 65°C. Place the bottle of Express-Hyb in the oven to warm, as well. This is important because the Express-Hyb contains elements that come out of solution at room temperature.
3. Boil 10 µg/mL of salmon testes DNA at 95–100°C for 10 min and then chill on ice for 5 min.
4. Add 0.15 mL of Express-Hyb per 1 $cm^2$ of membrane to a glass hybridization tube, and add the denatured salmon testes DNA to the Express-Hyb.
5. Carefully place the membrane in the hybridization tube DNA-side facing the inside of the tube, making sure that the edges do not overlap.
6. Allow the membrane to prehybridize for 30 min, rotating slowly at 65°C.
7. Boil the probe at 95–100°C for 10 min and then chill on ice for 5 min.
8. Pour out the prehybridization solution and add fresh Express-Hyb to the vial. Add the boiled probe to this new solution. Allow the probe to hybridize overnight at 65°C, rotating slowly.
9. Following the overnight hybridization, drain the probe and Express-Hyb into a radioactive waste container, and add 10 mL of wash solution 1. Allow this to wash for 30 min in the hybridization oven (65°C and rotating slowly). Repeat this wash twice. (Make sure to perform these steps behind a radioactivity shield, as the membrane and the washes will be radioactive.)
10. Add 10 mL of wash solution 2 and wash for 30 min in the hybridization oven, again at 65°C and 30 min. Repeat this wash twice.
11. Remove the membrane and carefully wrap it in plastic wrap, smoothing out the wrinkles. Place the wrapped membrane in a film cassette. Make sure to orient the membrane in such a way as to make analysis of the developed film simple.
12. In a dark room, place a piece of X-ray film over the membrane. Place the cassette at –80°C for 24 h.
13. Develop the film. If the film is very faint, put a new piece of film over the membrane, and store at –80°C for several more days.

## 4. Notes

1. DEPC treatment of glassware: Thoroughly wash the glassware with soapy water, and rinse with nondistilled water, then twice with distilled water. Soak the glass-

ware for 15 min in 0.1% DEPC (1 mL of DEPC/1 L of $dH_2O$). Drain the DECP-treated $dH_2O$ from the glassware, and cover the openings with aluminum foil. Finally, bake the glassware for at least 4 h at 180°C. To make DEPC-treated $dH_2O$ (0.1% DEPC), add 1 mL of DEPC per 1 L $dH_2O$. Mix thoroughly, and autoclave.

**Caution:** DEPC is a potential carcinogen, and should be handled at all times with gloves. In addition, if possible, perform the mixing of DEPC water in a chemical hood.

This is an important step in preparation, as RNases are extremely resistant to other methods of cleaning glassware, even autoclaving. RNase-contaminated glassware can leave a researcher at the end of their RNA purification with nothing but degraded RNA, so be sure to have DEPC-treated glassware, and perform all procedures involving RNA purification on a clean bench.

2. The components listed with an asterisk(*) can be purchased as part of an mRNA purification kit by Pharmacia Biotech.
3. When working with the membrane post-transfer, it is useful to mark the side of the membrane with a pencil. The lanes can be labeled without harming the membrane. This will ensure the correct placement of the gel in the hybridization tubes. In addition, it facilitates alignment and marker placement when labeling radiograms.
4. Because the size of restriction fragments may vary only slightly on a Southern blot, it is often useful to run longer gels in order to produce better resolution of fragments with very small size differences. A longer gel containing λ-HINDIII marker can be resolved by electrophoresis, and the gel cut to include the area in which the fragments are expected, and to fit a smaller size membrane onto which the fragments will be transferred.
5. The hybridization may produce either a very faint or a nonspecific dark exposure. If the exposure is too faint, try to hybridize the probe to the membrane at a lower temperature. If, in contrast, the exposure is very dark, and there is hybridization of the probe to bands are not specific, try raising the temperature. A gel that shows streaks or blotches may be fixed without having to reprobe the blot simply by washing the membrane three times with each solution, rather than twice.

## References

1. Killion, J. J. and Fidler, I. J. (1989) The biology of tumor metastasis. *Semin. Oncol.* **16,** 106–115.
2. Yoshida, B. A., Chekmareva, M. A., Wharam, J. F., Kadkhodaian, M., Stadler, W. M., Boyer, A., et al. (1998) Prostate cancer metastasis-suppressor genes: a current prospective. *In Vivo* **12,** 49–58.
3. Baylin, S. B., Herman, J. G., Graff, J. R., Vertino, P. M., and Issa, J. P. (1998) Alterations in DNA methylation: a fundamental aspect of neoplasia. *Adv. Cancer Res.* **72,** 141–196.
4. Southern, E. M. (1975) Detection of specific sequences among DNA fragments separated by gel electrophoresis. *J. Mol. Biol.* **98,** 503–517.
5. Bird, A. P. (1986) CpG-rich-islands and the function of DNA methylation. *Nature* **321,** 209–213.

6. Counts, J. L. and Goodman, J. I. (1995) Alterations in DNA methylation may play a variety of roles in carcinogenesis. *Cell* **83,** 13–15.
7. Gonzalez-Zulueta, M. G., Bender, C. M., Yang, A. S., Nguyen, T., Beart, R. W., Van Tornout, J. M, and Jones, P. A. (1995) Methylation of the 5' CpG island of the p16/CDKN2 tumor suppressor gene in normal and transformed tissues correlates with gene silencing. *Cancer Res.* **55,** 4531–4535.
8. Herman, J. G., Merlo, A., Mao, L., Lapidus, R. G., Issa, J. P. J, Davdson, N. E., et al. (1995) Inactivation of the CDKN2/p16/MTS1 gene is frequently associated with aberrant DNA methylation in all common cancers. *Cancer Res.* **55,** 4525–2530.
9. Alwine, J. C., Kemp, D. J., and Stark, G. R. (1977) Method for detection of specific RNAs in agarose gels by transfer to diazobenzyloxymethyl-paper and hybridization with DNA probes. *Proc. Natl. Acad. Sci. USA* **74,** 5350–5354.
10. Schroeder, M. and Mass, M. J. (1997) CpG methylation inactivates the transcriptional activity of the promoter of the human p53 tumor suppressor gene. *Biochem. Biophy. Res. Commun.* **235,** 403–406.
11. *Pharmacia-Biotech mRNA Purification Kit Instructions.* (1998) Revision 4.
12. *Amersham Life Science Megaprime DNA Labeling Systems Instruction Manual.* (1998) RPN 1604/5/6/7.
13. *Bio-Rad Zeta-Probe Membranes Instruction Manual.* (1998) pp. 2–4, 10–17, 25–26.

## Further Reading

Ausubel, F. M., Brent, R., Kingston, R. E., Moore, D. D., Seidman, J. G., Smith, J. A., and Struhl, K. (1998) *Current Protocols in Molecular Biology*, Vols. 1–3. John Wiley and Sons, pp. 2.1–3.15.

Bachman, K. E., Herman, J. G., Corn, P. G., Merlo, A., Costello, J. F., Cavanee, W. K., et al. (1999) Methylation-associated silencing of the tissue inhibitor of metalloproteinase-3 gene suggest a suppressor role in kidney, brain, and other human cancers. *Cancer Res.* **4,** 798–802.

# 18

## Methylation Analysis of CpG Islands

**Richard Lilischkis, Hermann Kneitz, and Hans Kreipe**

### 1. Introduction

The application of Southern blotting to determine the methylation status of a particular gene has already been alluded to in Chapter 17 by Tennant et al., and methodology for Southern blotting described. This chapter examines methylation analysis of CpG islands in more depth and describes a technique by which quantitative changes may be monitored with greater sensitivity than that achieved by Southern blotting, and by which multiple CpG sites can be monitored simultaneously.

In higher order eukaryotes, DNA is methylated primarily at cytosines that are located 5' to guanosines in a CpG dinucleotide. In mammalian species, 3–5% of the cytosine residues are modified to 5-methylcytosine **(Fig. 1A)** and there is now considerable evidence to show that this post-transcriptional modification plays an important role in gene function *(1,2)*. Some CpG dinucleotides are clustered together in 1–2-kb long stretches of DNA called "CpG islands" which account for approx 2% of the genome and have distinct properties when compared to the rest of the genome. CpG islands are often located in the promoter region or the first exon of expressed genes and show a high G + C content (60–70%), the remainder of the genomic DNA has a G + C content of 40% (*3* and references therein). Furthermore, bulk genomic DNA has only 25% of the CpG dinucleotides one would expect from random base composition, whereas CpG islands show the expected number. The "depletion" of CpG dinucleotides may be a result of spontaneous deamination of 5-methylcytosine to thymidine, leading to the mutation of CpG to TpG and CpA on the sense and the antisense strands, respectively. Indeed, when one calculates the percentage of these dinucleotides in the genomic DNA, it is clear that the deficiency of CpG dinucleotides is equivalent to the excess of TpG plus CpA *(3)*.

From: *Methods in Molecular Medicine, vol. 57:*
*Metastasis Research Protocols, Vol. 1: Analysis of Cells and Tissues*
Edited by: S. A. Brooks and U. Schumacher © Humana Press Inc., Totowa, NJ

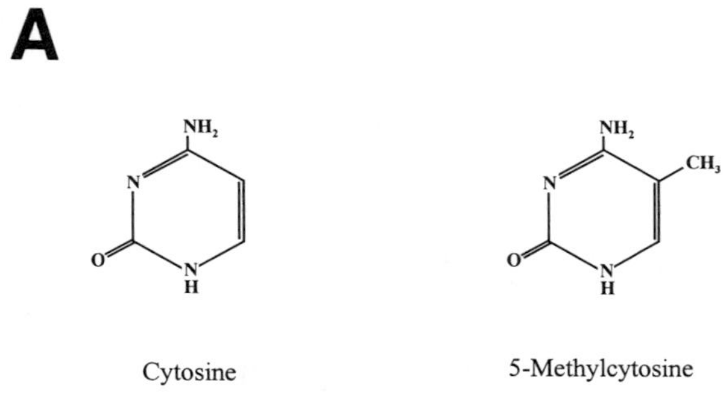


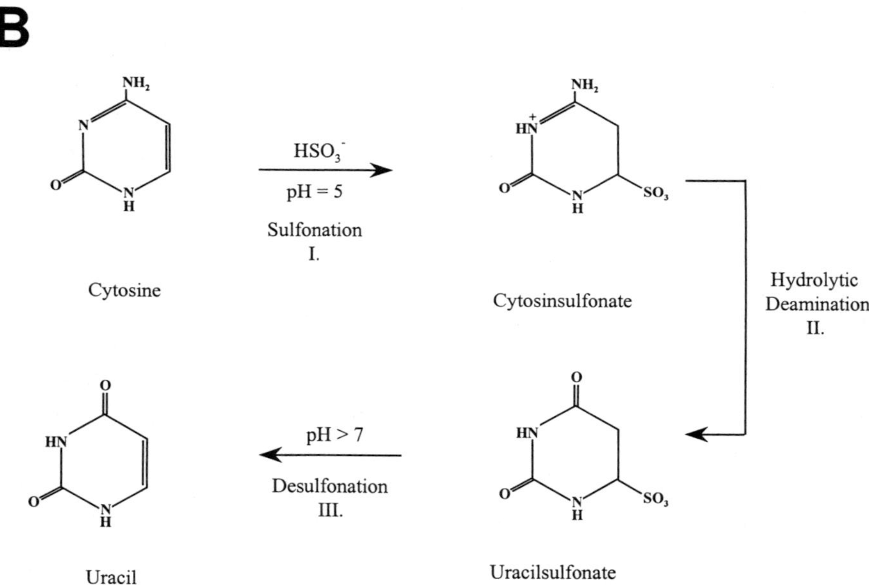


Fig. 1. **(A)** Cytosine is methylated at its 5-position to form 5-methylcytosine. **(B)** The chemical conversion of cytosine to uracil is achieved under the influence of high concentrations of bisulfite at low pH. Sulfonation of cytosine at its 6-position destabilizes the amino group in position 4, which is hydrolytically deaminated to form uracilsulfonate. Under alkaline conditions, the $SO_3^-$ group is split off again, resulting in a PCR-amplifiable uracil. Methylation of position 5, however, prevents sulfonation of cytosine and its conversion to uracil.

Whereas CpGs found in the bulk of genomic DNA are usually methylated, CpGs in CpG islands associated with expressed genes are usually protected from methylation. Of all known promoters, 60% are associated with these unmethylated CpG islands *(2)*. Methylation of CpG islands does occur, however, within the promoters of "imprinted" genes and genes lying on the X-chromosome, which is inactivated during female ontogenesis ***(4)***. In both cases the methylated allele is silent whereas the corresponding, unmethylated allele is expressed ***(3,5,6)***. This association of methylation with gene expression has emerged as a universal pattern and it is now understood that methylation functions to silence promoter activity ***(2,3,5–9)***. The mechanics of this "silencing" function remain to be elucidated; however, it has been suggested that methylation may render the DNA inaccessible to transcription, probably through a combination of direct inhibition of transcription factor binding and because methylation directs chromatin into an inactive structure ***(6,10,11)***.

Changes in the state of promoter methylation are associated with pathological changes. For example, *de novo* methylation of the estrogen receptor (*ER*) gene promoter and the switch from monoallelic to biallelic methylation of the normally imprinted insulin-like growth factor (*IGF2*) gene promoter have both been linked to aging and carcinogenesis ***(12,13)***. Furthermore, there is evidence that carcinogens target methylated CpGs directly and may hereby cause mutations at these positions ***(13,14)***.

Changes in the expression status, mediated by changes of promoter methylation, are conceivable for all genes known to play a pivotal role in carcinogenesis. Indeed, current evidence indicates that hypermethylation of the transcription start site of the cell adhesion molecule E-cadherin down-regulates the expression of E-cadherin in breast and prostate cancers and the loss of E-cadherin expression is related to tumor aggressiveness in these tumors ***(15)***. Loss of E-cadherin expression and the subsequent loss of specific cellular adhesiveness may be a critical step in the ability of epithelial tumor cells to invade and metastasize. In experimental tumor models, restored expression of E-cadherin inhibits invasion in vitro and in vivo, implicating E-cadherin as an "invasion suppressor gene" ***(16–18)***. Hypermethylation of the transcription start site of E-cadherin is an early event in thyroid carcinoma development ***(19)*** and might explain the aggressiveness and tendency to early metastasation in anaplastic thyroid carcinomas, which are generally E-cadherin negative (Mengel and Wasielewski, personal communication). In contrast to E-cadherin, where decreased expression leads to worse prognosis, it is an increased expression of the oncogenes c-*myc* and c-N-*ras* that promotes infiltration and tumor metastasis ***(20)***. This increased expression is mediated by DNA hypomethylation and is highly significant in cases with a tendency of tumor infiltration or metastasis.

The interest in cancer-associated changes in gene methylation has grown enormously in recent years with the conjecture that the promoter methylation status may provide an early marker for tumorigenesis. *De novo* methylation of tumor suppressor genes is often found in tumors and metastases, the archetype being $p16^{INK4a}$, which is commonly hypermethylated in a diverse range of tumors *(21–29)*. The contribution of $p16^{INK4a}$ promoter methylation to the multistep process of carcinogenesis could be shown in an in vitro model using normal human mammary epithelial cells (HMECs) *(30)*. In cultured HMECs, the proliferation is limited to a few passages owing to a proliferation block called M0 *(31)*. The few cells that are able to escape from this block show increased levels of $p16^{INK4a}$ promoter methylation and decreased $\text{p16}^{\text{INK4a}}$ protein *(30)*. This study proposes a link between the methylation status of $p16^{INK4a}$ and the loss of cell cycle control which is the hallmark of cancer. Furthermore, a recent in vivo study highlighted the potential importance of $p16^{INK4a}$ promoter methylation in early cancer diagnosis *(32)*. Of tumors that were induced in rats by the treatment of a tobacco-specific carcinogen, 94% showed increased methylation of the $p16^{INK4a}$ promoter. Most importantly, this methylation change was frequently detected in precursor lesions to the tumor, placing $p16^{INK4a}$ hypermethylation at an early stage of carcinogenesis, thus providing an essential marker for the early diagnosis of cancer *(32)*.

There are several methods that may be used to determine the methylation status of a gene promoter of interest. Earlier methods included restriction digests *(33,34)* and bisulfite genomic sequencing *(35)*. Restriction digest analysis relies on the fact that certain endonucleases are sensitive to sites of methylation and hence digestion followed by Southern blot analysis or polymerase chain reaction (PCR) may reveal sites of methylation. Methodology for this approach is given in Chapter 17 by Tennant et al. PCR amplification and sequencing of bisulfite-converted DNA will also expose sites of methylation, as bisulfite converts cytosine but not 5-methylated cytosine to uracil *(36)*. Bisulfite converted DNA may also be amplified with methylation-specific primers (MSPs) to differentiate between methylated and unmethylated samples *(37)*. MSP is considered to be the most sensitive method to date. A more recent method is referred to as methylation-sensitive single nucleotide primer extension (Ms-SnuPE) *(38)* and allows quantitation of single methylation sites by primer extension of bisulfite-converted DNA. This chapter describes a new method that combines endonuclease digestion with bisulfite modification to display methylated sites. This technique allows quantitative changes to be monitored and it is more sensitive than Southern blotting. In addition, multiple CpG sites can be monitored simultaneously.

In this method, the DNA is first treated with high bisulfite concentrations at low pH (**Fig. 1B**). Under these conditions, cytosine but not 5-methylcytosine is sulfonated, resulting in the unstable cytosinsulfonate (step I) and then hydrolytically deaminated to give uracilsulfonate (step II). After the DNA is purified from the bisulfite solution, the DNA is desulfonated in alkaline conditions (step III), resulting in a uracil residue. Because 5-methylcytosine is protected from sulfonation under these conditions *(36)*, every methylated cytosine will remain a cytosine after bisulfite treatment, whereas any unmethylated cytosine is converted

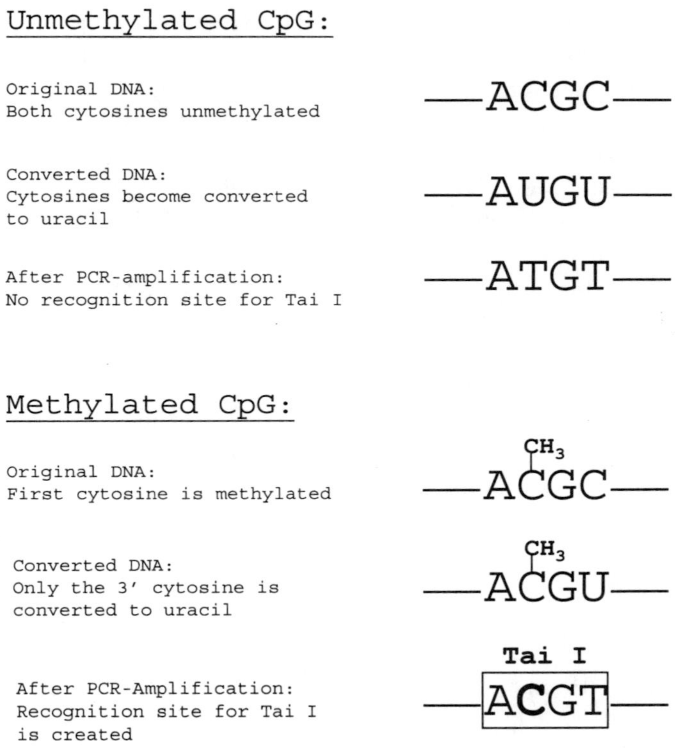

Unmethylated CpG:

Original DNA:
Both cytosines unmethylated —ACGC—

Converted DNA:
Cytosines become converted to uracil —AUGU—

After PCR-amplification:
No recognition site for Tai I —ATGT—

Methylated CpG:

Original DNA:
First cytosine is methylated —A$\overset{CH_3}{C}$GC—

Converted DNA:
Only the 3′ cytosine is converted to uracil —A$\overset{CH_3}{C}$GU—

After PCR-Amplification:
Recognition site for Tai I is created —ACGT— (Tai I)

Fig. 2. Creation of a restriction site by the bisulfite modification method. In the original DNA, the target site contains a potential CpG methylation site as well as a nonmethylatable cytosine. If both cytosines are unmethylated, they both become converted to uracil and are amplified as thymidines. This PCR product has no recognition site for *Tai*I. Only if the cytosine of the CpG dinucleotide is methylated its conversion prevented and a new recognition site for *Tai*I is created.

to uracil (**Fig. 1B**). The bisulfite-converted DNA provides the template for a subsequent PCR amplification step. The PCR primers are chosen to bind specifically to bisulfite converted DNA in stretches without CpG dinucleotides. This way only converted DNA will be amplified regardless of the methylation status of CpG dinucleotides between both primers. This PCR product can now be analyzed with simple endonuclease restriction digests. If, for example, the original DNA has an ACGC sequence in which both cytosines are unmethylated, then both become converted to uracils and are subsequently amplified as thymidines (**Fig. 2**, upper panel). The resulting sequence ATGT is not a recognition site for the restriction enzyme *Tai*I. In case of CpG methylation, however, only the 3' cytosine is converted whereas the methylated cytosine of the CpG dinucleotide remains stable (**Fig. 2**, lower panel). After PCR amplification, a recognition site for the restriction enzyme *Tai*I is created. In this manner, any methylated DNA will be visualized by virtue of its restriction fragments, even if unmethylated DNA dominates the DNA preparation. As shown in **Fig. 3**, the methylation status of 12 different cytosines can be monitored in one experiment utilizing the specificity of six restriction enzymes. An example of such an experiment is shown in **Fig. 4**. Genomic DNA derived from proliferating HeLa and Raji cells was bisulfite converted as described in the Methods section, nested PCRs were performed, and the radioactively labeled product was digested with the indicated restriction enzymes. In HeLa cells, all of the potential methylation sites under investigation were unmethylated and the PCR fragment resisted all of the restriction enzyme digests (**Fig. 4**, left panel). The PCR product derived from Raji DNA, however, could be digested by all of the restriction enzymes, indicating a 100% methylation at these sites. These cell lines can therefore serve as positive and negative controls in this assay.

## 2. Materials

### 2.1. Extraction

1. Lysis buffer: 50 m*M* KCl, 10 m*M* Tris-HCl, pH 8.4, 10 m*M* EDTA, 0.5% Tween-20; 0.5% Triton X-100.
2. Proteinase K: 10 mg/mL in water. Keep in small aliquots at –20°C and do not refreeze once thawed.
3. 6 *M* NaI solution: Add 1 µL of β-mercaptoethanol to every 10 mL of 6 *M* NaI.
4. Silica powder suspension "glass milk" (e.g., Oncor Appligene, Heidelberg).
5. 70% Ethanol.
6. TE; 10 m*M* Tris-HCl, pH 7.5, 1 m*M* EDTA.

### 2.2. Conversion

7. 3 *M* NaOH.
8. 3.6 *M* Sodium bisulfite, pH 5.0–1 m*M* hydroquinone, prepared immediately prior to use (*see* **Note 1**).
9. 20 m*M* NaOH–90% ethanol.
10. 70% and 90% ethanol.

### 2.3. PCR

1. Desalted oligonucleotides as indicated under **Subheading 3.3., step 1**.
2. PCR buffer and *Taq* polymerase as provided by the manufacturer, $MgCl_2$, and dNTP mix.
3. 10 m*M* dATP, dGTP, and dTTP.
4. Radioactive CTP ([α-$^{33}$P]dCTP, 370 MBq/mL, ≥ 92.5 TBq/mmol).
5. Restriction enzymes *Hinf*I, *Bsh*1236I, *Ita*I, *Mbo*I, *Tai*I, and *Mae*III
6. Acrylamide–*bis*-acrylamide, 19:1.
7. 1X TBE buffer: 90 m*M* Tris-borate, pH 8.0, 2 m*M* EDTA.

```
          primer E1                                        MboI/Bsh1236I
1    GAAGAAAGAG GAGGGGTTGG TTGGTTATTA GAGGGTGGGG CGGATCGCGT

                                                         ItaI
51   GCGTTCGGCG GTTGCGGAGA GGGGGAGAGT AGGTAGCGGG CGGCGGGGAG

            HinfI   ItaI
101 TAGTATGGAG TCGGCGGCGG GGAGTAGTAT GGAGTTTTCG GTTGATTGGT

       MaeIII   Bsh1236I
151 TGGTTACGGT CGCGGTTCGG GGTCGGGTAG AGGAGGTGCG GGCGTTGTTG

                             TaiI          MaeIII
201 GAGGCGGGGG CGTTGTTTAA CGTATCGAAT AGTTACGGTC GGAGGTCGAT

            primer E3
251 TTAGGTGGGT AGAGGGTTTG T
```

Fig. 3. Amplified fragment of the p16$^{INK4a}$ promoter. After conversion and amplification of the p16$^{INK4a}$ promoter DNA, recognition sites for six different restriction enzymes are created. Potentially methylated cytosines are underlined, primer binding sites for the second amplification step are double underlined. Recognition sites of the indicated restriction enzymes are given in boldface. Digesting the DNA with these enzymes results in two or three fragments of 42–229 base pairs in length **(Table 1)**. Note that *Ita*I and *Bsh*1236I will cut the DNA only if both of the CpG sites within the restriction enzyme recognition site are methylated.

## 3. Methods

### *3.1. Extraction of DNA from Frozen Tissue Samples*

1. Using a Cryostat, cut 5–10 sections of 10 µm of the frozen tissue sample and place into the cap of an Eppendorf tube.
2. Add 300 µL of lysis buffer and 3 µL of Proteinase K solution. Mix well.
3. Incubate at 55°C for 3 h or overnight.
4. Add 300 µL of phenol–chloroform, mix, centrifuge for 2 min (*see* **Note 2**) and transfer aqueous (upper) phase to a fresh tube.
5. Add 1 mL of NaI and 20 µL of glass milk; mix well.
6. Incubate at room temperature for 10 min and mix by inverting the tube every 2 min.
7. Pellet the glass milk (*see* **Note 2**) for 2 min and discard supernatant.
8. Add 1 mL of 70% ethanol, resuspend the pellet completely and centrifuge for 10 s.
9. Discard supernatant and repeat **step 8**.
10. Discard supernatant and air dry the pellet for 5–10 min. Prevent overdrying (i.e., do not use a Speed-Vac), as this can reduce DNA recovery dramatically.
11. To elute the DNA, resuspend pellet in 50 µL of TE or water and incubate at 55°C for 5–10 min.
12. Centrifuge for 2 min and transfer the supernatant to a fresh tube, being careful not to disturb the glass milk pellet (*see* **Note 3**).
13. Measure $OD_{260}$ and determine concentration.

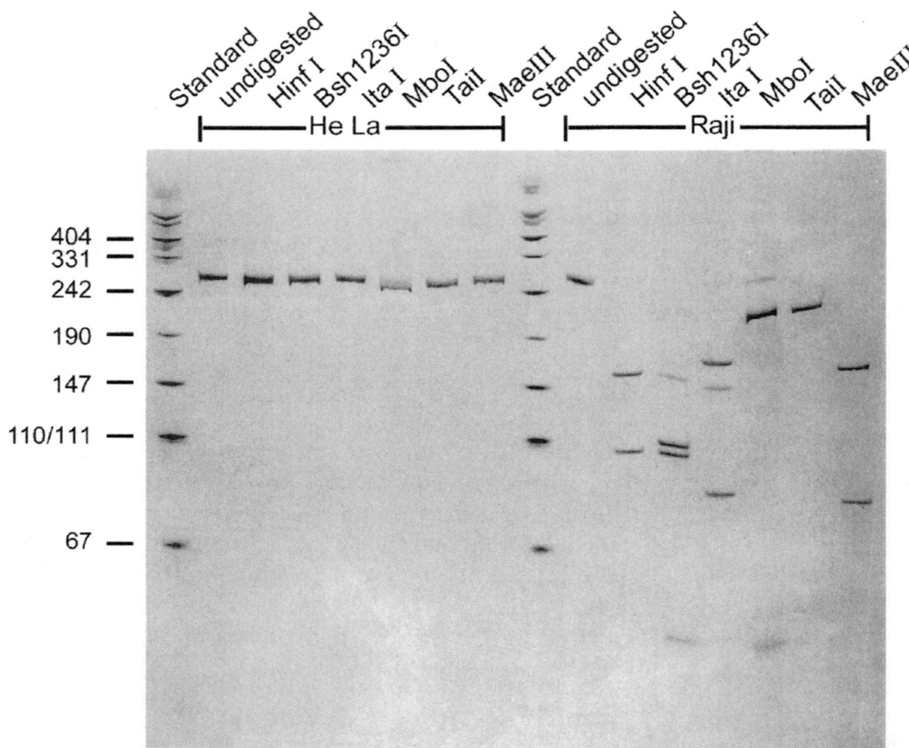


Fig. 4. Restriction digest pattern of cell line DNA. DNA derived from HeLa or Raji cell lines was converted, amplified, and digested as described. In HeLa DNA, the PCR product is not cut by any of the selected enzymes, indicating no detectable methylation of the target sites. The amplification product derived from Raji DNA, however, is cut by all of the selected enzymes, resulting in the predicted restriction fragments, indicating fully methylated DNA at the target sites.

### 3.2. Bisulfite Conversion of DNA

1. Take 500–1000 ng of extracted chromosomal DNA and dilute to 100 µL with water.
2. Add 7 µL of freshly prepared 3 *M* NaOH.
3. Incubate at 37°C for 10 min.
4. Add 550 µL of bisulfite–hydroquinone solution (*see* **Note 1**) to each tube and mix well.
5. Incubate at 55°C for 16 h *in the dark* (*see* **Note 4**).
6. Add 750 µL of NaI solution and 5 µL of glass milk to each tube and mix well (*see* **Note 5**).
7. Incubate at room temperature for 10 min and mix by inverting the tube every 2 min.
8. Pellet the glass milk (*see* **Note 2**) for 30 s and discard the supernatant.

**Table 1**
**Restriction Fragments from a p16$^{INK4a}$ PCR Product Derived from Fully Methylated DNA**

| Enzyme | Recognition site | No. of cuts | Fragment sizes |
|---|---|---|---|
| *Hinf*I | GANTC | 1 | 107 bp + 164 bp |
| *Mbo*I | GATC | 1 | 42 bp + 229 bp |
| *Tai*I | ACGT | 1 | 219 bp + 52 bp |
| *Ita*I | GCNGC | 2 | 89 bp + 24 bp + 158 bp |
| *Mae*III | GTNAC | 2 | 152 bp + 79 bp + 40 bp |
| *Bsh*1236 I | CGCG | 2 | 45 bp + 115 bp + 111 bp |

9. Wash the pellet by resuspending it in 1 mL of 70% ethanol and centrifuging for 10 s.
10. Discard supernatant and wash twice more.
11. Resuspend pellet in 100 µL of 20 m*M* NaOH–90% ethanol.
12. Incubate for 5 min at room temperature.
13. Centrifuge and carefully discard supernatant.
14. Wash the pellet twice in 1 mL of 90% ethanol.
15. Air-dry pellet for a few minutes on the bench, resuspend in 25 µL of TE, and incubate at 55°C for 5–10 min.
16. Centrifuge for 2 min, then transfer the supernatant to a fresh tube. This solution contains the bisulfite-converted and purified DNA (*see* **Note 6**).

### 3.3. p16$^{INK4a}$ PCR and Restriction Digest

1. The PCR reaction is carried out as nested PCR using the following primers:

   p16E0 5'TTAGAGGATTTGAGGGATAGGGT3',
   p16E1 5'GAAGAAAGAGGAGGGGTTG GTT3',
   p16E3 5'ACAAACCCTCTACCCACCTAAAT3',
   p16E4 5'CTACCTAATT CCAATTCCCCTACA3'.

2. For the first ("outer") PCR, assemble the PCR cocktail on ice as follows: 2 µL of bisulfite converted DNA, 10 pmol of each primer p16E0 and p16E4, 5 µL of 10X PCR buffer (Quiagen, Valencia, CA), 2.75 m*M* $MgCl_2$ (final concentration), 0.2 m*M* each dNTP and 1 U of *Taq* polymerase (Quiagen), and $H_2O$ to a final volume of 50 µL.
3. Insert the PCR tubes into the preheated thermocycler and after a 2-min denaturing step at 94°C perform 35 cycles of 30 s each 94°C denaturation, 60°C annealing, and 72°C extension. As a last step, add 8 min of 72°C final extension.
4. For the second ("nested") PCR, use 1 µL of the first PCR reaction and 10 pmol of each primer p16E1 and p16E3. PCR buffer and cycling conditions are the same as described previosly. The size of the amplified fragment should be 271 bp.
5. Check the purity of the PCR product on a 1.5% agarose–TBE gel.
6. For radioactive labeling of the PCR fragment, add 10 µL of the nested PCR fragment to 10 pmol of each primer p16E1 and p16E3; 5 µL of 10X PCR buffer

(Quiagen, Valencia, CA); 2.75 m*M* $MgCl_2$; 0.2 m*M* each of dATP, dGTP; and dTTP, 2 µL of [α-$^{33}$P] dCTP (370 MBq/mL, ≥92.5 TBq/mmol); 1.5 U of *Taq* polymerase (Quiagen), and $H_2O$ to a final volume of 50 µL.

7. Perform 10 cycles of PCR as described previously including both initial denaturation and final extension steps.
8. Digest 5 µL of the labeled PCR product with 2.5–5 U of the restriction enzymes *Hin*fI, *Bsh*1236I, *Ita*I, *Mbo*I, *Tai*I, and *Mae*III, respectively, for 2 h in a reaction volume of 10 µL with the restriction enzyme buffer recommended by the manufacturer.
9. Resolve the restriction fragments on a 12% nondenaturing polyacrylamide gel (*see* **Note 7**) in 1X TBE buffer.
10. After completion of the run, dry the gel under vacuum and expose to X-ray film for 4–16 h without enhancing screens at room temperature.

## 4. Notes

1. Most preparations of sodium bisulfite are usually mixtures of sodium bisulfite ($NaHSO_3$) and sodium metabisulfite ($Na_2S_2O_5$). The latter dissolves resulting in two $HSO_3^-$ ions such that 2.5 *M* metabisulfite is equivalent to 5 *M* bisulfite. Assuming a 1:1 ratio of both components in the preparation, 3.6 *M* bisulfite solution is prepared by dissolving 8.1 g of bisulfite in 15 mL of water. Adjust the pH to 5.0 with 10 *M* NaOH (approx 1 mL) and heat the solution to 50°C if necessary to dissolve all of the bisulfite, add 1 mL of freshly prepared 20 m*M* hydroquinone and fill up to a final volume of 20 mL.
2. All centrifugation steps are carried out at maximum speed in a microcentrifuge at top speed (>10,000*g*).
3. Because glass milk absorbs UV, the quantitation of the DNA will be perturbed by any glass milk transferred into the new tube.
4. For detailed discussion of variables influencing the bisulfite conversion, *see* **refs. *39–42***.
5. For both of the DNA purifications (extracting DNA from tissues and purification out of bisulfite solution) DNA extraction kits may be used that are designed to extract DNA out of agarose gel slices. In our hands, DNA purification kits based on silica powder using either NaI or $NaClO_4$ as a chaotropic salt work equally well. With kits based on DNA binding to glass filters, however, the recovery seems to be much lower and they tend to lead to the precipitation of bisulfite, which can interfere with the subsequent PCR.
6. Bisulfite-converted DNA can be stored at –20°C in the dark for at least 3 mo.
7. To obtain maximum resolution, use a 20 × 20 cm gel apparatus and do not exceed 1 mm thickness of the gel.

## Acknowledgment

We would like to thank E. Barczak for technical help with the artwork and K. Sweeney for critical reading of the manuscript.

## References


1. Holliday, R. and Grigg, G. W. (1993) DNA methylation and mutation. *Mutat. Res.* **285,** 61–67.
2. Issa, J.-P. J., Baylin, S. B., and Herman, J. G. (1997) DNA methylation changes in hematologic malignancies: biologic and clinical implications. *Leukemia* **11,** 7–11.
3. Cross, S. H. and Bird, A. P. (1995) CpG islands and genes. *Genes Dev.* **5,** 309–314.
4. Li, E., Beard, C., and Jaenisch, R. (1993) Role for DNA methylation in genomic imprinting. *Nature* **366,** 362–365.
5. Pfeifer, G. P., Steigerwald, S. D., Müller, P. R., Wold, B., and Riggs, A. D. (1989) Genomic sequencing and methylation analysis by ligation mediated PCR. *Science* **246,** 810–813.
6. Riggs, A. D. and Pfeifer, G. P. (1992) X-chromosome inactivation and cell memory. *Trends Genet.* **8,** 169–174.
7. Bird, A. P. (1986) CpG-rich islands and the function of DNA methylation. *Nature* **321,** 209–213.
8. Bird, A. (1992) The essentials of DNA methylation. *Cell* **70,** 5–8.
9. Merlo, A., Herman, J. G., Mao, L., Lee, D. J., Gabrielson, E., Burger, P. C., et al. (1995) 5' CpG island methylation is associated with transcriptional silencing of the tumor suppressor p16/CDKN2/MTS1 in human cancers. *Nat. Med.* **1,** 686–692.
10. Jost, J.-P., Saluz, H.-P., and Pawlak, A. (1991) Estradiol down regulates the binding activity of an avian vitellogenin gene repressor (MDBP-2) and triggers a gradual demethylation of the mCpG pair of its DNA binding site. *Nucleic Acids Res.* **19,** 5771–5775.
11. Meehan, R., Lewis, J., Cross, S., Nan, X., Jeppesen, P., and Bird, A. (1992) Transcriptional repression by methylation of CpG. *J. Cell Sci.* **16** (Suppl.), 9–14.
12. Vertino, P. M., Issa, J.-P., Pereira-Smith, O. M., and Baylin, S. B. (1994) Stabilization of DNA methyltransferase levels and CpG island hypermethylation precede SV40-induced immortalization of human fibroblasts. *Cell Growth Differentiat.* **5,** 1395–1402.
13. Issa, J.-P. J., Vertino, P. M., Boehm, C. D., Newsham, I. F., and Baylin, S. B. (1996) Switch from monoallelic to biallelic human IGF2 promoter methylation during ageing an carcinogenesis. *Proc. Natl. Acad. Sci. USA* **93,** 11757–11762.
14. Chen, J. X., Zheng, Y., West, M., and Tang, M. (1998) Carcinogens preferentially bind at methylated CpG in the p53 mutational hot spots. *Cancer Res.* **58,** 2070–2075.
15. Graff, J. R., Herman, J. G., Lapidus, R. G., Chopra, H., Xu, R., Jarrard, D. F., et al. (1995) E-Cadherin expression is silenced by DNA hypermethylation in human breast and prostate carcinomas. *Cancer Res.* **55,** 5195–5199.
16. Frixen, U. H., Behrens, J., Sachs, M., Eberle, G., Voss, B., Warda, A., et al. (1991) E-cadherin-mediated cell-cell adhesion prevents invasiveness of human carcinoma cells. *J. Cell Biol.* **113,** 173–185.
17. Navarro, P., Gomez, M., Pizarro, A., Gamallo, C., Quintanilla, M., and Cano, A. (1991) A role for the E-cadherin cell–cell adhesion molecule during tumor progression of mouse epidermal carcinogenesis. *J. Cell Biol.* **115,** 517–533.

18. Vleminckx, K., Vakaet, L., Mareel, M., Fiers, W., and Van Roy, F. (1991) Genetic manipulation of E-cadherin expression by epithelial tumor cells reveals an invasion suppressor role. *Cell* **66,** 107–119.
19. Graff, J. R., Greenberg, V. E., Herman, J. G., Westra, W. H., Boghaert, E. R., Ain, K. B., et al. (1998) Distinct patterns of E-Cadherin CpG island methylation in papillary, follicular, Hurthle's cell, and poorly differentiated human thyroid carcinoma. *Cancer Res.* **58,** 2063–2066.
20. Shen, L., Fang, J., Qiu, D., Zhang, T., Yang, J., Chen, S., and Xiao, S. (1998) Correlation between DNA methylation and pathological changes in human hepatocellular carcinoma. *Hepato-Gastroenterol.* **45,** 1753–1759.
21. Herman, J. G., Merlo, A., Mao, L., Lapidus, R. G., Issa, J.-P. J., Davidson, N. E., et al. (1995) Inactivation of the CDKN2/p16/MTS1 gene is frequently associated with aberrant DNA methylation in all common human cancers. *Cancer Res.* **55,** 4525–4530.
22. Gonzalez-Zuleta, M., Bender, C. M., Yang, A. S., Nguyen, T., Beart, R. W., VanTornout, J. M., and Jones, P. A. (1995) Methylation of the 5' CpG island of the p16/CDKN2 tumor suppressor gene in normal and transformed human tissues correlates with gene silencing. *Cancer Res.* **55,** 4531–4535.
23. Fueyo, J., Gomez-Manzano, C., Bruner, J. M., Saito, Y., Zhang, B., Zhang, W., et al. (1996) Hypermethylation of the CpG island of p16/CDKN2 correlates with gene inactivation in gliomas. *Oncogene* **13,** 1615–1619.
24. Reed, A. L., Califano, J., Cairns, P., Westra, W. H., Jones, R. M., Koch, W., et al. (1996) High frequency of 16 (CDKN2/MTS-1/INK4A) inactivation in head and neck squamous cell carcinoma. *Cancer Res.* **56,** 3630–3633.
25. El-Naggar, A. K., Lai, S., Clayman, G., Lee, J.-K. J., Luna, M. A., Goepfert, H., and Batsakis, J. G. (1997) Methylation, a major mechanism of p16/CDKN2 gene inactivation in head and neck squamous carcinoma. *Am. J. Pathol.* **151,** 1767–1774.
26. Swafford, D. S., Middleton, S. K., Palmisano, W. A., Nikula, K. J., Tesfaigzi, J., Baylin, S. B., et al. (1997) Frequent aberrant methylation of p16INK4A in primary rat lung tumors. *Mol. Cell. Biol.* **17,** 1366–1374.
27. Wong, D. J., Barret, M. T., Stöger, R., Emond, M. J., and Reid, B. J. (1997) p16INK4A promoter is hypermethylated at high frequency in esophageal adenocarcinomas. *Cancer Res.* **57,** 2619–2622.
28. Martinez-Delgado, B., Fermandez-Piqueras, J., Garcia, M. J., Arranz, E., Gallego, J., Robledo, M., and Benitez, J. (1997) hypermethylation of a 5' CpG island of p16 is a frequent event in non-Hodgkin's lymphoma. *Leukemia* **11,** 425–428.
29. Gonzalgo, M. L., Hayashida, T., Bender, C. M., Pao, M. M., Tsai, Y. C., Gonzales, F. A., et al. (1998) The role of DNA methylation in expression of the p19/p16 locus in human bladder cancer cell lines. *Cancer Res.* **58,** 1245–1252.
30. Foster, S. A., Wong, D. J., Barrett, M. T., and Galloway, D. A. (1998) Inactivation of p16 in human mammary epithelial cells by CpG island methylation. *Mol. Cell. Biol.* **18,** 1793–1801.
31. Foster, S. A. and Galloway, D. A. (1996) Human papillomavirus type 16 E7 alleviates a proliferation block in early passage human mammary epithelial cells. *Oncogene* **12,** 1773–1779.

32. Belinsky, S. A., Nikula, K. J., Palmisano, W. A., Michels, R., Saccomanno, G., Gabrielson, E., et al. (1998) Aberrant methylation of p16INK4a is an early event in lung cancer and a potential biomarker for early diagnosis. *Proc. Natl. Acad. Sci. USA* **95,** 11891–11896.
33. Razin, A. and Cedar, H. (1991) DNA methylation and gene expression. *Microbiol. Rev.* **55,** 451–458.
34. Stöger, R., Kubicka, P., Liu, C.-G., Kafri, T., Razin, A., Cedar, H., and Barlow, D. P. (1993) Maternal-specific methylation of the imprinted mouse Igf2r locus identifies the expressed locus as carrying the imprinting signal. *Cell* **73,** 61–71.
35. Frommer, M., McDonald, L. E., Millar, D. S., Collis, C. M., Watt, F., Grigg, G. W., et al. (1992) A genomic sequencing protocol that yields a positive display of 5-methylcytosine residues in individual DNA strands. *Proc. Natl. Acad. Sci. USA* **89,** 1827–1831.
36. Wang, R. Y.-H., Gehrke, C. W., and Ehrlich, M. (1980) Comparison of bisulfite modification of 5-methyldeoxycytidine and deoxycytidine residues. *Nucleic Acids Res.* **8,** 4777–4790.
37. Herman, J. G., Graf, J. R., Myöhänen, S., Nelkin, B. D., and Baylin, S. B. (1996) Methylation-specific PCR: a novel PCR assay for methylation status of CpG islands. *Proc. Natl. Acad. Sci. USA* **93,** 9821–9826.
38. Gonzalgo, M. L. and Jones, P. A. (1997) Rapid quantitation of methylation differences at specific sites using methylation-sensitive single nucleotide primer extension (Ms-SNuPE). *Nucleic Acids Res.* **25,** 2529–2531.
39. Feil, R., Charlton, J., Bird, A. P., Walter, J., and Reik, W. (1994) Methylation analysis on individual chromosomes: improved protocol for bisulphite genomic sequencing. *Nucleic Acids Res.* **22,** 695–696.
40. Schanke, J. T., Quam, L. M., and Van Ness, B. G. (1994) An improved method for detection of 5-methylcytosine by PCR-based genomic sequencing. *BioTechniques* **16,** 416–417.
41. Raizis, A. M., Schmitt, F., and Jost, J.-P. (1995) A bisulfite method of 5-methylcytosine mapping that minimizes template degradation. *Analyt. Biochem.* **226,** 161–166.
42. Grigg, G. W. (1996) Sequencing 5-methylcytosine residues by the bisulfite method. *DNA Sequence* **6,** 189–198.

# 19

# Detection of Metastatic Tumor Cells in Blood by Reverse Transcriptase-Polymerase Chain Reaction

**Katja Haack, Krishna Soondrum, and Michael Neumaier**

## 1. Introduction

In oncology, the correct assessment of tumor stage is crucial information, as it relates to disease-free interval and prognosis. Routine staging modalities use histopathology to evaluate the presence or absence of metastasis mainly in tissues such as lymph nodes, as described in Chapter 1 by Roskell and Buley. Retrospective analysis using fine sectioning techniques has shown that approx 30% of tumor patients may be understaged at the time of diagnosis *(1)*. Since the early 1980s, immunohistochemical and -cytochemical methods have been devised to increase the sensitivity in detecting small tumor loads or single tumor cells *(2)*. Clinical studies using monoclonal antibodies against cytokeratins, intermediate filament proteins characterizing epithelial cells and thus carcinomas, have shown that the presence of such cells in the bone marrow is an independent prognostic marker of disease-free and overall survival *(3–5)*. The methodology for this approach is given in Chapter 5 by Braun and Pantel.

With the advent of extremely powerful nucleic acid amplification and detection methods a significant increase in analytical and diagnostic sensitivity of tumor cell detection has been achieved. Such micrometastasis assays detect messenger RNA (mRNA) molecules transcribed from genes active in the cancer cell, but not the non-malignant cellular components otherwise present in the sample material. In this technique, a first strand cDNA copy is first produced from the RNA molecules using a reverse transcriptase (RT). The RT step is then followed by an enzymatic exponential amplification reaction, most commonly performed using the polymerase chain reaction (PCR). In principle, any protein target that can be used for micrometastasis detection by

From: *Methods in Molecular Medicine, vol. 57: Metastasis Research Protocols, Vol. 1: Analysis of Cells and Tissues*
Edited by: S. A. Brooks and U. Schumacher © Humana Press Inc., Totowa, NJ

immunocytochemistry is a candidate for analysis at the level of the mRNA precursor. In contrast to detection of the gene product by antibodies, the logarithmic amplification power makes it possible to obtain specific amplification signals starting from very few mRNA molecules. This clearly puts a single circulating tumor cell well within reach of diagnostic scrutiny. An increasing body of literature has evolved during the recent years describing RT-PCR tests for targeting mRNAs, such as tyrosinase ***(6)***; prostate-specific antigen (PSA) ***(7)***; prostate-specific membrane antigen (PSM) ***(8)***; carcinoembryonic antigen (CEA) ***(9)***; a number of cytokeratins (CKs) such as CK18 ***(10)***, CK19 ***(11)***, and CK20 ***(12,13)***; and others. RT-PCR assays show an exquisite analytic sensitivity and specificity in model systems that use limiting dilutions of tumor cells seeded into normal bone marrow or peripheral blood nucleated cells. In contrast, the diagnostic sensitivities and specificities in clinical samples are, for a number of reasons, still disputed. In our experience, there are three major problem areas that are related, but need separate consideration, when designing a RT-PCR assay or interpreting its results.

First, with decreasing numbers of target tumor cells in a clinical sample, the probability of obtaining a positive result from a single test will be increasingly governed by stochastic effects that can be predicted by Poisson distribution calculation. This applies to all assays that analyze a fraction of the specimen in a single round of testing. In general, the statistical dependence of the single result is inverse proportional to the size of the sample fraction investigated in a single test.

Second, the tumor specificity of the target mRNA expression investigated may not be warranted under all circumstances. For example, expression of the PSA has been observed in tumor cells of women suffering from breast cancer on both the mRNA and the protein levels ***(14)***. Also, nonmalignant PSA-expressing prostate cells can be shed into the circulation upon digital rectal investigation/manipulation of the prostate gland ***(15)***. Other examples include CK18, for which mRNA expression has been shown to occur in normal blood nucleated cells at a low level of constitutive background transcription ***(10)***. Other targets may be inducible in cells in which transcription of the respective gene is normally not observed. For example, the human CEA is induced specifically in granulocytes by interferon (IFN), but not by other inflammatory cytokines ***(16)***. Finally, the diagnostic specificity of a RT-PCR test may be compromised due to processed pseudogenes in the human genome. This has been demonstrated for CK18 ***(10,17)*** and also CK19 ***(11)*** and others ***(18,19)***.

Third, and most importantly, methodological differences between studies may lead to different test results and clinical significance even in cases of comparable tumor type, clinical stage, and mRNA target. These differences include procedures in sample stabilization, optional tumor cell enrichment, RNA preparation procedures, the use of different reagents and enzymes, number of amplification cycles, read-out format, and so forth, making it difficult to interpret a finding, as has been concluded recently for the PSA *(20)*.

Taken together, RT-PCR systems must be thoroughly characterized to make use of their potential in detection of tissue-specific gene expression for minimal residual disease detection. In the following subheadings, we present the protocols for nucleic acids preparation that have worked most consistently in our laboratory for a number of different mRNA targets in both standard and nested RT-PCR. As an example of the detection of circulating tumor cells we describe the protocols established for RT-PCR of CEA *(9,16)*. This assay has also been used by independent groups with good success *(21–23)*. In our experience the details given in the following subheadings have been helpful in designing RT-PCR assays for other mRNA targets used for micrometastasis detection.

## 2. Materials

### 2.1. Isolation of RNA

1. Stabilization solution: 70.1 g guanidinium isothiocyanate (GITC), 3.3 mL of 0.75 *M* sodium citrate, 5 mL of 10% *N*-laurylsarcosine sodium salt. Adjust the volume to 100 mL with distilled water. Add 36 µL of β-mercaptoethanol per 5 mL of GITC solution and store at room temperature.
2. Organic solvents: Water-saturated phenol, pH 4.2. Chloroform–isoamylalcohol: Add 24 vol of chloroform to 1 vol of isoamylalcohol prior to each use, fresh for each extraction.
3. 2 *M* Sodium acetate, pH 4.0.
4. 8 *M* Lithium chloride: Dissolve 17 g of solid lithium chloride in a 50-mL volume of RNase-free distilled water (necessary only when specimens have been anticoagulated with heparin).
5. TE8 buffer: 10 m*M* Tris-HCl, pH 8.0, and 1 m*M* EDTA.
6. Random hexamer primers: $pdN_6$ (Pharmacia-Upjohn, Freiburg, Germany), Superscript (murine Moloney leukemia virus reverse transcriptase) (Gibco-BRL, Paisley, UK). Thermostable DNA polymerases are available from various manufacturers and are supplied together with their respective reaction buffers.
7. Loading buffer: 10 m*M* EDTA, 0.25% w/v bromophenol blue, and 50% v/v glycerol.
8. TAE electrophoresis buffer: 4 m*M* Tris-acetate, pH 7.8, 1 m*M* EDTA.
9. PCR primer:

10a. CEA RT-PCR: Primer A: 5' TCT GGA ACT TCT CCT GGT CTC TCA GCT GG 3'. primer B: 5' TGT AGC TGT TGC AAA TGC TTT AAG GAA GAA GC 3'; primer C: 5' GGG CCA CTG TCG GCA TCA TGA TTG G 3'. Product length for primers A and B: 160 bp; for primers B and C: 131 bp *(2)*; CEAos 5' GGCCTCTAACCCATGCCCGCAGTAT 3'; CEAoa 5' AAGCCCAGCTCATTTTTGTATTTTT 3'; CEAis 5' AGTCTCTGCATCTGGAACTTCTCC 3'; CEAia 5' TTTAGACTGTAGCTGTTGCAAATGCTTTAAGG 3'; Product length for CEAos and CEA oa: 370 bp; for primers CEAis and CEAia: 172 bp *(16)*.

10b. $\beta_2$-Microglobulin RT-PCR: Primer M1: 5' CCT GAA TTG CTA TGT GTC TGG GTT TCA TCC A 3'; primer M2: 5' GGA GCA ACC TGC TCA GAT ACA TCA AAC ATG G 3'. Product length for primers M1 and M2: 441 bp.

## 3. Methods

The following subheadings relate to the schematic representation shown in **Fig. 1**. Selected aspects depicted in the flow sheet are also referred to in the Notes section.

### 3.1. Sampling, Additives, Lysis, and Storage

1. When sampling peripheral blood, always discard the first draw after phlebotomy. Prior to bone marrow sampling, perform a skin incision.
2. Add 1 mL of EDTA–blood to 5 mL of 6 *M* GITC solution and thoroughly mix by vortexing. RNA is stable at room temperature for up to 3 d and is thus suitable for sending by regular surface mail. Long-term storage of the blood lysate requires temperatures of –80°C or below, preferably in liquid nitrogen (*see* **Note 1**).

### 3.2. Preparation of Total RNA

1. Transfer 600 µL of the GITC hemolysate (room temperature) to a fresh 2-mL Eppendorf tube. Add 60 µL of 2 *M* sodium acetate pH 4.0; 500 µL of water-saturated phenol, pH 4.2 (*see* **Note 2**); and 300 µL of chloroform/isoamylalcohol. Vortex-mix vigorously for 30 s.
2. Incubate for 30 min at 4°C and centrifuge at 16,000*g* for 30 min at 4°C.
3. Carefully transfer the upper, aqueous phase to a fresh tube with 600 µL of 2-propanol and precipitate overnight at –20°C.
4. Collect the precipitate after centrifugation at 16,000*g* for 30 min at 4°C.
5. Discard supernatant, wash pellet by adding 1 mL of ice-cold (–20°C) 70% ethanol, and centrifuge again as in **step 4**. Repeat wash step once.
6. Carefully remove supernatant with a pipet without disturbing the pellet. Dry pellet for 5–15 min in a vacuum centrifuge until all traces of ethanol have evaporated. Do not overdry the pellet.
7. Resuspend in 100 µL of TE8 buffer and store at –80°C.
8. If the original blood sample contained heparin, precipitate the RNA sample with lithium chloride (*see* **Note 3**).

This method will provide total RNA that can be used for RT-PCR without further manipulation. Consistently good results are obtained when 1/10 of the volume of this RNA sample is used for further analysis (*see* **Note 4**). Increasing the volume of RNA in the following reverse transcription/amplification step may result in a net loss of sensitivity, probably caused by residual inhibitors in the RNA preparation becoming critical. If required, a more highly purified RNA preparation can be generated using commercially available reagents. We use the following protocol adapted from the manufacturer's recommendations.

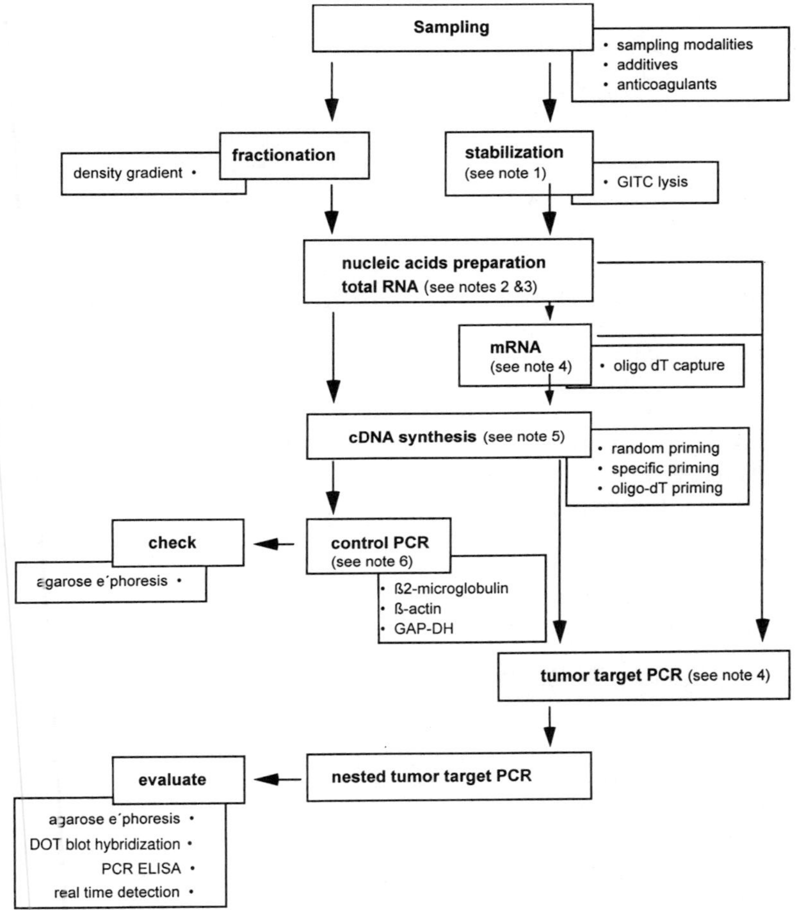


Fig. 1. Flow sheet of RT-PCR for micrometastasis detection.

### 3.3. Affinity Adsorption Method for Preparation of mRNA

1. Dissolve five tablets of solid glass magnetic particles (GMPs) (Roche Molecular Biochemicals) in a 50-mL polypropylene tube by adding 300 μL of RNase-free redistilled water by vortex-mixing and add 30 mL of RNA/DNA stabilization reagent.
2. Add 3 mL of the GITC hemolysate and incubate the mixture for 30 min at room temperature on a roller incubator to bind the total nucleic acids to the GMPs.
3. Centrifuge at 1100*g* for 2 min at room temperature.
4. Discard the supernatant and drain the inverted tube on filter paper for 30 s.
5. Resuspend the pelleted GMPs by thorough pipeting in 1 mL of GMP washing buffer. Transfer suspension to a fresh 2-mL Eppendorf tube and vortex-mix briefly.

6. Separate the GMPs using a magnetic separator and discard the supernatant. Repeat the wash step twice.
7. Centrifuge the tube briefly at 2000*g* and remove all traces of liquid with a drawn-out pipet tip (sequence gel loading tips are ideally suited for that purpose).
8. Elute nucleic acids in 600 µL of elution buffer by incubation for 5 min at 70°C. A suitable shaker is important. We use a heatable vibro shaker.
9. Following magnetic separation of the GMPs, transfer supernatant instantly to a clean 1.5-mL Eppendorf tube and keep on ice.
10. Homogenize streptavidin-coated magnetic particles (SMPs) by vortex-mixing and transfer a 60-µL aliquot to a fresh 1.5-mL Eppendorf tube. Separate the liquid from the SMPs using magnetic separation.
11. Charge the streptavidin-coated SMPs with 6 µL of biotinylated Oligo$(dT)_{20}$ probe in 300 µL of hybridization reagent included in the kit.
12. To the charged SMPs add the sample **(step 9)** and incubate for 5 min at 37°C. The mRNA is captured from the solution by the oligo(dT)-loaded SMPs.
13. Magnetically separate the captured mRNA from the solution and discard the supernatant. It is advisable to repeat the washing/separation step twice.
14. Centrifuge the tube briefly at 2000*g* and remove supernatant completely as in **step 7**.
15. Resuspend the dried SMPs in 1 m*M* Tris-HCl, pH 8.0, and elute the mRNA by incubating at 70°C for 2 min. Store the mRNA sample at –80°C.

### 3.4. First-Strand cDNA Synthesis

1. To 10 µL of RNA add 2 µL of 100 ng/µL of random hexamer primer; alternatively, 500 ng of oligo$(dT)_{12–18}$ or 2 pmol of antisense primer can be used, if full length or specifically primed first strand cDNA synthesis is required, respectively (*see* **Note 5**).
2. Cover with mineral oil, incubate in a thermocycler for 5 min at 70°C, and chill on ice to eliminate mRNA secondary structures.
3. Add RT buffer sufficient for a 20-µL reaction volume, 2 µL of 100 m*M* DTT, 1 µL of 10 m*M* dNTPs, and RT.
4. Incubate 30–60 min at 42°C. Inactivation of the reaction by heating at 90°C for 1 min is optional. If secondary structure is suspected to limit reverse transcription, the first-strand cDNA snythesis reaction can be carried out at temperatures up to 55°C. However, the stability of the enzyme is diminished at this temperature.

### 3.5. PCR

To verify that cDNA reverse transcription has been successful, it is advisable to choose as a control a target mRNA ubiquitously expressed. We prefer the $\beta_2$ microglobulin transcribed in all somatic cells (*see* **Note 6**). For this purpose, a small fraction of the cDNA is sufficient.

1. Use 1–2 µL of the cDNA in a separate tube for an RT control reaction with primers specific for $\beta_2$-microglobulin mRNA. Add 5 µL of 10X PCR buffer, 1 µL of 10 m*M* dNTPs, 1 µL of 25 µ*M* 5'-primer M1, 1 µL 25 µ*M* 3'-primer

M2, 1–3 U of *Taq* DNA polymerase, and distilled water to a final volume of 50 μL (work on ice and proceed with thermocycling). Thermocycling profile: 5 min denaturation at 95°C followed by 30–35 rounds of amplification performed at 95°C for 30 s, 60 °C for 30 s, and 72°C for 1 min. Append a 10-min 72°C final extension step after the last cycle.

The control reaction is evaluated most conveniently by agarose gel electrophoresis (*see* **Subheading 3.6.**) prior to further testing. A positive result will establish a successful RT reaction. To test for tumor-associated mRNA expression, for example, of CEA, a so-called nested PCR reaction is carried out to enhance sensitivity and also specificity. Specifically, following the first round of amplification, a small fraction of the first-round reaction is transferred to a fresh tube and subjected to a second round of PCR. This reaction employs primers amplifying internal sequences of the first round's product. As an example, the nested PCR protocol for the detection of CEA mRNA-expressing cells is given below.

2. The remaining 18 μL of the first strand cDNA synthesis (**step 1**) are used directly for the first round of CEA amplification. Add 5 μL of 10X PCR buffer, 1 μL of 10 m*M* dNTPs, 1 μL of 25 μ*M* 5'-primer A, 1 μL of 25 μ*M* 3'-primer B, 1–3 U of *Taq* DNA polymerase, and distilled water to a final volume of 50 μL (work on ice and proceed with thermocycling). Thermocycling profile: Denature the reaction for 5 min at 95°C. Run a two-step protocol (omitting annealing time) for 20 cycles consisting of 95°C for 1 min and 72°C for 2 min . At the end of this round of amplification append a 72°C final extension step for 10 min.
3. Prepare on ice a new reaction containing all reagents listed in **step 2** with the exception that primer A is replaced by primer C. Add 5 μL of the first-round CEA amplification reaction and proceed with the second round of amplification. Thermocycling profile: 30 cycles at 95°C for 1 min, 69 °C for 1 min, and 72°C for 1 min with a final 10-min extension step at 72°C.

### 3.6. Electrophoresis

PCR reactions are evaluated most conveniently by agarose gel electrophoresis. A positive result in this test will establish a successful reverse transcription. Mix 9 μL of the PCR product with 1 μL of gel loading buffer and load onto a submarine horizontal 1–2% agarose gel, containing 4 μL of 1% ethidium bromide solution per 100 mL (**Caution:** Ethidium bromide is a mutagen, so wear gloves at all times when handling the gel.) Run electrophoresis at 6 V/cm using TAE running buffer. When the bromphenol dye has migrated halfway down the gel (after approx 1 h), the bands can be visualized on an ultraviolet (UV) transilluminator.

## 4. Notes

1. Two modalities have been devised to handle clinical specimens for micrometastasis detection, that is, fractionation of nucleated cells or stabilization using chaotropic solutions. Traditionally, samples are fractionated using density gradient centrifugation. This results in (1) depletion of erythrocytes, that is, hemoglobin that, even in traces, inhibits the enzymatic reaction steps in the

subsequent RT-PCR and (2) in enrichment of the mononuclear cell fraction for analysis. However, keep in mind that the sedimentation of the tumor cells with this fraction is not warranted. Also, the processing of a specimen cannot be performed without delay in some situations, for example, the specimen is in longer transit to the laboratory. In this case, RNA may be lost due to degradation. We therefore prefer the stabilization of the sample using chaotropic reagents to effectively protect the mRNA. The price to pay for the stability is the massive hemoglobin contamination caused by the instant complete hemolysis that must be completely removed during the purification protocol. Note that, in contrast to the original protocol using a 4 *M* final concentration of chaotropic salts ***(25)***, the final concentration given for stabilization of RNA in our protocol is 5 *M,* the concentration necessary for irreversibly denaturing RNases ***(26)***.

GITC solution is stable for months and can be stored at room temperature. If GITC solution crystallizes due to low temperature, it can be redissolved by gently warming with no loss of performance. Once β-mercaptoethanol has been added, the shelf-life is approx 8 wk, after which we have observed rapid deterioration of RNA yields and quality.

2. For isolation of RNA do not use Tris-buffered phenol. Instead, use water-saturated phenol, pH 4.2. Under the conditions of acid pH, the DNA will partition into organic–anorganic interphase, while RNA will stay in solution, thus depleting the sample of DNA.
3. Additives for anticoagulation: Heparin should be avoided for specimen anticoagulation. If a heparin-containing sample is processed using a protocol employing any nucleic acid precipitation steps, be aware that the mucopolysaccharide heparin will coprecipitate/copurify and inhibit the subsequent amplification reaction even at low concentrations. The use of lithium chloride for precipitation completely restores the RT-PCR sensitivity, because only the RNA is precipitated selectively ***(24)***. This method can be easily incorporated into the RNA preparation protocol as follows: Following **step 6** in the preparation protocol, resuspend the RNA in 200 μL of TE8 buffer. Add lithium chloride to a final concentration of 1.8 *M* and incubate overnight at 4°C. Centrifuge, wash, and dry the sample as detailed in **Subheading 3.2.**

   Some targets of interest, for example, cytokeratins or CEA, are expressed in normal skin. Discarding the first draw effectively avoids contamination of the specimen with epithelial cells. For bone marrow sampling, the skin incision is mandatory.
4. Note that in the protocol given in **Subheading 3.2.**, total RNA is prepared from 1/10 of the specimen. Hereof, 1/10 is used for each single RT-PCR reaction, that is, equivalent to 1% of the total specimen. To recover mRNA directly, nucleic acid adsorption methods as detailed in **Subheading 3.3.** can be used. As mRNA represents 1–5% of the total RNA, the amount of nucleic acids that can be analyzed in a single reaction can be considerably increased. Although we routinely detect few circulating tumor cells/mL of peripheral blood with the regular protocol, the more expensive adsorption technique may be of advantage for very low copy numbers of target molecules and also to save time.

5. The choice of primers for the reverse transcriptase reaction should be considered carefully (*see* **Subheading 3.2.**). Through the use of random oligonucleotide primers all mRNA in the reaction is, in principle, copied into cDNA. This allows to use one transcription reaction for both control mRNA and tumor target mRNA amplification. In addition, random primed cDNA can be stored for other targets to be amplified later from the same cDNA. This decreases inter-assay variance and may be of significant advantage, when comparing copy numbers of target mRNAs in quantitative assay systems *(27)*. In contrast, specific priming has, in our experience, been more sensitive by a factor of 2–3. The lowest sensitivity in subsequent PCR amplification reactions has been found with oligo-dT-primed cDNA, particularly when amplification is performed in far upstream region of the target mRNA. In these cases the efficiency to generate full-length first-strand cDNA copies is limiting.
6. The mRNA dependency of the amplification signal has to be verified for each target. The occurrence of specific PCR fragments in reactions in which the RT step has been omitted may be indicative of processed pseudogene(s), if all possibilities of contamination have been excluded. Pseudogene sequences have been demonstrated for housekeeping genes such as β-actin *(18)* or GAPDH *(19)* as well as targets for micrometastasis detection *(11,17)*. We use $\beta_2$-microglobulin as a reverse transcription control, because no processed pseudogenes have been found for this target.

All PCR assays are prone to contamination. This is particularly true for tests designed for high-sensitivity RT-PCR or nested PCR reactions that require pipetting of amplification products to set up the second round of amplification. It is far beyond the scope of this manuscript to address all possibilities of contamination avoidance. Aspects of good laboratory practice have been described elsewhere *(28)*. However, the main sources of contamination of PCR reagents and PCR reactions are aerosols forming during the use of automatic pipettors or when opening the snap-top reaction tubes. Nanoliter droplets leaving the liquid will dry to dustlike DNA/salt crystals while traveling through the air. Accordingly, it is a good idea to keep pipetting to a minimum and to protect contact surfaces in the pipettors using aerosol-tight disposable tips. The second useful measure is the use of mineral oil known from times when the thermocycler did not usually feature heatable lids and evaporation was prevented by overlaying the reaction mixture with light mineral oil. The oil layer will seal the aqueous PCR reaction volume and thus is a good protection against aerosol formation. If necessary, the oil can be removed almost completely prior to further analysis: Place the reaction tubes into –20°C for approx 30 min. After the reaction mixture is frozen, the oil can be pipetted off before the aqueous phase starts thawing.

An alternative to RT-PCR in two separate steps, that is, reverse transcription/amplification is the one-step RT-PCR (*see also* **Fig. 1**), in which both the cDNA synthesis reaction and the subsequent PCR amplification are performed in a single reaction mix without intermediate pipetting steps. Several optimized buffer and enzyme systems are commercially available using combinations of reverse transcriptases together with mixtures of nonproofreading and proofreading thermostable DNA polymerases. In our experience, these systems work well with the optimized reagents. However, their use is considerably more expensive. Also, we see a disadvantage in using an all-included multi-enzyme system in troubleshooting. Particularly when setting up and testing a new assay, we find it more convenient to work with separate reactions.

## References


1. Gusterson, B. (1991) Are micrometastases clinically relevant? *Br. J. Hosp. Med.* **47,** 247–248.
2. Dearnaley, D. P., Sloane, J. P., Ormerod, M. G., Steele, K., Coombes, R. C., Clink, H. M., et al. (1981) Increased detection of mammary carcinoma cells in marrow smears using antisera to epithelial membrane antigen. *Br. J. Cancer* **44,** 85–90.
3. Pantel, K., Izbicki, J. R., Angstwurm, M., Braun, S., Passlick, B., Karg, O., et al. (1993) Immunocytological detection of bone marrow micrometastasis in operable non-small lung cancer. *Cancer Res.* **53,** 1027–1031.
4. Diel, I. J., Kaufmann, M., Görner, R., Costa, S. D., Kaul, S., and Bastert, G. (1992) Detection of tumor cells in bone marrow of patients with primary breast cancer: a prognostic factor for distant metastasis. *J. Clin. Oncol.* **10,** 1–5.
5. Lindemann, F., Schlimok, G., Dirschedl, P., Witte, J., and Riethmuller, G. (1992) Prognostic significance of micrometastatic tumor cells in bone marrow of colorectal cancer patients. *Lancet* **340,** 685–689.
6. Smith, B., Selby, P., Southgate, J., Pittman, K., Bradley, C., and Blair, G. E. (1991) Detection of melanoma cells in peripheral blood by means of reverse transcriptase and polymerase chain reaction. *Lancet* **338,** 1227–1229.
7. Moreno, J. G., Croce, C. M., Fischer, R., Monne, M., Vihko, P., Mulholland, S. G., and Gomella, L. G. (1992) Detection of hematogenous micrometastasis in patients with prostate cancer. *Cancer Res.* **52,** 6110–6112
8. Israeli, R. S., Powell, C. T., Corr, J. G., Fair, W. R., and Heston, W. D. W. (1994) Expression of the prostate-specific membrane antigen. *Cancer Res.* **54,** 1807–1811.
9. Gerhard, M., Juhl, H., Kalthoff, H., Schreiber, H. W., Wagener, C., and Neumaier, M. (1994) Specific detection of carcinoembryonic antigen-expressing tumor cells in bone marrow aspirates by polymerase chain reaction. *J. Clin. Oncol.* **12,** 725–729.
10. Tschentscher, P., Wagener, C., and Neumaier, M. (1997) Sensitive and specific cytokeratin 18 RT-PCR excluding the amplification of processed pseudogenes from contaminating genomic DNA. *Clin. Chem.* **43,** 2244–2250.
11. Datta, Y. H., Adams, P. T., Drobyski, W. R., Ethier, S. P., Terry, V. H., and Roth, M. S. (1994) Sensitive detection of occult breast cancer by the reverse transcriptase polymerase chain reaction. *J. Clin. Oncol.* **12,** 475–482.
12. Burchill, S. A., Bradbury, M. F., Pittman, K., Southgate, J., Smith, B., and Selby, P. (1995) Detection of epithelial cancer cells in peripheral blood by reverse transcriptase-polymerase chain reaction. *Br. J. Cancer* **71,** 278–281.
13. Soeth, E., Vogel, I., Roder, C., Juhl, H., Marxsen, J., Kruger, U., Henne-Bruns, D., et al. (1997) Comparative analysis of bone marrow and venous blood isolates from gastrointestinal cancer patients for the detection of disseminated tumor cells using reverse transcription PCR. *Cancer Res.* **57,** 3106–3110.

14. Yu, H., Diamandis, E. P., Levesque, M., Sismondi, P., Zola, P., and Katsaros, D. (1994) Ectopic production of prostate specific antigen by a breast tumor metastatic to the ovary. *Clin. Lab. Anal.* **8,** 251–253.
15. Moreno, J. G., O'Hara, S. M., Long, J. P., Veltri, R. W., Ning, X., Alexander, A. A., and Gomella, L. G. (1997) Transrectal ultrasound-guided biopsy causes hematogenous dissemination of prostate cells as determined by RT-PCR. *Urology* **49,** 515–520.
16. Jung, R., Krüger, W., Hosch, S., Holweg, M., Kröger, N., Wagener, C., et al. (1998) Specificity of RT-PCR assays designed for the detection of circulating cancer cells is influenced by cytokines in-vivo and in-vitro. *Br. J. Cancer* **78,** 1194–1198.
17. Neumaier, M., Gerhard, M., and Wagener, C. (1995) Diagnosis of micrometastases by the amplification of tissue-specific genes. *Gene* **159,** 43–47.
18. Raff, T., van der Giet, M., Endemann, D., Wiederholt, T., and Paul, M. (1997) Design and testing of beta-actin primers for RT-PCR that do not co-amplify processed pseudogenes. *BioTechniques* **23,** 456–460.
19. Ercolani, L., Florence, B., Denaro, M., and Alexander, M. (1988) Isolation and complete sequence of a functional human glyceraldehyde-3-phosphate dehydrogenase gene. *J. Biol. Chem.* **263,** 15335–15341.
20. Gomella, L. G., Ganesh, V. R., and Moreno, J. G. (1997) Reverse transcriptase polymerase chain reaction for prostate specific antigen in the managment of prostate cancer *J. Urol.* **158,** 326–337.
21. Mori, M., Mimori, K., Ueo, H., Tsuji, K., Shiraishi, T., Barnard, G. F., et al. (1998) Clinical significance of molecular detection of carcinoma cells in lymph nodes and peripheral blood by reverse transcription-polymerase chain reaction in patients with gastrointestinal or breast carcinomas. *Clin. Oncol.* **16,** 128–132.
22. Castaldo, G., Tomaiuolo, R., Sanduzzi, A., Bocchino, M. L., Ponticiello, A., Barra, E., et al. (1997) Lung cancer metastatic cells detected in blood by reverse transcriptase-polymerase chain reaction and dot-blot analysis. *Clin. Oncol.* **15,** 3388–3393.
23. Liefers, G. J., Cleton-Jansen, A. M., van de Velde, C. J., Hermans, J., van Krieken, J. H., Cornelisse, C. J., and Tollenaar, R. A. (1998) Micrometastases and survival in stage II colorectal cancer. *N. Engl. J. Med.* **339,** 223–228.
24. Jung, R., Lübcke, C., Wagener, C., and Neumaier, M. (1997) Reversal of RT-PCR inhibition observed in heparinized clinical specimens. *Biotechniques* **23,** 24–28.
25. Chomczynski, P. and Sacchi, N. (1987) Single step method of RNA isolation by acid guanidinium thiocyanate-phenol-chloroform extraction. *Analyt. Biochem.* **162,** 156–159.
26. Gillespie, H. H. (1994) Dissolve and capture: a strategy for analysing of RNA in blood *Nature* **367,** 390–391.
27. Jung, P., Ahmad-Nejad, P., Wimmer, M. , Gerhard, M., Scheike, K., Wagener, C., and Neumaier, M. (1997) Quality management and influential factors for the detection of single metastatic cancer cells by reverse transcriptase PCR. *Eur. J. Clin. Chem. Clin. Biochem.* **35,** 3–10.
28. Neumaier, M., Braun, A., and Wagner, C. (1998) Fundamentals of quality assessment of molecular amplification methods in clinical diagnostics. *Clin. Chem.* **44,** 12–26.

# 20

# Differential Display

**Ruey Ho Kao and Giulio Francia**

## 1. Introduction

The study of cancer genetics has focused primarily, until now, on changes at the genomic level. However, activating, or inactivating, mutations/deletions/rearrangements in DNA are not the only way in which alterations in gene expression can occur. Since the development of subtractive hybridization, which screens for differentially expressed genes between related populations at the RNA level, evidence has accrued showing that alterations in downstream gene expression might play a significant role in the biology and evolution of cancer.

Establishment of the novel technique of differential display (DD) *(1,2)* has facilitated studies in this new field that has been termed "expression genetics" *(3)*. Compared with traditional methods of gene expression analysis DD possesses the advantages of simplicity, sensitivity, reproducibility, and speed. It is the current method of choice in expression genetics not only in the cancer biology field but also in such diverse areas as embryology, developmental biology, and neurobiology. This novel technique involves the reverse transcription (RT) of mRNAs with oligo-dT primers anchored to the beginning of the poly(A) tail, followed by a polymerase chain reaction (PCR) in the presence of a second arbitrary random primer. The amplified cDNAs of 3' termini of mRNAs obtained (with the same primers) from different samples can then be run side by side on a sequencing gel to allow differentially expressed genes, either upregulated or downregulated, to be visualized (*see* **Fig. 1**). By changing primer pair combinations approx 15,000 individual mRNA species from a mammalian cell may be analyzed reasonably quickly, thus providing a "fingerprint" of the mRNA in any particular cell type. Differential cDNA fragments can easily be recovered and reamplified from the gel and cloned into vectors for future screening and characterization work. The DD technique can be subdivided into five steps:

From: *Methods in Molecular Medicine, vol. 57: Metastasis Research Protocols, Vol. 1: Analysis of Cells and Tissues*
Edited by: S. A. Brooks and U. Schumacher © Humana Press Inc., Totowa, NJ

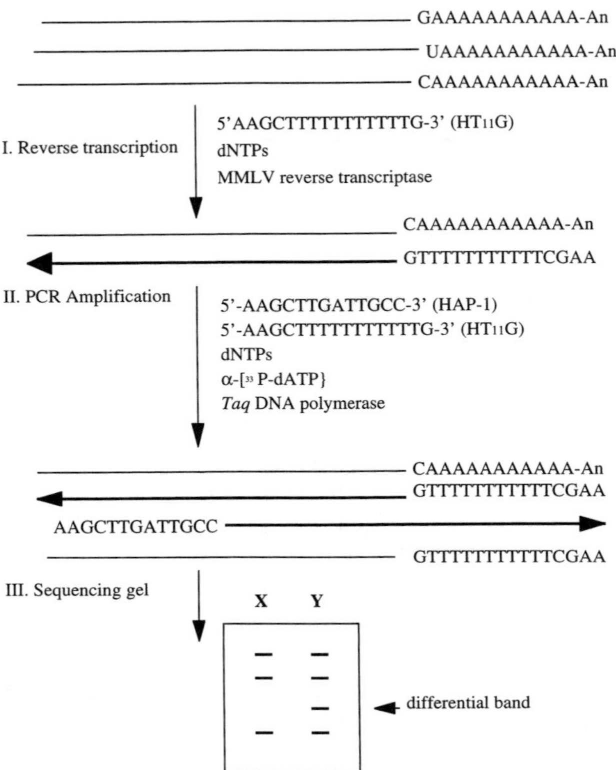


Fig. 1. Principal procedures of differential display technique.

1. RNA preparation, including DNase treatment.
2. Reverse transcription.
3. PCR.
4. Sequencing gel analysis.
5. cDNA recovery and reamplification.

The following protocol is based on the third generation of DD methodology using the RNAimage™ Kit (GeneHunter, Brookline, MA) in which three one-base-anchored oligo-dT primers are used ***(4)***.

## 2. Materials

### *2.1. RNA Preparation*

1. Autoclaved distilled $H_2O$ ($dH_2O$).
2. RNase-free DNase (1000 U/mL; Promega).
3. RNase inhibitor (33 U/µL; Promega).
4. 0.5-mL Microcentrifuge tube.

### 2.2. Reverse Transcription

1. One-base anchor primers (2 $\mu M$ ): $HT_{11}A$, $HT_{11}G$, $HT_{11}C$ with a *Hin*dIII restriction site (5'-AAGCTT-3') at the 5' end (GeneHunter, Brookline, MA).
2. dNTPs mixture (dATP, dTTP, dCTP, dGTP; 250 $\mu M$ each).
3. 10X RT buffer: 0.5 *M* Tris-HCl, pH 8.3, 0.75 *M* KCl, 0.03 *M* $MgCl_2$ (Stratagene).
4. 0.1 *M* Dithiothreitol (DTT).
5. Moloney Murine Leukemia Virus (MMLV) reverse transcriptase (50 U/µL; Stratagene).
6. 0.5-mL Microcentrifuge tube.

### 2.3. PCR

1. 10X PCR buffer: 0.1 *M* Tris-HCl, pH 8.3, 0.5 *M* KCl (Perkin-Elmer).
2. $MgCl_2$ (25 m*M;* Perkin-Elmer).
3. dNTPs (dATP, dTTP, dCTP, dGTP; 25 $\mu M$ each).
4. Anchor primer: $HT_{11}A$, $HT_{11}G$, $HT_{11}C$ (2 $\mu M$ ).
5. Arbitrary primer HAP1-40 (2 $\mu M$ ): 13-mer random primers with a *Hin*dIII restriction site (5'-AAGCTT-3') at the 5' end.
6. [a-$^{33}$P] dATP (2000 Ci/mmol).
7. *Taq* DNA polymerase (ICRF).
8. Perkin-Elmer thermocycler.
9. 0.5 mL of GeneAmp PCR reaction tube (Perkin-Elmer).
10. Mineral oil.

### 2.4. Sequencing Gel Analysis

1. 5X TBE buffer: 0.45 *M* Tris-borate, 0.01 *M* EDTA.
2. Protogel: 30% w/v acrylamide, 0.8% w/v *bis*-acrylamide.
3. 10% Ammonium persulfate, freshly made.
4. *N*, *N*, *N'*, *N'*-Tetramethylethylenediamine (TEMED).
5. 50% Glycerol.
6. DNA loading dye: 95% formamide, 10 m*M* EDTA, pH 8.0, 0.09% xylene cyanole FF, 0.09% bromophenol blue.
7. 3MM chromatography paper (Whatman).
8. Sequencing cell.
9. Long flat-tipped pipet.
10. Plastic wrap.
11. Gel dryer with vacuum.
12. Kodak X-ray film.
13. X-ray intensifying cassette.

### 2.5. cDNA Recovery and Reamplification

1. 3 *M* Sodium acetate, pH 5.2.
2. 35 mg/mL of glycogen.
3. Absolute ethanol.
4. 85% Ethanol, ice cold.

5. Parafilm.
6. The same PCR components as in **Subheading 2.3.** except dNTP (250 µ*M*) and without radioisotope.
7. Agarose.
8. 10 mg/mL ethidium bromide in water.
9. 1.5-mL Microcentrifuge tube.
10. Refrigerated microcentrifuge.

## 3. Methods

### *3.1. RNA Preparation (see Note 1)*

1. Add 1 µL of DNase and 0.1 µL of RNase inhibitor in 20 µL of total RNA, and incubate at 37°C for 10 min to remove DNA contamination.
2. Dilute RNA to 1 µg/µL concentration and aliquot 2 µL each into 0.5-mL microcentifuge tubes; store at –70°C.

### *3.2. Reverse Transcription*

1. Set up 3 RT reactions, one for each of the anchor primers ($HT_{11}A$, $HT_{11}G$, $HT_{11}C$), in each of the RNA samples to be compared.
2. Take one RNA aliquot from –70°C and add 18 µL of $dH_2O$ to obtain 0.1 µg/µL concentration.
3. Add together the following components in a 0.5 mL microcentrifuge tube:

| | |
|---|---|
| RNA (0.1 µg/µL) | 2.0 µL |
| 10X RT buffer | 2.0 µL |
| dNTPs (250 µ*M* ) | 1.6 µL |
| HT11M (2 µ*M* ) | 2.0 µL |
| $dH_20$ | 9.4 µL |

   Mix well and centrifuge briefly.
4. Place tubes in the thermocycler and program as follows: 65°C /5 min, 37°C /1 h, 75°C /5 min (*see* **Note 2**).
5. After 5 min at 37°C add 1 mL of 0.1 *M* DTT to the reaction. After another 5 min add 2 µL of MMLV reverse transcriptase (50 U/mL), mix well, and proceed with the reaction.
6. At the end of the reaction, centrifuge briefly to collect any condensation and store at –20°C until the next step.

### *3.3. PCR*

1. Set up the required number of PCR reactions, for each anchor primer RT reaction, in each RNA sample.
2. Add 2 µL of RT reaction mix and 2 µL of each individual arbitrary primer in a 0.5-mL GeneAmp PCR reaction tube (*see* **Note 3**).
3. Make up the core reaction mix consisting of the following components:

| | |
|---|---|
| 10X PCR buffer | 2.0 µL |
| $MgCl_2$ (25 m*M*) | 1.2 µL |

| | |
|---|---|
| dNTPs (25 μ*M*) | 1.6 μL |
| $HT_{11}M$ (2 μ*M*) (the same as in RT) | 2.0 μL |
| [α-$^{33}$P]dATP (2000 Ci/mmol) | 0.04 μL |
| *Taq* polymerase | 0.2 μL (*see* **Note 4**) |
| $dH^2O$ | 9.0 μL |

Make two more reaction volumes than required in case of any pipetting error, for example, 12 vol for 10 reactions. Mix well.

4. Pipet 16 μL of core reaction mix into each PCR reaction tube to make up 20 μL of total volume. Mix well. Overlay with one drop of mineral oil and centrifuge briefly.
5. Place PCR tubes in thermocycler (*see* **Note 5**) and program as follows:

| | | |
|---|---|---|
| Step 1 | 95°C, 5 min | 1 cycle |
| Step 2 | 94°C, 30 s (denature) | |
| | 40°C, 2 min (annealing) | |
| | 72°C, 30 s (extension) | 40 cycles |
| Step 3 | 72°C, 5 min (elongation) | 1 cycle |

6. Store at –20°C until the next step (*see* **Note 6**).

### 3.4. Sequencing Gel Analysis

1. Set up the sequencing cell apparatus. Prepare 50 mL of sequencing gel by mixing 10 mL of 5X TBE, 10 mL of Protogel, and 30 mL of $dH_2O$. Add 760 μL of 10% ammonium persulfate (freshly made) and 37 μL of TEMED immediately before pouring the gel into the sequencing cell apparatus. Insert a 20-well forming comb 1 cm into the gel mixture and allow to polymerize for 1 h. Make up 1X TBE as running buffer. Prerun the gel at 60 W for 20 min before removing the comb and loading the samples.
2. Take 7.0 μL from each PCR reaction into a 0.5-mL tube and add 1 μL of 50% glycerol and 1.5 μL of DNA loading dye. Heat at 80°C in a thermocycler for 2 min and quench in ice for 30 s.
3. Remove comb, and use long flat-tipped pipet to load samples. Restart running the gel at 55 W for approx 2.5 h until the slower moving light blue dye (xylene dye) reaches the bottom of the gel.
4. Blot the gel onto Whatman 3MM chromatography paper, cover with plastic wrap, and dry it under vacuum on a gel dryer for approx 1 h until temperature reaches 70°C.
5. Remove plastic wrap. Place dry gel on its mounting of 3MM paper in an X-ray intensifying cassette. In the dark room cut a Kodak X-ray film into a size smaller than the 3MM paper but larger than the dry gel. Place it onto 3MM paper covering the whole dry gel. Use a marker pen to draw a line crossing the film and the 3MM paper at four corners for the purpose of orientation. Expose the dry gel to a film at room temperature from overnight for up to 3 d, then develop the film (*see* **Fig. 2**).
6. Repeat **Subheadings 3.3.** and **3.4.** in independent experiments. Select reproducibly differentially expressed bands to take forward to the next step (*see* **Note 7**).

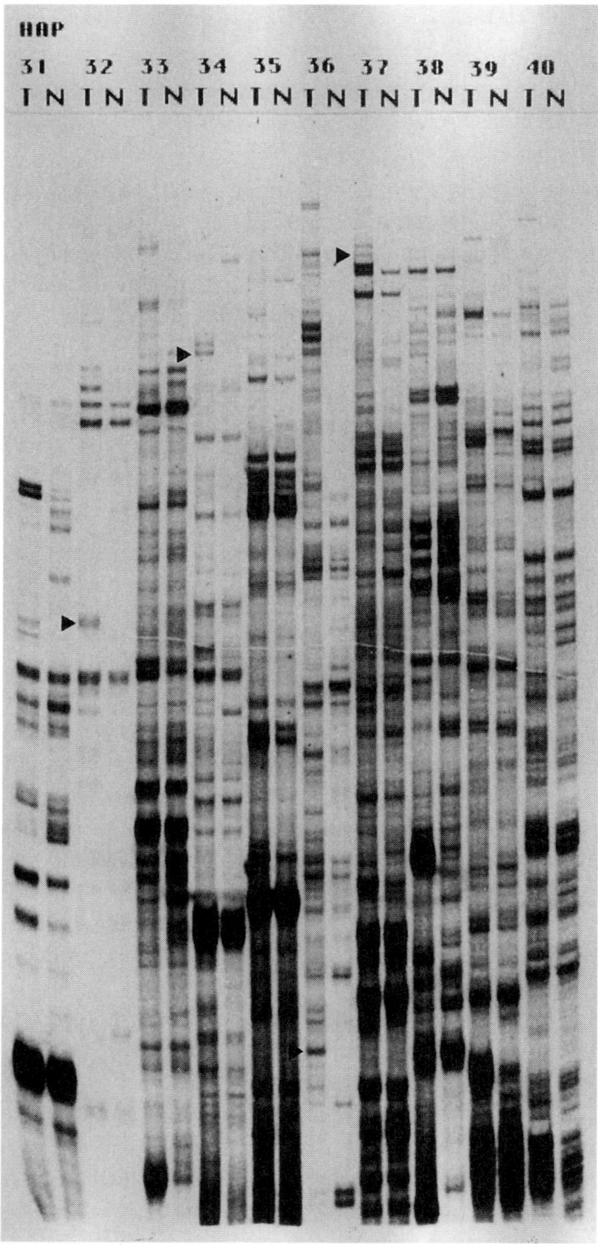


Fig. 2. Photograph of a differential display gel. This photograph compares tumor tissue with normal tissue from a single breast cancer patient using the combination of anchor primer $HT_{11}A$ and arbitrary primer HAP31-40. T indicates tumor tissue and N indicates normal tissue. Most of the band patterns are identical between T and N although there are a number of differentially expressed bands (arrowhead).

### 3.5. cDNA Recovery and Reamplification

1. After developing the film, orient the autoradiogram with the gel by matching the mark lines. Locate bands of interest by punching through the film with a needle at both sides and cut through the dry gel at this location with a clean scalpel.
2. Soak the gel slice, along with the 3MM paper, in 100 µL of $dH_2O$ in a 1.5-mL microcentrifuge tube for 10 min. Boil the tube, with the cap tightly closed by parafilm, for 15 min. Centrifuge briefly to collect condensation and pellet the gel and paper debris.
3. Transfer the supernatant to a new tube. Add 10 µL of 3 *M* sodium acetate, pH 5.2, 2 mL of glycogen (35 mg/mL, for visibility of pellet); and 450 µL of absolute ethanol. Store at –70°C for 30 min.
4. Centrifuge at 8700*g* at 4°C for 10 min to pellet cDNA. Remove supernatant carefully. Avoid disturbing the pellet. Rinse pellet with 200 µL of ice-cold 85% ethanol. Centrifuge briefly and remove ethanol. Air-dry for 5–10 min. Resuspend the pellet with 10 µL of $dH_2O$.
5. Perform the reamplification reaction by adding the following components into a 0.5-mL GeneAmp PCR reaction tube:

| | |
|---|---|
| Recovered cDNA | 4.0 µL |
| 10X PCR buffer | 4.0 µL |
| dNTPs (250 µ*M* ) | 3.2 µL |
| $MgCl_2$ (25 m*M* ) | 2.4 µL |
| $HT_{11}M$ (2 µ*M* ) | 4.0 µL |
| HAP (2 µ*M* ) | 4.0 µL |
| *Taq* polymerase | 0.4 µL |
| $dH_2O$ | 18.0 µL |
| | 40.0 µL |

   Mix well. The primer pair and PCR conditions should be the same as those used originally in the differential display PCR.
6. Run 10 µL of the reamplification reaction on a 1.5% agarose gel with ethidium bromide (2.5 µL of 10 mg/mL ethidium bromide in 100 mL of 1.5% agarose gel) to see if the reamplification worked well, and to check if the sizes of the reamplified PCR products were consistent with their sizes on the sequencing gel (*see* **Fig. 3**) (*see* **Notes 8** and **9**).

## 4. Notes

1. The success of the DD technique depends largely on the integrity of the RNA. You should run a portion of the starting RNA on an agarose formaldehyde gel to check its integrity. Store RNA at a concentration of >1 µg/µL at –70°C at all times. Do not repeat thawing and freezing more than three times. Dilute the RNA just before use. We do not use DEPC-treated $H_2O$; if you handle the RNA as described previously and wear gloves at all times then regular, autoclaved distilled $H_2O$ is adequate. It also is essential that through the whole course of the protocol the RNA should come from the same extraction batch. Ensure there is sufficient RNA before starting. A 0.2 µg quantity of total RNA is sufficient for 10 PCR reactions. Calculate the minimum

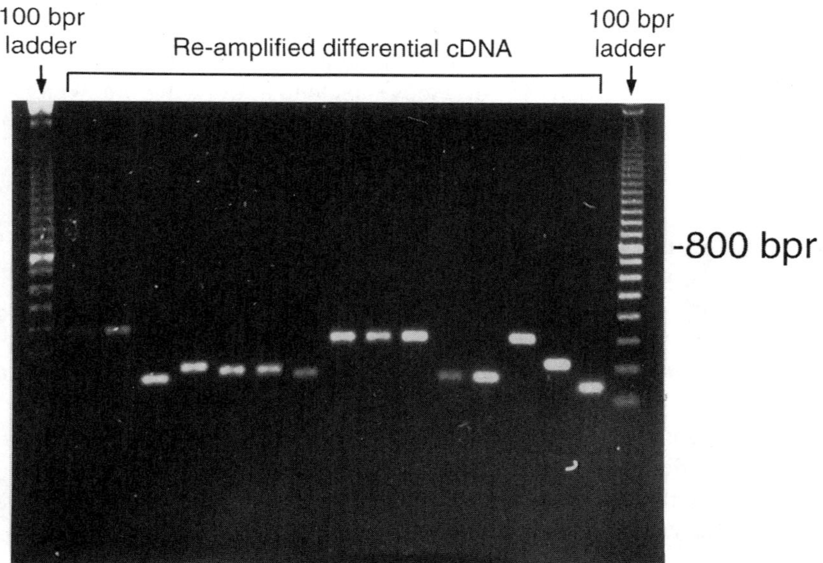


Fig. 3. Reamplification reactions. Reamplification products are running on an ethidium bromide agarose gel. The sizes of the reamplified cDNAs should be consistent with those obtained in the differential display gel.

amount of RNA required by estimating the number of intended reactions you will perform. There is no necessity to use mRNA as the initial material, as this does not improve results.

2. Temperatures of 65°C for 5 min not only denature RNA but also inactivate DNase. We find it unnecessary to do phenol–chloroform extractions after treatment with DNase.
3. Some makes and types of tubes have been reported as being less than optimal for DD *(5)*; we find these tubes to give acceptable results.
4. Use the same *Taq* polymerase through the entire experimental series. Different *Taq*s have been shown to give different band patterns.
5. Use one thermocycler only. Different thermocyclers, even the same model from the same company, may give different band patterns.
6. Contaminating chromosomal DNA in the RNA sample is one of the major sources of false-positives. We suggest that in **Subheading 3.3.** you perform a negative control in which RNA, instead of the RT product, is used as a template and then run it alongside the other PCR reactions in the sequencing gel. Any occurrence of a band pattern in this negative control indicates DNA contamination. You should prepare the RNA again if this happens.
7. This reduces the number of false-positives due to the variability of the PCR reactions.
8. Be sure the size of the reamplifed cDNA corresponds to the size of the differential band in the sequencing gel. If you obtain more than one band in a reamplifi-

cation reaction, it most likely is due to the inclusion of a nearby contaminant band in your gel cutting. To confirm this, expose the cut dry gel to another film to see whether the correct bands are cut out. Any doubt then repeat the recovery and reamplification procedures again from another duplicate gel. On rare occasions, there may be more than one differential band of the same size. This will be proved in the following confirmation test.

9. At this point, you may have a number of candidates for differential cDNA. Because of the possibility of chromosomal DNA contamination and the variability of each PCR reaction it is probable that many of these are false-positives. You must do at least one confirmation test to rule out such false-positives. Traditionally, the standard confirmation test is Northern blot analysis. Sometimes a more sensitive method, such as the nuclease protection assay or even RT-PCR, is required to detect low-abundance messenger RNA. This possible difficulty arises because of the high sensitivity of DD. If you use tissue material as the source of the initial RNA input then you may have to do *in situ* hybridization, instead of Northern blot analysis, as the confirmatory procedure. We have successfully applied *in situ* hybridization to confirm differential cell expression of isolated cDNA in breast cancer tissue, a procedure that permits identification of the cell types expressing the particular gene under study. For a large number of differential cDNA fragments it may prove necessary to use the reverse Northern blot method ***(6)***. Alternatively, you may wish to clone these PCR products into a suitable vector, using any PCR-based cloning method, and then sequence them. Following GeneBank searching, select the best potential candidates for the confirmation test.

## Acknowledgment

We thank Ian Hart for the help on preparation of the manuscript. This work was supported by the Tzu-Chi Buddhist Compassion Relief Foundation.

## References

1. Liang, P. and Pardee, A. B. (1992) Differential display of eukaryotic messenger RNA by means of the polymerase chain reaction. *Science* **257,** 967–971.
2. Liang, P., Averboukh, L., and Pardee, A. B. (1993) Distribution and cloning of eukaryotic mRNAs by means of differential display: refinement and optimization. *Nucleic Acids Res.* **21,** 3269–3275.
3. Sager, R. (1997) Expression genetics in cancer: Shifting the focus from DNA to RNA. *Proc. Natl. Acad. Sci. USA* **94,** 952–955.
4. Liang, P., Zhu, W., Zhang, X., Guo, Z., O'Connell, R. P., Averboukh, L., et al. (1994) Differential display using one-base anchored oligo-dT primers. *Nucleic Acids Res.* **22,** 5763–5764.
5. Chen, Z., Swisshelm, K., and Sager, R. (1994) A cautionary note on the reaction tubes for differntial display and cDNA amplification in thermal cycling. *BioTechniques* **16,** 1003–1006.
6. Francia, G., Mitchell, S. D., Moss, S. E., Hanby, A. M., Marshall, J. F., and Hart, I. R. (1996) Identification by differential display of annexin-VI, a gene differentially expressed during melanoma progression. *Cancer Res.* **56,** 3855–3858.

# III

# Mathematical Modeling of Metastasis

21

# Predictive Mathematical Modeling in Metastasis

Jonathan A. Sherratt

## 1. Introduction

Mathematical modeling is emerging as a powerful predictive tool in many areas of biology and medicine, with applications to cancer metastasis increasingly widespread and effective. This type of modeling involves quantitatively accurate representations of specific cellular activities, and is quite different from more traditional applications of mathematics to cancer, such as the simple fitting of experimental data to Gompertzian growth curves. It is made possible by the twin revolutions in molecular biology and nonlinear mathematics over the last two decades, and involves using experimental data at the cell and molecular level to construct mathematical models, which can then be used to predict the macroscopic implications of this data.

Mathematical models have been in use for biological prediction since the early part of the century, initially in ecology and embryology. These early models were phenomenological, that is, they acted as a convenient way to express and explore theories, but did not represent particular postulated mechanisms. Establishment of mathematical biology as a recognized scientific field was achieved by a number of major successes for these early models. Most notable amongst these is the work of Hodgkin and Huxley on electrical signaling in nerve axons, which underlies much of neurophysiology, and for which they were awarded the Nobel Prize for Physiology and Medicine in 1963. More recently, the ability to isolate biological mechanisms at the molecular level has led to a new type of mathematical model, which represents specific low-level mechanisms, either known or hypothesised. This use of mathematical modeling was pioneered during the late 1970s and early 1980s by James Murray at the University of Oxford, mainly in applications to developmental biology. An

From: *Methods in Molecular Medicine, vol. 57:*
*Metastasis Research Protocols, Vol. 1: Analysis of Cells and Tissues*
Edited by: S. A. Brooks and U. Schumacher 

example of a success story from this work is the determination of the mechanism of aggregation in the starvation response of cellular slime molds. These normally live as individual cells, but in starvation conditions the individual cells aggregate to form a multicellular slug, several millimeters long, which crawls in a coordinated manner and then redistributes the individual cells in a new location. Mathematical models showed that chemicals secreted by the cells when starved form spiral signaling waves that cause the initial cellular aggregation *(1)*.

The ability that now exists to mathematically model specific cellular activities in medically important contexts is a direct progression from this earlier work. Within cancer biology, the range of potential applications is very wide. The tumor–immune system interaction has been extensively modeled, addressing the role of the immune system in tumor composition and morphology, and the potential for different immunotherapeutic strategies *(2,3)*. The growth constraints on tumors prior to vascularization have also been modelled widely, leading to predicted relationships between cellular parameters and size limitation *(4,5)*. In the area of metastasis, angiogenesis and invasion have both been studied using mathematical modeling. In the former case, work initially focussed on the basic mechanisms via which angiogenic factors stimulate vascular ingrowth *(6)*, and recent achievements include prediction of the ways in which the extracellular matrix can modulate the cellular response to angiogenic factors *(7)*, and simulation of antiangiogenesis strategies *(8)*. Cancer invasion is an ideal topic for mathematical modeling because of the recent discovery of many underlying molecular mechanisms: Models are able to predict the larger scale implications of these mechanisms. This approach has been used to study the delicate balance between proteolytic enzymes and their inhibitors *(9)* and the role of pH gradients in the invasive process *(10)*.

## 2. Materials

### 2.1. Computer Hardware

The basic materials requirement is an appropriate computer; PC, Apple Macintosh and Unix Workstation frameworks are all suitable. Many modeling studies do not require levels of processing power that are problematic for most modern machines. However, investigation of behavior in two or three space dimensions does often demand high level computing, in terms of processing power, RAM, and graphics. Before embarking on such a study, it is advisable to model the behavior concerned in one spatial dimension.

### 2.2. Computer Software

Some types of mathematical model can easily be simulated directly using commercially available computer packages such as Maple *(11)* or Mathematica *(12)*; this includes models involving variation in time but not in space, and some discrete-cell models such as cellular automata. Other types of model require preprocessing before they can easily be simulated: This is particularly true for the "continuum models" discussed in **Subheading 3.2.** For those familiar with the computer languages Fortran or C, writing source code is often the most efficient method of simulation in these cases, as there are a number of relevant subroutines available free of charge, notably from the Higher Education National Software Archive (ftp://unix.hensa.ac.uk/mirrors/netlib/). Another useful software source is the Clinical and Biomedical Computing Unit at the University of Cambridge, UK, which has produced a number of software packages specifically designed for medical problems (*see* http://www.cbcu.cam.ac.uk/software.htm for details). If none of these approaches is suitable, those unfamiliar with numerical simulation are best advised to consult a local applied mathematician.

### 2.3. Mathematical Expertise

In practice, the majority of mathematical modeling studies occur as a collaboration between an experimental or clinical scientist and a mathematician. A major advantage of this approach is that it enables traditional "pen and paper" analysis of the model. This can often give significant insights into the underlying biology, yielding formulae for predicted behavior as a function of cell and molecular parameters. However, the basic processes of model development and simulation do not require high levels of mathematical expertise.

## 3. Methods

### 3.1. Suitability of Mathematical Modeling

The essential first step in developing a mathematical model is to identify a problem that is suitable for a theoretical approach. In general terms, such problems fall into one of two categories: (1) Quantitative predictions, for which the key criterion is that reliable quantitative data exists for all the model inputs. Here the objective of the work will be to predict a macroscopic feature of the system, such as the amount of an anti-angiogenic factor required to reduce the angiogenic response by a given percentage. (2) Qualitative hypothesis testing. It is a common misconception that mathematical modeling is effective only for quantitative prediction. In fact, some of the most effective modeling in cancer biology has been of a qualitative nature. In this case, the objective is to test the feasibility of one or more qualitative hypotheses. For example, it is known that in breast carcinomas, tumor-associated macrophages, which are an important regulator of angiogenesis, are found in clusters distinct from vascular hot spots. One hypothesis is that the macrophage clustering might arise prior to vascularization; a model can be used to test this, determining the precise conditions under which prevascularization patterning arises. (For details of this example, *see* **refs.** *3* and *13*).

It is important to stress that the objective of mathematical modeling work is not to generate a large-scale computer simulation of the whole metastatic cascade; although such a model is feasible in principle, its complexity would make it so sensitive to underlying assumptions as to be of no practical value. Rather, models are used in a very focused way, to study in detail particular aspects of metastasis, which may be as limited as a single cell–cytokine interaction, for example. Thus the philosophy of using a mathematical model is similar to that underlying the use of an in vitro experiment to help in the understanding of an in vivo system.

### 3.2. Choice of Modeling Framework

The most fundamental choice when developing a mathematical model is the choice of mathematical framework used to represent the biological problem. The approach with the longest history is continuum modeling, in which the discreteness of individual cells is neglected, and local cell densities are used as model variables. This is very well established, and gives mathematical equations closest to those used in more traditional areas of applied mathematics (such as fluid dynamics and solid mechanics). A number of good reference books are available on models of this type in biology and medicine, for example ***(14,15)***.

The alternative to a continuum model is the representation of cells as discrete objects. Discrete models are very well established in some other areas of biology, most notably ecology; the articles ***(16,17)*** compare various different discrete and continuous models for the same ecological phenomena, and are useful background reading when choosing a model formulation. However, in cell biology, discrete models are relatively new and there is not yet a coordinated body of expertise. Successful approaches include direct computational models that track cell boundaries ***(18)***, cellular automata ***(19)***, and discrete cells studied in a continuous extracellular matrix ***(20)***. There is currently no reference book describing these different approaches, and thus the choice of an appropriate discrete formulation requires the reading of original research papers, making the modeling procedure rather more involved. The great advantage of discrete models is their ability to predict phenomena at the individual cell level, for example, the paths taken by individual cells in an invasion assay.

### 3.3. Procedure for Model Development

1. Selection of variables is a key step in the modeling process. Variables can include cell types, chemical regulators, extracellular matrix components, and cell surface receptors. The most effective models usually have only a small of variables, because this makes it much easier to draw clear precise conclusions. However, in some cases it is helpful to begin developing a larger model as an aid to the derivation of an appropriate simpler form.

2. Word equations are a valuable first step before attempting a mathematical formulation. These should indicate the interdependencies between the various model variables, and it is also helpful at this stage to summarize the various data inputs available.
3. Mathematical equations are simply a rewrite of the word equations, but in terms of mathematical symbols. Typically all the terms are relatively standard, with mathematical novelty coming from the way in which they are combined.
4. Parameter identification is important for qualitative as well as quantitative modeling. The values of some parameters can be obtained directly from the scientific literature: Kinetic rate constants and chemical diffusion coefficients often fall into this category. Others will typically be determined using experimental data, often from in vitro experiments that led to the modeling study. It is often the case that some parameters cannot be estimated accurately, but that approximate or order of magnitude values are known. This is in no sense a barrier to effective use of the model, but indicates that it is important to perform a parameter sensitivity analysis, described in **item 6**.
5. Numerical simulation is a key step in the use of a mathematical model. A variety of approaches are possible, as discussed in **Subheading 2.** The best practice is to initially consider one particular set of biological parameters, and vary numerical parameters (such as spatial discretization) to confirm accuracy. The simulations can then be used as a "mathematical experiment," in which biological parameters can be varied to make quantitative predictions or test qualitative hypotheses, as appropriate. When model simulations run quickly on the computer, there is a temptation to rapidly generate a large volume of simulations that are difficult to handle; this can be effectively dealt with by identifying a particular series of "experiments" to do initially.
6. Parameter sensitivity analysis is not necessary in all cases but is often very instructive. The basic procedure is to vary individual parameters by a given amount (say 10%) and determine the resulting percentage change on a particular aspect of the model prediction. Ideally one should vary parameters in combinations as well as individually, but this is feasible only if the total number of parameters is relatively small. The purpose of this calculation is to determine whether the model predictions are particularly sensitive to one or two model inputs; if so, a more detailed study of these inputs may be appropriate.
7. Mathematical analysis is a powerful tool for the investigation of continuum models, as one can use analytical techniques developed previously in other areas of applied mathematics. Murray's book ***(14)*** provides a wide-ranging review of these methods. For discrete models, analytical tools are much less well developed, and one is usually limited to numerical investigation in these cases.

## 4. Notes

1. Numerical problems are usually manifested by simulation results that are clearly nonsensical (e.g., variables becoming extremely large, or negative), corresponding to the instability of the numerical scheme. Often this requires nothing more

than a change in numerical parameters, such as a reduction in the time step. Results that are biologically reasonable but whose precise form is sensitive to numerical parameters usually correspond to a numerical scheme that is stable but has not yet converged to the actual solution. A useful rule of thumb is that time/space steps should be reduced until they no longer visibly affect the solution; detailed numerical tests for convergence are described in numerical analysis textbooks, for example *(21)*.

2. Refinement of the model in the light of initial simulations is very common. For instance, the results may indicate that one particular term in the model is central to the predictions, in which case it may be appropriate to include a more detailed representation of this term. An example of this is provided by work on cell chemotaxis. This is conventially modeled as a movement of cells up gradients of chemoattractant, at a rate dependent on the local chemoattractant concentration. However, in some cases, the predicted behavior depends crucially on the details of this concentration dependence, and in such cases more detailed representations are used in which cell surface receptors are added as explicit variables in the model *(22)*.
3. Reformulation of the model is occasionally appropriate; this means alteration of a basic aspect of the model, rather than simply addition of a new feature as in **Note 2**. Reformulation is required when the model predictions are inconsistent with a key modeling assumption. For example, if a continuum model predicts a spatial pattern on the scale of one or two cell diameters, then the prediction will not be reliable and should be checked using a discrete model.

## References


1. Tyson, J. J. and Murray, J. D. (1989) Cyclic AMP waves during aggregation of *Dictyostelium* amoebae. *Development* **106,** 421–426.
2. Adam, J. A. and Bellomo, N. (1997) *A Survey of Models for Tumor-Immune System Dynamics*. Birkhäuser, Boston.
3. Owen, M. R. and Sherratt, J. A. (1997) Pattern formation and spatiotemporal irregularity in a model for macrophage-tumor interaction. *J. Theor. Biol.* **189,** 63–80.
4. Ward, J. P. and King, J. R. (1997) Mathematical modeling of avascular tumor growth. *IMA J. Math. Appl. Med. Biol.* **14,** 39–69.
5. Byrne, H. M. and Chaplain, M. A. J. (1995) Growth of nonnecrotic tumors in the presence and absence of inhibitors. *Math. Biosci.* **130,** 151–181.
6. Chaplain, M. A. J. and Stuart, A. M. (1993) A model mechanism for the chemotactic response of endothelial cells to tumor angiogenesis factor. *IMA J. Math. Appl. Med. Biol.* **10,** 149–168.
7. Sleeman, B. D. (1997) Mathematical modeling of tumor growth and angiogenesis. *Adv. Exp. Med. Biol.* **428,** 671–677.
8. Orme, M. E. and Chaplain, M. A. J. (1996) A mathematical model of the first steps of tumor-related angiogenesis: capillary sprout formation and secondary branching. *IMA J. Math. Appl. Med. Biol.* **13,** 73–98.
9. Perumpanani, A. J., Sherratt, J. A., Norbury, J., and Byrne, H. M. (1996) Biological inferences from a mathematical model for malignant invasion. *Invas. Metastas.* **16,** 209–221.

10. Gatenby, R. A. and Gawlinski, E. T. (1996) A reaction-diffusion model of cancer invasion. *Cancer Res.* **55,** 4151–4156.
11. Redfern, D. (1996) *The Maple Handbook: Maple V* release 4. Springer-Verlag, New York.
12. Wolfram, S. (1993) *Mathematica: A System for Doing Mathematics by Computer.* Addison-Welsey, Reading, MA.
13. Leek, R. D., Lewis, C. E. Whitehouse, R., Greenall, M., Clarke, J., and Harris, A. L. (1996) Association of macrophage infiltration with angiogenesis and prognosis in invasive breast carcinoma. *Cancer Res.* **56,** 4625–4629.
14. Murray, J. D. (1989) *Mathematical Biology*. Springer-Verlag, Berlin.
15. Edelstein-Keshet, L. (1988) *Mathematical Models in Biology*. McGraw-Hill, New York.
16. Sherratt, J. A., Eagan, B. T., and Lewis, M. A. (1997) Oscillations and chaos behind predator-prey invasion: mathematical artifact or ecological reality? *Philos. Trans. R. Soc. B* **352,** 21–38.
17. Durrett, R. and Levin, S. A. (1994) Stochastic spatial models—a user's guide to ecological applications. *Philos. Trans. R. Soc. B* **343,** 329–350.
18. Weliky, M. and Oster, G. (1990) The mechanical basis of cell rearrangement. 1. Epithelial morphogenesis during fundulus epiboly. *Development* **109,** 373.
19. Ermentrout, G. B. and Edelstein-Keshet, L. (1993) Cellular automata approaches to biological modeling. *J. Theor. Biol.* **160,** 97–133.
20. Dallon, J. C. and Othmer, H. G. (1997) A discrete cell model with adaptive signalling for aggregation of *Dictyostelium discoideum. Philos. Trans. R. Soc. Lond. B* **352,** 391–417.
21. Morton, K. W. and Mayers, D. F. (1994) *Numerical Solution of Partial Differential Equations*. Cambridge University Press, Cambridge, UK.
22. Höfer, T. Maini, P. K., Sherratt, J. A., Chaplain, M. A. J., Chauvet, P., Metevier, D., et al. (1994) A resolution of the chemotactic wave paradox. *Appl. Math. Lett.* **7,** 1–5.

# Index